Modern Concepts of Plant Biology

Modern Concepts of Plant Biology

Editor: Cristiano Shepherd

R CALLISTO REFERENCE

www.callistoreference.com

Callisto Reference,
118-35 Queens Blvd., Suite 400,
Forest Hills, NY 11375, USA

Visit us on the World Wide Web at:
www.callistoreference.com

ISBN: 978-1-64116-124-4 (Hardback)

Cataloging-in-Publication Data

Modern concepts of plant biology / edited by Cristiano Shepherd.
     p. cm.
Includes bibliographical references and index.
ISBN 978-1-64116-124-4
1. Botany. 2. Plants. 3. Biology. I. Shepherd, Cristiano.
QK45.2 .M63 2019
580--dc23

Table of Contents

Preface

In my initial years as a student, I used to run to the library at every possible instance to grab a book and learn something new. Books were my primary source of knowledge and I would not have come such a long way without all that I learnt from them. Thus, when I was approached to edit this book; I became understandably nostalgic. It was an absolute honor to be considered worthy of guiding the current generation as well as those to come. I put all my knowledge and hard work into making this book most beneficial for its readers.

Plant biology is the sub-discipline of biology that is concerned with the scientific study of plants and their biological processes. It is a multidisciplinary subject integrating principles of different areas of science and technology. It explores the study of plant structure, growth, differentiation, reproduction, diseases, evolution, taxonomy, etc. Such studies are made using multiple techniques of optical microscopy, live cell imaging, plant chemistry and chromosome number analysis, besides others. Botany has wide applications in agriculture, forestry and horticulture. It is also crucial for the production of commercial products like staple foods, fiber, drugs, rubber, etc. It also has relevance in energy production, environmental management and in the maintenance of biodiversity. This book strives to provide a fair idea about this discipline and to help develop a better understanding of the latest advances within this field. The aim of this book is to present researches that have transformed this discipline and aided its advancement. Those who want to broaden the expanse of their knowledge will be immensely assisted by it.

I wish to thank my publisher for supporting me at every step. I would also like to thank all the authors who have contributed their researches in this book. I hope this book will be a valuable contribution to the progress of the field.

Editor

1

In vitro evaluation of antidermatophytic activity of five lichens

Ashutosh Pathak[1], Rohit K. Mishra[2], Shashi Kant Shukla[1], Rajesh Kumar[1], Madhu Pandey[1], Manisha Pandey[1], Afifa Qidwai[1] and Anupam Dikshit[1]*

*Corresponding author: Anupam Dikshit, Biological Product Lab, Department of Botany, University of Allahabad, Allahabad 211012, Uttar Pradesh, India
E-mail: anupambplau@rediffmail.com
Reviewing editor: Raffaele Dello Ioio, Universita degli Studi di Roma La Sapienza, Italy

Additional information is available at the end of the article

Abstract: Lichens (a composite organism) are known for their secondary metabolites and have several properties as photoprotection, allelopathy, antioxidant, antimicrobial, and antiviral. In this study, based on alarming situation of prevalence and developing resistance in dermatophytes, the new biological source in the form of lichens was screened for their antidermatophytic potential. Three dermatophytes viz. *Microsporum canis*, *Trichophyton mentagrophytes*, and *Trichophyton rubrum* were procured from Microbial Type Cell Culture, Chandigarh, India and susceptibility of aforementioned pathogens were tested via Clinical Laboratory and Standard Institute recommended broth microdilution procedure for filamentous fungi. Five lichens viz. *Bulbothrix setschwanensis*, *Myelochroa aurulenta*, *Parmotrema nilgherrense*, *Parmotrema reticulatum*, and *Ramalina conduplicans* were tested for their antidermatophytic activity (fungistatic and fungicidal concentrations) in the form of MIC and MFC values. *M. aurulenta* exhibited most promising MIC and MFC values against all dermatophytes and provides new leads in the form of secalonic acid A and leucotylic acid for future investigations.

Subjects: Dermatology; Mycology; Natural Products

Keywords: dermatophytes; lichens; secalonic acid A; leucotylic acid

ABOUT THE AUTHORS

The main focus area of our research group is the bioprospection of new biological resources viz. plants, lichens and microbes, available in India and we are working in this particular field since last three decades. The other areas of interest of our group are microbiology of human pathogen and agricultural microbes, taxonomy of angiosperms and lichens. We are trying to develop new biological products for human welfare and have succeeded in some which have been patented by us. Recently, the group is also working with synthesis, characterization, formulation of various metal, metal oxide and polymeric nanoparticles with their application in human as well as plant health.

PUBLIC INTEREST STATEMENT

The developing resistance against first line clinical drug in pathogens is an evolutionary process for survival. Due to developing resistance in pathogens, we have to search for new drugs or compounds or herbals for our backup. We humans always rely on nature for our needs and mostly nature provides us solution in the form of any biological source. In present study, one of the less explored biological sources i.e. lichens were investigated. Lichens are a composite organism (composed of a fungus and a photobiont (algae and cyanobacteria or both)), slow growing in nature and acts as a pollution indicator. Lichens are used since long back in human civilization for their several benefits viz. antimicrobial, perfume, edible and many more. The present study enhances the existing knowledge and provides new leads for the cure of prevalent disease dermatophytosis (fungal infection of skin).

1. Introduction

Filamentous keratinophilic fungi causing cutaneous mycoses are called dermatophytes and are classified into three genera, *Trichophyton*, *Microsporum*, and *Epidermophyton* (Peres, Maranhão, Rossi, & Martinez-Rossi, 2010). So far, about 30 species of dermatophytes have been identified among human pathogens (White, Oliver, Graser, & Henn, 2008). Dermatophytes are cosmopolitan in distribution. A study involving 16 European countries showed that 35–40% of the analyzed individuals had infection of the foot (tinea pedis) caused by dermatophytes (Burzykowski et al., 2003). A study involving children's in US revealed that between 22 and 55% were having hair scalp infection of dermatophytes (Abdel-Rahman, Simon, Wright, Ndjountche, & Gaedigk, 2006). Although, the prevalence of drug resistance in dermatophytes is rare but resistance cases have been reported for griseofulvin, terbinafine, and fluconazole (Orozco et al., 1998; Peres et al., 2010; Smith et al., 1986; Stephenson, 1997; Wingfield, Fernandez-Obregon, Wignall, & Greer, 2004). In present study, based on the alarming situation documented in aforementioned literatures, the less explored biological source in the form of lichens was screened for their antidermatophytic potential. Lichen (a composite organism) thallus is a consortium of mycobiont and photobiont in a mutualistic relationship (Hawksworth, 2000). Apart from photobiont and mycobiont the lichen thallus is comprised of several other lichenicolous fungi, endophytic fungi, and bacteria (Grube & Berg, 2009). Lichens are known for their secondary metabolites which are quite unique to them and have several properties as photoprotection, allelopathy, antioxidant, antimicrobial, and antiviral (Molnar & Farkas, 2010).

2. Results and discussion

2.1. Percent yield

Percent yield obtained for *Bulbothrix setschwanensis* was highest i.e. 5.5%; *Parmotrema nilgherrense* was 4.5%; *Parmotrema reticulatum* was 4.45%; *Myelochroa aurulenta* was 3.75%; and least for *Ramalina conduplicans* i.e. 3.5%.

2.2. Antidermatophytic activity of lichens

The antidermatophytic activity of various lichen extracts were tabulated in Table 1 and graphically represented in Figure 1 in the form of IC_{50} (concentration of drug/extract required to inhibit the pathogens growth up to 50 percent) and MIC (minimum concentration required to inhibit the growth of pathogens) values along with the standard drug used i.e. Sertaconazole. Sertaconazole nitrate exhibited very low MIC values in other studies against *Trichophyton rubrum* (1 μg ml^{-1}) and *Trichophyton mentagrophytes* (8 μg ml^{-1}) (Carrillo-Munoz, Tur-Tur, Cardenes, Estivill, & Giusiano, 2011). The difference in the MIC values is due to the size of inoculum. In the previous study, the initial inoculum size was 4.7×10^3 to 1.5×10^4 CFU/ml whereas in present study the inoculum size was ca

Table 1. Antidermatophytic activity of lichen extracts and reference standard i.e. sertaconazole						
	M. canis		*T. mentagrophytes*		*T. rubrum*	
	IC_{50}	MIC	IC_{50}	MIC	IC_{50}	MIC
Sertaconazole	0.040 ± 0.03*	0.043 ± 0.03*	0.037 ± 0.05*	0.090 ± 0.05*	0.033 ± 0.05*	0.064 ± 0.05*
B. setschwanensis	0.076 ± 0.16*		0.706 ± 0.04**	–	0.639 ± 0.21	
M. aurulenta	0.013 ± 0.02*	0.098 ± 0.02*	0.428 ± 0.03**	1.318 ± 0.03**	0.463 ± 0.24*	0.921 ± 0.24*
P. nilgherrense	0.037 ± 0.02**					
P. reticulatum	1.201 ± 0.08		0.339 ± 0.04**			
R. conduplicans			0.478 ± 0.02**		0.107 ± 0.15	

Note: All the values are in mg ml^{-1}.

*Level of significance is less than or equal to 0.05.

**Level of significance is less than or equal to 0.01.

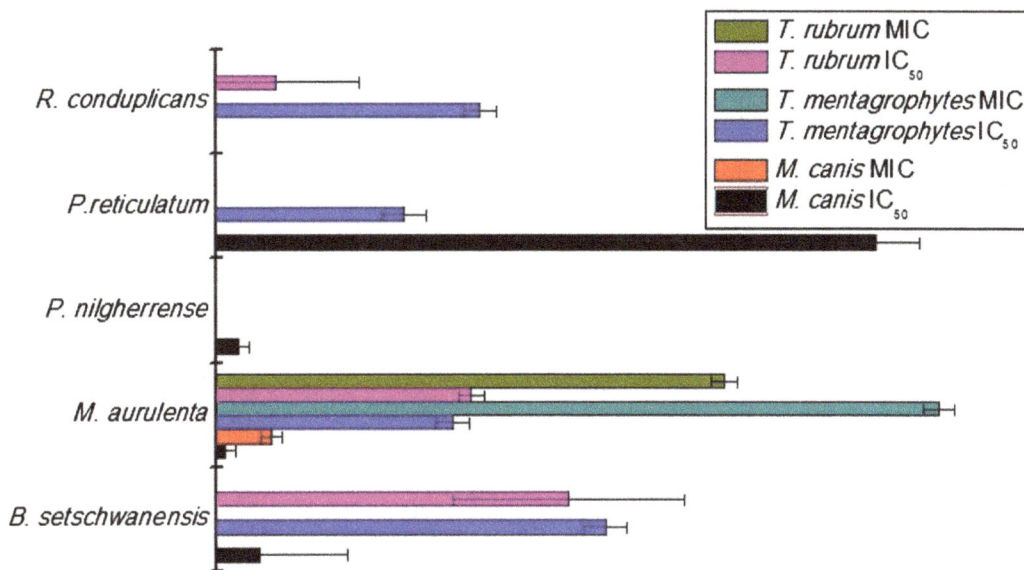

Figure 1. Antidermatophytic activity of sertaconazole nitrate and lichen extracts.

0.5×10^6 CFU/ml. Sertaconazole nitrate a highly active chemical drug having low fingistatic and fungicidal activities but it cause inflammation and itching to the patients (Liebel, Lyte, Garay, Babad, & Southall, 2006). Lichen extracts possess antidermatophytic activity but the most potent lichen extract was of *M. aurulenta* which had MIC values against all three pathogens present in this study. *B. setschwanensis* exhibited activity against all three pathogens but only 50% inhibition was achieved. *R. conduplicans* showed activity against *Trichophyton* spp. whereas; *P. nilgherrense* exhibited activity against *Microsporum* sp. and was least effective amongst all; *P. reticulatum* exhibited activity against *Microsporum canis* and *T. mentagrophytes*.

Fungicidal activities were reported in the form of minimum fungicidal concentration (MFC) in Sertaconazole and *M. aurulenta*. Sertaconazole exhibited fungicidal activity against *M. canis* and *T. rubrum* at 0.078 mg ml^{-1}; 0.156 mg ml^{-1} for *T. mentagrophytes*; while *M. aurulenta* exhibited fungicidal concentration for *M. canis* at 0.156 mg ml^{-1} and 1.25 mg ml^{-1} for *T. rubrum*.

2.3. Statistical analysis

The significance level in the mean difference of control and treated columns in independent samples *t*-test are: Sertaconazole were 0.011 (*M. canis*), 0.047 (*T. mentagrophytes*), 0.025 (*T. rubrum*); *B. setschwanensis* were 0.011 (*M. canis*), 0.001 (*T. mentagrophytes*), 0.898 (*T. rubrum*); *M. aurulenta* were 0.029 (*M. canis*), 0.001 (*T. mentagrophytes*), 0.043 (*T. rubrum*); *P. nilgherrense* were 0.002 (*M. canis*); *P. reticulatum* were 0.277 (*M. canis*), 0.001 (*T. mentagrophytes*); *R. conduplicans* were 0.001 (*T. mentagrophytes*), 0.890 (*T. rubrum*).

Secondary metabolites reported from *M. aurulenta* are zeorin, atranorin, secalonic acid A, and leucotylic acid (Singh & Sinha, 2010). Among aforementioned compounds atranorin and zeorin exhibits weak antifungal properties. Atranorin doesn't exhibit any activity against filamentous fungi but was found active against *Candida albicans* and *Candida glabrata* with MIC values of 500 µg for each (Turk, Yilmaz, Tay, Turk, & Kivanc, 2006; Yilmaz, Turk, Tay, & Kivanc, 2004). On the other hand, zeorin exhibited MIC values above 3 mg ml^{-1} against filamentous fungi (Marijana, Branislav, & Slobodan, 2010). Both the aforementioned compound either possess no or very weak antifungal activity. Thus, secalonic acid A and leucotylic acid need to be tested against dermatophytes or it might be possible that there was a synergy in between aforementioned compounds is the future leads of this present investigation.

3. Material and methods

3.1. Collection of lichens
Lichen thalli were collected from Chakrata district, Uttarakhand, India and identified with the help of relevant key (Awasthi, 2007). The voucher specimens were submitted in Central Regional Circle, Allahabad, Botanical Survey of India viz. *B. setschwanensis* (BSA-8763), *M. aurulenta* (BSA-8761), *P. nilgherrense* (BSA-8757), *P. reticulatum* (BSA-8762), and *R. conduplicans* (BSA-8759).

3.2. Preparation of lichen extracts
Two grams of lichen thalli were taken for cold extraction in 50% acetone (50 ml) (the solvent is a mixture of polar and non-polar solvent so that maximum amount of secondary as well as primary metabolites can diffuse into the solvent). Subsequently, the extracts were filtered by Whatman No. 1 filter paper and crude extract was obtained using rotary evaporator and weighed. The weight of crude extracts obtained were 0.11 g, 0.075 g, 0.09 g, 0.089 g, and 0.07 g for *B. setschwanensis, M. aurulenta, P. nilgherrense, P. reticulatum,* and *R. conduplicans,* respectively. Percent yield of crude extracts were calculated according to below mentioned formula:

$$Percent\ yield(\%) = \left((Dry\ weight\ of\ extract)/(Dry\ weight\ of\ sample)\right) \times 100$$

Stock solutions (50 mg/ml) of crude extracts were prepared in dimethyl sulphoxide (DMSO) for antifungal susceptibility test.

3.3. Test pathogens and inocula preparation
Fungal cultures of *M. canis* (MTCC No. 3270), *T. mentagrophytes* (MTCC No. 7687), and *T. rubrum* (MTCC No. 296) were procured from Microbial Type Culture Collection and Gene Bank (MTCC), Chandigarh, India, which were subcultured on SDA medium under laminar flow cabinet (Laminar flow ultra clean air unit, Micro-Filt, India). Inocula were prepared in saline media and then adjusted to a 0.5 McFarland standard, corresponding to ca 0.5×10^6 CFU/ml and transmittance of inoculum prepared was 70–72% at 520 nm for each culture (Santos, Barros, & Hamdan, 2006).

3.4. Antidermatophytic susceptibility test
Antidermatophytic susceptibility test was performed according to the Clinical Laboratory Standard Institute (CLSI) recommended broth microdilution method for filamentous fungi in RPMI-1640 medium HEPES modification (Sigma Aldrich) supplemented with MOPS buffer (3-morphollinopropane-1-sulfonic acid) (Qualigens Fine Chemicals) (Rex et al., 2008). Six 96-well plates were used separately i.e. five for lichen extracts and one for sertaconazole. Brief steps involved per plate were as follows: inocula prepared was diluted 1:50 times in testing media i.e. RPMI 1640; test was performed in 96-well flat-bottom microtitre plates; column 1 named as negative control consist of 100 μl RPMI-1640 broth media and 100 μl of inocula prepared and formaldehyde less than 1%; column 2 named as broth control consist of 200 μl of media; columns 3 and 4, 6 and 7, 9 and 10 were vertically diluted with extract having final concentration of 1.25, 0.625, 0.313, 0.156, 0.078, 0.039, 0.019, and 0.009 mg/ml and named as treated (The final concentration of DMSO is equal to and less than 2.5% which does not interfere the growth of tested pathogens); column 5, 8, and 11 were taken as positive control and contains only 100 μl inocula and 100 μl of RPMI-1640 broth media; column 12 named as extract control and contains vertically diluted extract/drug in the aforementioned concentrations. To encounter the color of extract, optical density (O.D.) of extract control was subtracted from treated columns corresponds to extract treated (Pathak et al., 2015). Percent inhibition was calculated using following equation:

$$Perccent\ inhibition\ (\%) = \left((O.D.\ positive\ control - O.D.\ extract\ treated)/(O.D.\ positive\ control)\right) \times 100$$

Minimum Inhibition Concentrations (MICs) were calculated based on optical density recorded with a spectrophotometer (SpectraMax Plus384, Molecular Devices Corporation, USA) at 530 nm after 96 h incubated at 30 ± 2°C (Table 1).

The antifungal activity of lichen acetone extracts was performed along with Sertaconazole Nitrate BP (Glenmark Pharmaceuticals, Nasik, India) as standard chemical. 50 mg ml^{-1} of stock solution of sertaconazole was prepared and tested at same concentrations as lichen extracts.

MFC was determined by classic method for microbial MBC determination method but with slight modification. 20 µl from treated columns from above MIC wells were transferred into 7 ml tubes of fresh RPMI 1640 medium. Tubes were incubated at 30 ± 2°C for 4 weeks and checked for the turbidity. Aforementioned procedure was performed with sertaconazole nitrate BP and the concentration at which no turbidity has been achieved was defined as MFC (Veinović et al., 2013).

3.5. Statistical analysis
Independent sample *t*-test was performed between the control and treated dermatophytes for the measure of Levene's Test for Equality of Variances and *t*-test for equality of means via SPSS v20.

4. Conclusion
M. aurulenta exhibited antidermatophytic activity and provides new leads for the exploration of new compounds viz. secalonic acid A and leucotylic acid.

Acknowledgements
Thanks are due to Head, Department of Botany, University of Allahabad, Allahabad, India for providing library facilities; Dr D.K. Upreti, Scientist, National Botanical Research Institute-CSIR, Lucknow, India for determining the lichens; Dr G.P. Sinha, Head of Office, Central Regional Circle, Botanical Survey of India, Allahabad, India for providing accession numbers.

Funding
This work was financially supported by DST Fast Track Young Scientist Scheme [grant number SB/YS/LS-44/2014]; the CSIR-UGC Senior Research Fellowship Scheme.

Competing Interests
The authors declare no competing interest.

Author details
Ashutosh Pathak[1]
E-mail: ashupathaks@yahoo.in
ORCID ID: http://orcid.org/0000-0001-5404-1690
Rohit K. Mishra[2]
E-mail: rohit_ernet@yahoo.co.in
Shashi Kant Shukla[1]
E-mail: shashibplau@rediffmail.com
Rajesh Kumar[1]
E-mail: rajeshdubey.au@gmail.com
Madhu Pandey[1]
E-mail: madhubpl1989@gmail.com
Manisha Pandey[1]
E-mail: pandeymanisha81@gmail.com
Afifa Qidwai[1]
E-mail: afifaqidwai04@gmail.com
Anupam Dikshit[1]
E-mail: anupambplau@rediffmail.com
[1] Biological Product Lab, Department of Botany, University of Allahabad, Allahabad 211012, Uttar Pradesh, India.
[2] CMDR, Motilal Nehru National Institute of Technology, Allahabad 211004, Uttar Pradesh, India.

Cover image
Source: Authors.

References
Abdel-Rahman, S. M., Simon, S., Wright, K. J., Ndjountche, L., & Gaedigk, A. (2006). Tracking *Trichophyton tonsurans* through a large urban child care center: Defining infection prevalence and transmission patterns by molecular strain typing. *Pediatrics, 118*, 2365–2373. http://dx.doi.org/10.1542/peds.2006-2065
Awasthi, D. D. (2007). *A compendium of the macrolichens from India, Nepal and Sri Lanka.* Dehra Dun: Bishen Singh Mahendra Pal Singh. ISBN 978-81-211-0600-9.
Burzykowski, T., Molenberghs, G., Abeck, D., Haneke, E., Hay, R., Katsambas, A., ... Marynissen, G. (2003). High prevalence of foot diseases in Europe: Results of the Achilles Project. *Mycoses, 46*, 496–505. http://dx.doi.org/10.1046/j.0933-7407.2003.00933.x
Carrillo-Munoz, A. J., Tur-Tur, C., Cardenes, D. C., Estivill, D., & Giusiano, G. (2011). Sertaconazole nitrate shows fungicidal and fungistatic activities against *Trichophyton rubrum, Trichophyton mentagrophytes,* and *Epidermophyton floccosum,* causative agents of tinea pedis. *Antimicrobial Agents and Chemotherapy, 55*, 4420–4421. http://dx.doi.org/10.1128/AAC.00219-11
Grube, M., & Berg, G. (2009). Microbial consortia of bacteria and fungi with focus on the lichen symbiosis. *Fungal Biology Reviews, 23*, 72–85. http://dx.doi.org/10.1016/j.fbr.2009.10.001
Hawksworth, D. L. (2000). Freshwater and marine lichen-forming. In K. D. Hyde, W. H. Ho, & S. B. Pointing (Eds.), *Aquatic mycology across the millennium,* (Vol. 5, pp. 1–7). Hong Kong: Fungal Diversity.
Liebel, F., Lyte, P., Garay, M., Babad, J., & Southall, M. D. (2006). Anti-inflammatory and anti-itch activity of sertaconazole nitrate. *Archives of Dermatological Research, 298*, 191–199. http://dx.doi.org/10.1007/s00403-006-0679-8
Marijana, K., Branislav, R., & Slobodan, S. (2010). Antimicrobial activity of the lichen *Lecanora frustulosa* and *Parmeliopsis hyperopta* and their divaricatic acid and zeorin constituents. *African Journal of Microbiology Research, 4*, 885–890.
Molnar, K., & Farkas, E. (2010). Current results on biological activities of lichen secondary metabolites: A Review. *Zeitschrift für Naturforschung C, 65*, 157–173.
Orozco, A., Higginbotham, L., Hitchcock, C., Parkinson, T., Falconer, D., Ibrahim, A., ... Filler, S. G. (1998). Mechanism

of fluconazole resistance in *Candida krusei. Antimicrobial Agents and Chemotherapy, 42*, 2645–2649.

Pathak, A., Shukla, S. K., Pandey, A., Mishra, R. K., Kumar, R., & Dikshit, A. (2015). In vitro antibacterial activity of ethno medicinally used lichens against three wound infecting genera of enterobacteriaceae. *Proceedings of National Academy of Sciences India Section B.* doi:10.1007/ s40011-015-0540-y

Peres, N. T. A., Maranhão, F. C. A., Rossi, A., & Martinez-Rossi, N. M. (2010). Dermatophytes: Host-pathogen interaction and antifungal resistance. *Anais Brasileiros de Dermatologia, 85*, 657–667. doi:10.1590/S0365-05962010000500009

Rex, J. H., Alexander, B. D., Andes, D., Arthington-Skaggs, B., Brown, S. D., Chaturveli, V., ... Walsh, T. J. (2008). *Reference method for broth dilution antifungal susceptibility testing of filamentous fungi* (Approved Standard-2nd ed., Vol. 28(16)). Clinical and Laboratory Standard Institute (CLSI), M38-A2.

Santos, D. A., Barros, M. E. S., & Hamdan, J. S. (2006). Establishing a method of inoculum preparation for susceptibility testing of *Trichophyton rubrum* and *Trichophyton mentagrophytes. Journal of Clinical Microbiology, 44*, 98–101. http://dx.doi.org/10.1128/JCM.44.1.98-101.2006

Singh, K. P., & Sinha, G. P. (2010). *Indian lichens: An annotated checklist.* Kolkata: Botanical Survey of India. ISBN 978-81-8177-036-3.

Smith, K. J., Warnock, D. W., Kennedy, C. T. C., Johnson, E. M., Hopwood, V., van Cutsem, J., & Vanden Bossche, H. (1986). Azole resistance in *Candida albicans. Medical Mycology, 24*, 133–144. http://dx.doi.org/10.1080/02681218680000201

Stephenson, J. (1997). Investigators seeking new ways to stem rising tide of resistant fungi. *JAMA: The Journal of the American Medical Association, 277*, 5–6. http://dx.doi.org/10.1001/jama.1997.03540250013006

Turk, H., Yilmaz, M., Tay, T., Turk, A. O., & Kivanc, M. (2006). Antimicrobial activity of extracts of chemical races of the lichen *Pseudoevernia furfuracea* and their physodic acid, chloroatranorin, atranorin, and olivetoric acid constituents. *Zeitschrift für Naturforschung C, 61*, 499–507.

Veinović, G., Cerar, T., Strle, F., Lotrič-Furlan, S., Maraspin, V., Cimperman, J., & Ružić-Sabljić, E. (2013). *In vitro* susceptibility of European human *Borrelia burgdorferi* sensu stricto strains to antimicrobial agents. *International Journal of Antimicrobial Agents, 41*, 288–291. http://dx.doi.org/10.1016/j.ijantimicag.2012.11.016

White, T. C., Oliver, B. G., Graser, Y., & Henn, M. R. (2008). Generating and testing molecular hypotheses in the dermatophytes. *Eukaryotic Cell, 7*, 1238–1245. http://dx.doi.org/10.1128/EC.00100-08

Wingfield, A. B., Fernandez-Obregon, A. C., Wignall, F. S., & Greer, D. L. (2004). Treatment of tinea imbricata: A randomized clinical trial using griseofulvin, terbinafine, itraconazole and fluconazole. *British Journal of Dermatology, 150*, 119–126. http://dx.doi.org/10.1111/bjd.2004.150.issue-1

Yilmaz, M., Turk, A. O., Tay, T., & Kivanc, M. (2004). The antimicrobial activity of extracts of the lichen *Cladonia foliacea* and its (-)-Usnic acid, Atranorin, and Fumaroprotocetraric acid constituents. *Zeitschrift für Naturforschung C, 59*, 249–254.

2

In vitro allelopathic effects of extracts and fractions of five plants on tomato seed germination and vigor index

Komguem Tagne Pélagie Michelin[1], Aghofack-Nguemezi Jean[1], Gatsing Donatien[2], Lunga Paul Keilah[3], Lacmata Tamekou Stephen[2] and Kuiate Jules-Roger[2]*

*Corresponding author: Kuiate Jules-Roger, Laboratory of Microbiology and Antimicrobial Substances, Faculty of Science, Department of Biochemistry, University of Dschang, P.O. Box 67, Dschang, Cameroon

E-mail: jrkuiate@yahoo.com

Reviewing editor: Sabrina Sabatini, Sapienza University of Rome, Italy

Additional information is available at the end of the article

Abstract: In nature, allelopathic substances are likely to influence the germination of seeds. *Callistemon viminalis*, *Tephrosia vogelii*, *Senna spectabilis*, *Cupressus lusitanica*, and *Polyscias fulva* are found around some tomato culture fields in Cameroon. These plants may produce allelochemicals that can influence tomato seed germination and seedling characteristics. Methanol extracts and their fractions were prepared from leaves of these plants and tested on seed germination rate, shoot diameter, root and stem lengths, and vigor index of seedlings. The tested substances did not significantly ($p > 0.05$) affect the seed germination rate. Aqueous extracts of *T. vogelii*, *C. lusitanica*, and *C. viminalis* exerted the highest inhibition on seed germination while methanol extract, hexane, and ethyl acetate fractions of all the plants induced significant increases in diameters of seedlings. Aqueous extracts were effective on the stem length. All treatments generally reduced the root length. Methanol extracts induced significantly higher vigor indices ($p < 0.05$).

Subjects: Bioscience; Environment & Agriculture; Food Science & Technology

Keywords: Allelopathy; germination; vigor index; extract; fractions

ABOUT THE AUTHORS

Our main research domain is the physiology of growth and development of the plant, ripening and senescence of fruits. Events occurring in the field greatly influence these processes. The conventional cultivation methods of plant like tomato require the use of important quantities of chemical inputs, which can impair the nutritional virtues of fruits and harm the environment. Thus, an integrated management of the production of fruits with known health-promoting properties should involve pre-harvest treatments based on biological material. In this regard, many of our research activities in the past years have been focused on bioregulators/biostimulants, emerging notions that refer to any biological organism, its powder or extracts that can induce positive alterations in plant growth and development parameters. The use of allelochemicals in the cultivation of tomatoes is one of pragmatic approaches to reduce biotic and abiotic stresses effects, enhance growth and yield, and preserve fruit quality.

PUBLIC INTEREST STATEMENT

With increasing global population and decreasing arable soil superficies, achieving food security has become a challenge for most of governments. The major threats to crop production are weeds, insect pests, diseases, and drought. Unfavorable but unavoidable climatic changes are other important threats to food security. In order to face some of these challenges, there is use of great quantities of chemical inputs whose excessive use has negative impacts of the quality of agricultural commodities, environment, and human health. That is why aware and well-adviced consumers increasingly prefer produces which are not susceptible to negatively affect their health. Allelopathy has emerged as one of the sustainable and integrated approaches (e.g. crop rotations, cover crops, intercropping, use of powder, or extracts of biological material) to tackle these problems. The present findings, that some plant extracts led to the improvement in tomato seedling growth parameters, are in straight line with this approach.

1. Introduction

Tomato (*Solanum lycopersicum* L.) is one of the most produced and consumed fruits in world, and Cameroon in particular. It is a rich source of vitamins A and C, folic acid, α-lipoic acid, lycopene, choline, β-carotene, and lutein (Afzal et al., 2009). Its production in Cameroon has significantly increased, passing from 25,000 t in 1960 to 954,000 t in 2013 (FAOSTAT, 2013). However, this production is unequally distributed in the country, with the western highlands among the high production areas. According to Tonfack et al. (2009), tomato cultures in the western highlands of Cameroon produce very weak outputs despite the use of large quantities of fertilizers. It is known that the production yield of tomato and other plants depends on many factors including soil nature, pest management, the variety used, seed germination, and seedling problems. It can also be affected by neighboring plants through allelopathic effects.

Indeed, plants interfere with each other in their environments in different ways, through secondary metabolites released into the soil as exudates or resulting from the decomposition of plant residues (Kuzyakov, Hill, & Jones, 2007). Besides the classic competition for water, nutrients and light, it has been demonstrated that many plant species synthesize allelopathic molecules capable of inhibiting or stimulating the germination and growth of neighboring plants (Dayan, Romagni, & Duke, 2000). Allelopathy is then considered as an ecological and chemical interaction characterized by stimulatory or inhibitory effects among different plant families (Badmus & Afolayan, 2012). This is particularly evident in the context of integrated and organic agricultural production, as an alternative to chemical treatments against weeds that are currently scarce and very expensive, with the allelopathy phenomenon being highly selective.

In the west Region of Cameroon, tomato is very often grown together with other plant species such as *Senna spectabilis*, *Tephrosia vogelii*, *Polyscias fulva*, *Cupressus lusitanica* and *Callistemon viminalis*, some of which are ornamental and others used as fertilizers. Sometimes, the leaves of *S. spectabilis* or *T. vogelii* are wrinkled and spread on the soil before the sowing. The potential allelopathic effect of *C. lusitanica* (used to materialize the limits of a field) on seed germination, radicle, and seedling growth was investigated on chickpea, maize, pea, and teff (Lisanework & Michelsen, 1993). *Cupressus arizonica* extracts also showed an inhibitory effect on the seed germination and seedling growth of *Lolium perenne* and *Poa pratensis*. The author found that the inhibition rates of the two studied grasses increased with extract concentration (Aliloo, 2012). Bioassay showed that the aqueous extract from fresh leaves of *T. vogelii* significantly inhibited seed germination and seedling growth of *Festuca arundinacea*, *Cynodon dactylon*, and *Digitaria sanguinalis* (Ruilong et al., 2011).

Based on the above facts, we hypothesized that *C. lusitanica*, *S. spectabilis*, *T. vogelii*, *C. viminalis*, and *P. fulva* may contain substances that could affect the germination of tomato and subsequently plant vigor index in the field. This can be verified *in vitro* in a preliminary study before field investigation. Thus, this work was designed to investigate the *in vitro* allelopathic effects of extracts and their fractions from these five plant species on tomato seed germination and vigor.

2. Material and methods

2.1. Plant material

Plant material consisted of leaves from the following species: *C. viminalis*, *T. vogelii*, *S. spectabilis*, *C. lusitanica*, collected in January 2014 at the campus of the University of Dschang, as well as *P. fulva* collected in May 2014 in Bafou, all situated in the western administrative Region of Cameroon. The plant materials were identified at the Cameroon National Herbarium in Yaoundé where a voucher specimen was kept, respectively, under the reference numbers 49872/HNC, 43546/HNC, 45740/HNC, 35436/HNC, and 47801/HNC. The leaves of each plant species were dried in the shade for three weeks at 22 ± 2°C indoor. They were finely crushed in a mechanical mill and the resulting powders were used to prepare plant extracts.

Tomato seeds (Rio Grande, lot N° 58480, packaging date September 2013. Vikima Seed A/S from Denmark) respecting the EC Standard Norms, were purchased from Holland Farming Cmr Sarl, Cameroon. All experiments were conducted in a randomized complete block design. Treatments of each experiment were replicated three times and all experiments were repeated thrice. The initial seed germination count was carried out after two days of incubation. The final germination, percentage of inhibition, seedling diameter, and lengths were evaluated after 10 days.

2.2. Preparation of plant extracts and fractions
Aqueous extracts were prepared by macerating, respectively, 2.5, 1.25, and 0.625 g of plant powder in 100 ml of distilled water for 24 h. These extracts were then filtered using Whatman paper No. 1.

Methanolic extracts were prepared by macerating each plant powder (1,000 g) in 6 L methanol for two days, and then filtered using Whatman paper No. 1. The evaporation of solvent was carried out using a rotary evaporator (Buchi R-200) under vacuum at 40°C. The extracts obtained were placed in an oven at 40°C for 24 h to remove residual solvent.

The methanol extracts were successively and separately partitioned using n-hexane and ethyl acetate. For this, 80 g of each methanol extract were dissolved in 200 ml of methanol. To this solution, 200 ml of hexane were added and the mixture was gently shaken and the two phases were separated using separating funnel. The upper phase, the hexane fraction, was kept aside while to the lower portion, 200 ml of water and 200 ml of ethyl acetate were added. After shaking, the upper phase constituted the ethyl acetate fraction while the lower phase was the residual fraction. The solvents were evaporated under vacuum at 40°C to give the hexane fraction (Hex), the ethyl acetate fraction (EA), and the residual fraction (Res) (Figure 1).

2.3. Treatment and seed germination
Stock solutions of methanolic extracts and their fractions were prepared by dissolving separately 10 mg of each extract in 800 µl of Tween 80 (surfactant) and the total volume adjusted to 8 ml with sterile distilled water for a final concentration of 1.25 mg/ml. Tested concentrations (0.625, 0.312, and 0.156 mg/ml) were obtained by serial dilution of the stock solutions. Positive control consisted of a 0.025 µg/ml naphthalene acetic acid (NAA) from a stock solution of 10 µg/ml in sterile distilled water, while distilled water was used as negative control.

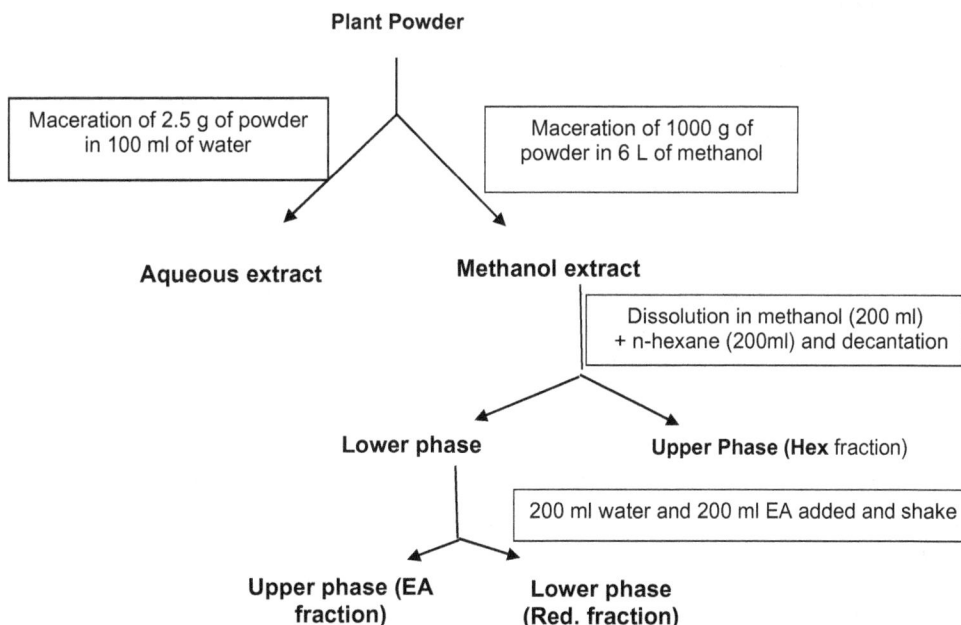

Figure 1. General procedure of plant extracts and fractions preparation.

The tomato seed germination experiment was carried out using 90 mm Petri dishes containing two layers of Whatman papers No. 2. Ten milliliters of tested solutions were used to moisten this paper. Seed germination was monitored by looking for a visible protrusion of the radicle from day 1 of experiment onwards. All sets of treatments and controls were prepared in triplicate of 10 seeds each and the experiment was repeated three times. The experiment was carried out at 25 ± 2°C and 12 h photoperiod. The germination percentage was calculated using the following formula:

$$\% \text{ Germination} = \frac{\text{Number of germinated seeds}}{\text{Total number of seeds sown}} \times 100$$

2.4. Evaluation of physiological growth parameters

The measurements of the diameter, length of stems and roots were made on the fifteenth day of culture. The diameter of the plants was measured using a micrometer brand. The stem and root length was measured using a straight edge in cm.

The vigor index was calculated by the formula of Patel and Manhad (2014).

Vigor index = Germination percentage × total length of the plant

The data from the repeated experiments were pooled before being subjected to ANOVA and treatment means separated by Duncan's Multiple Range Test at 5% probability level.

3. Results and discussion

3.1. Results

3.1.1. Effect of extract/fraction on the germination kinetics of tomato seeds
The kinetics of tomato seed germination in the presence or absence of plant extracts are presented in Figure 2. The extracts or fractions used did not have any effect on the time taken for the seeds to germinate. However, the positive control accelerated seed germination, with a germination time of 2 days, instead of 3 days observed in the negative control or extract/fraction-treated groups. Moreover, the percentage of germination generally increased with time up to day 6. The plant extracts and their fractions significantly ($p < 0.05$) and differentially affected the percentage of seed germination. The aqueous extracts of T. vogelii, C. lusitanica, and C. viminalis exerted the highest inhibition on seed germination. Globally, treated seeds responded differentially and gradually to germination. Forty percent of seeds quickly responded to the treatment with NAA and the number of germinated seeds increased with time up to 90%; with up to 10% of seed not germinating. Also, considering NAA and negative control, the germination pattern was the same in each experiment with up to 90 and 83% of seeds germinating, respectively.

3.1.2. Effects of plant extracts and fractions on the tomato seedling diameter
For all plant species, the methanol extract, hexane and ethyl acetate fractions induced significant increases in diameters of seedlings, while the negative control, aqueous extract and the residual fraction had comparable effects ($p > 0.05$) (Figure 3). However, the positive control significantly increased ($p < 0.05$) diameter of seedlings compared to all the above. There was a significant increase in the diameter of seedlings by the methanol extract and hexane fractions of C. lusitanica and S. spectabilis.

3.1.3. Effects of plant extracts and fractions on the tomato seedling stem lenght
The activation or inhibition effect of the plant extracts and fractions on the seedling stem length is illustrated in Figure 4. Generally, the aqueous extracts which could not increase the diameter of seedlings were the most effective on the stem length. Although the methanol extracts generally induced a reduction in the stem length of seedlings relative to the negative control, the treatment of tomato seedlings with ethyl acetate fractions of P. fulva, T. vogelii, and C. lusitanica and residual fraction of P. fulva and C. lusitanica residue fractions generally led to increases in the stem length of seedlings.

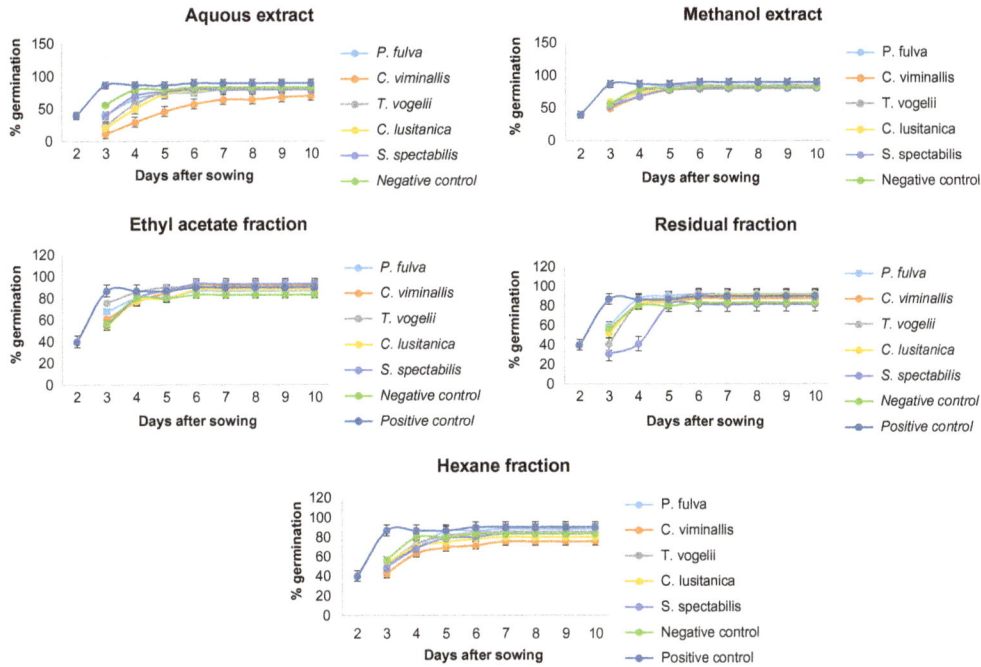

Figure 2. Germination kinetics of tomato seeds in presence and absence of plant extract.

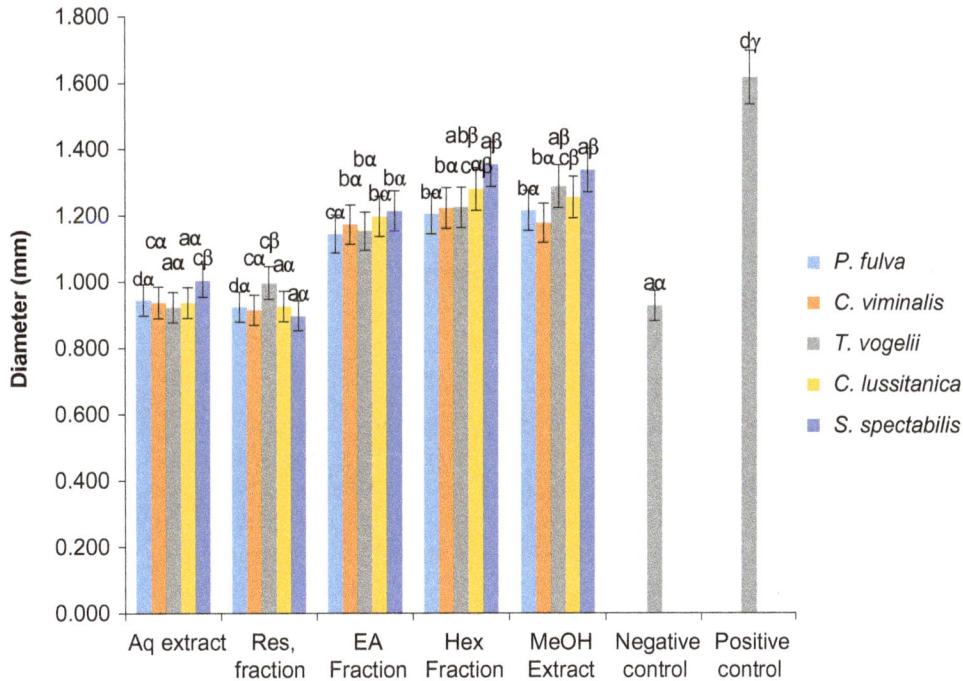

Figure 3. Effects of plants extracts and fractions on the tomato seedling diameter at the end of the experiment (10 days).

Notes: Aq: Aqueous; Res: Residual; EA: Ethyl acetate; Hex: Hexane. a, b, c: for the same plant, extracts or fractions carrying different letters induced significant differences in seedlings diameters (Waller-duncan test, $p < 0.05$). α, β, γ: for the same type of extracts or fractions, plant species carrying different letters induced significant differences in seedlings diameters (Waller-duncan test, $p < 0.05$).

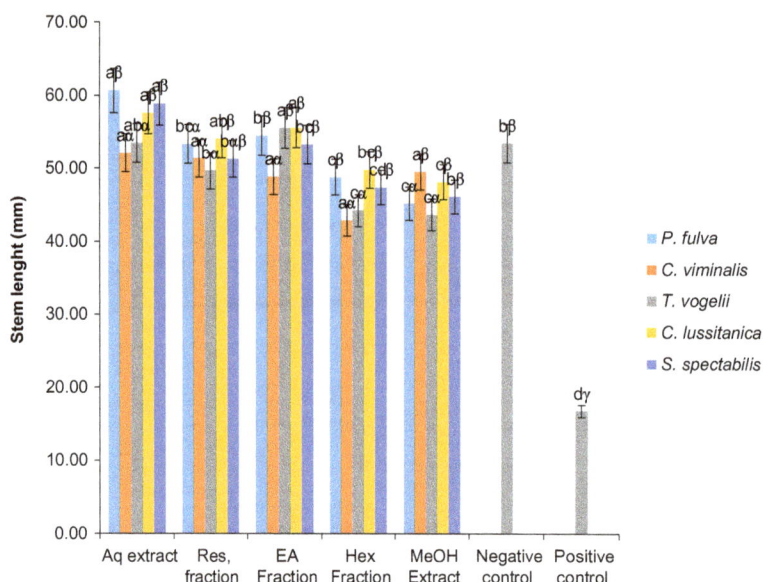

Figure 4. Effects of plants extracts and fractions on the tomato seedling plant stem lenght at the end of the experiment (10 days).

Notes: Aq: Aqueous; Res: Residual; EA: Ethyl acetate; Hex: Hexane. a, b, c: for the same plant, extracts or fractions carrying different letters induced significant differences in seedlings stem lenght (Waller-duncan test, $p < 0.05$). α, β, γ: for the same type of extracts or fractions, plant species carrying different letters induced significant differences in seedlings stem lenght (Waller-duncan test, $p < 0.05$).

3.1.4. Effect of extracts and fractions on the seedling root lengths
The effect of the plant extracts/fractions on the root length of tomato seedlings after germination is presented in Figure 5. Treatments generally reduced the root length compared to negative control. For the methanol extracts, only those from C. viminalis and S. spectabilis preserved the root length, to a dimension comparable to that of the negative control. Hexane fractions, especially from T. vogelii significantly led to increases in the root length, though to a value below that of the negative control. The aqueous extracts from T. vogelii had the strongest activity on root length, very close to the negative control and significantly ($p < 0.05$) higher than those of T. vogelii methanol extract and fractions, but comparable to the effect of the Hexane fraction. The methanol extract of T. vogelii induced the strongest reduction in the root length but less than the positive control which completely inhibited the formation of roots.

3.1.5. Effect of extracts and fractions on tomato seedling total length
The seedlings treated with the aqueous extracts of T. vogelii and S. spectabilis as well as the methanol extracts of C. viminalis and S. spectabilis did not show significant ($p > 0.05$) variation in total length compared to the negative control (Figure 6). The resulting seedling lengths were close to 140 mm, while those of seedlings treated with the other extracts were significantly reduced. The fractions from T. vogelii, P. fulva and C. lusitanica methanol extracts had reduced total seedlings length compared to the control.

3.1.6. Effects of plants extracts and fractions on tomato seedling vigor index
The plant extracts and fractions exerted variable effects on tomato seedling plant vigor index (Figure 7). Apart from T. vogelii, the vigor index of seedlings treated with methanol extracts were significantly ($p < 0.05$) higher than those treated with their corresponding aqueous extracts. The aqueous and methanol extracts from S. spectabilis were the most effective, inducing seedling vigor index of about 10,200 and 12,100 units, respectively. The methanol extract of S. spectabilis induced vigor index comparable to that of the negative control, while the rest of the extracts induced a reduction in the vigor index. Treatment of seedlings with ethyl acetate fractions led to significant increases in the vigor index.

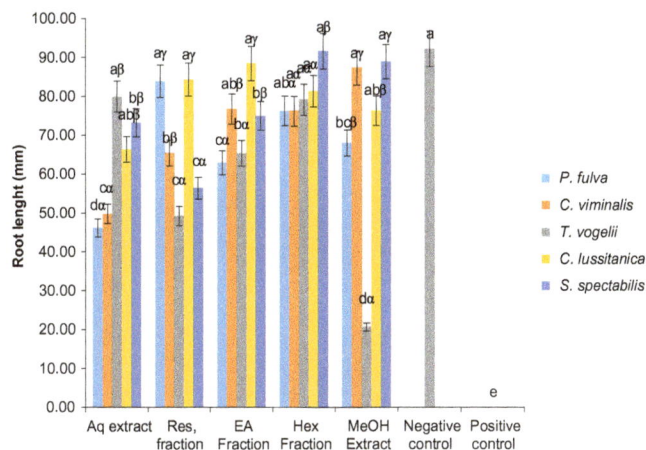

Figure 5. Effects of plants extracts and fractions on the tomato seeding plant root lenght at the end of the experiment (10 days).

Notes: Aq: Aqueous; Res: Residual; EA: Ethyl acetate; Hex: Hexane. a, b, c: for the same plant, extracts or fractions carrying different letters induced significant differences in seedlings root lenght (Waller-duncan test, $p < 0.05$). α, β, γ: for the same type of extracts or fractions, plant species carrying different letters induced significant differences in seedlings root lenght (Waller-duncan test, $p < 0.05$).

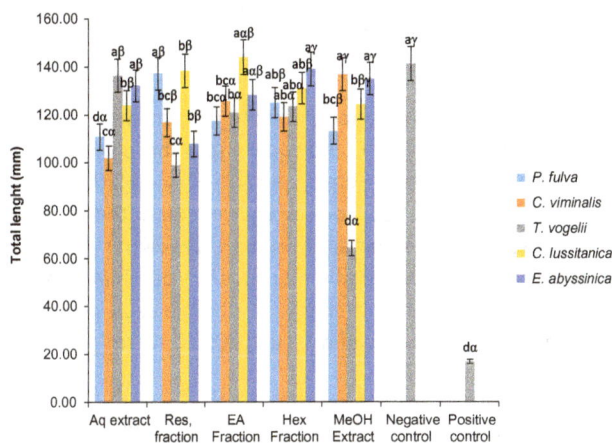

Figure 6. Effects of plants extracts and fractions on the tomato seedling plant total lenght at the end of the experiment (10 days).

Notes: Aq: Aqueous; Res: Residual; EA: Ethyl acetate; Hex: Hexane. a, b, c: for the same plant, extracts or fractions carrying different letters induced significant differences in seedlings total lenght (Waller-duncan test, $p < 0.05$). α, β, γ: for the same type of extracts or fractions, plant species carrying different letters induced significant differences in seedlings total lenght (Waller-duncan test, $p < 0.05$).

3.2. Discussion

Some plant extracts, particularly *C. viminalis* and *S. spectabilis* reduced the overall seed germination percentage indicating the presence of potential allelopatic inhibitory substances in such extracts. Reduction in germination is the most obvious allelopathic effect observed in bioassays (Alagesa Boopathi, 2011; Samreen, Hussain, & Sher, 2009) and in environment (ZhongQun, Junan, HaoRu, & Zhi, 2012). Thus, previous studies showed that different leaf aqueous extracts of *C. arizonica* showed significant inhibitory effects on germination percentage of tomato seeds (Aliloo, 2012). Likewise, germination efficiency of tomato seeds decreased after treatment with *Eucalyptus camaldulensis* extracts (Fikreyesus, Kebebew, Nebiyu, Zeleke, & Bogale, 2011).

In this study, solvent polarity effect was highly pronounced and the reduction in germination percentage was intensified with increasing extract polarity, suggesting an increase in active principles with polarity. The ethyl acetate fraction, less polar than aqueous and methanol fractions exhibited

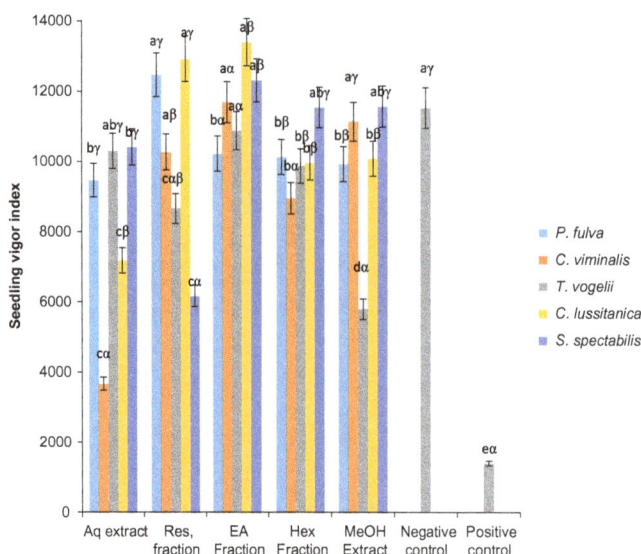

Figure 7. Effects of plants extracts and fractions on the tomato seedling plant vigor index at the end of the experiment (10 days).

Notes: Aq: Aqueous; Res: Residual; EA: Ethyl acetate; Hex: Hexane. a, b, c: for the same plant, extracts or fractions carrying different letters induced significant differences in seedlings vigor index (Waller-duncan test, $p < 0.05$). α, β, γ: for the same type of extracts or fractions, plant species carrying different letters induced significant differences in seedlings vigor index (Waller-duncan test, $p < 0.05$).

the highest stimulatory activities leading to a higher germination percentage than the negative control. Kiran, Lalitha, and Raveesha (2011) showed that extract obtained using polar solvent could have phytotoxic effect on tomato seeds while Al-Wakeel, Gabr, Hamid, and Abu-El-Soud (2007) suggested that it may possess allelochemicals with inhibitory effects on seed germination. This may be the case for the ethyl acetate fractions of the tested plants of this study, since it is known that allelochemicals present in some leaf extracts can prevent growth of embryo or cause its death (Anjana & Pramod, 2010). Different other mechanisms may explain the observed reduction in seed germination, including obstruction of water imbibition by the seeds (Tawaha & Turk, 2003), alteration in the synthesis or activities of gibberellic acid (Olofsdotter, 2001).

- The result on the seedling diameter indicates that stem-widening chemical compounds are concentrated in the hexane fractions, as well as in the ethyl acetate fractions to a lesser extent. Considering this active extract and its active fractions, *S. spectabilis*, *C. lusitanica*, and *T. vogelii* are more likely to induce such an effect. The same groups of compounds could be responsible for the stem-lengthening effects of the ethyl acetate fractions of *P. fulva*, *T. vogelii*, and *C. lusitanica* as well as *P. fulva* and *C. lusitanica* residue fractions. The increase in shoot length by these extracts/fractions corroborates the work of Perelló, Gruhlke, and Slusarenko (2013), who demonstrated the same effect for garlic juice treatment on wheat cultivar as a function of garlic juice concentration tested. In a previous study, aqueous leaf extracts of *C. lusitanica* was shown to significantly reduce both germination and radicle growth of chickpea, maize, pea, and teff (Lisanework & Michelsen, 1993).

Overall, no plant extract accelerated seedling growth in length since all the values were less than or comparable to that of the negative control. Apart from the aqueous extract of *T. vogelii* and the MeOH extracts of *C. viminalis* and *S. spectabilis*, the rest of the extracts may possess plant length-inhibiting factors. Distilled water and methanol extracts are expected to contain polar substances, known to reduce the growth of seedling by inhibiting cell division or cell enlargement in the meristematic areas of shoot and root of plants and may also interfere with the rearrangement of microtubules during cell division (Singh, Pandey, & Singh, 2009). Consequently, these extracts, together with their residual fractions, caused the plants to have shorter shoots and roots, and thinner diameters. This may explain why the hexane fractions containing non-polar substances were generally the best fractions regarding increase in shoot and root lengths, and diameters. These fractions therefore have non polar growth-accelerating compounds (Sornchai, Saithong, Srichompoo, Unartngam, & Iamtham, 2014), probably in appreciable amounts to provoke visible length increase.

All the extracts, apart from the MeOH extract of *S. spectabilis* which induced comparable vigor index to that of the negative control, induced a reduction in the vigor indices. This shows that the vigor index-reducing extracts may contain inhibiting factors. It could be observed that most of the vigor index inhibitors were distributed, as a function of the plant extract, in the residue and hexane fraction, while the ethyl acetate fraction contained mostly vigor index activators.

4. Conclusion

The overall results of the present study show that the MeOH extracts of *C. viminalis*, *T. vogelii*, *S. spectabilis*, *C. lusitanica*, and *P. fulva*, as well as their ethyl acetate fractions, might contain compounds that can differentially affect positively or negatively tomato seed germination as well as seedling length and vigor index. Extract fractions could also contain both stimulatory and inhibitory allelochemicals, as related to the parameters measured. The stimulatory allelochemicals present in the ethyl acetate extract may have good biotechnological application potential in agriculture. It would be very interesting to identify such compounds for further investigations.

Funding
The authors received no direct funding for this research.

Competing Interest
The authors declare no competing interest.

Author details
Komguem Tagne Pélagie Michelin[1]
E-mail: komguemprm@yahoo.com
Aghofack-Nguemezi Jean[1]
E-mail: aghofack@yahoo.fr
Gatsing Donatien[2]
E-mail: gatsingd@yahoo.com
Lunga Paul Keilah[3]
E-mail: lungapaul@yahoo.ca
Lacmata Tamekou Stephen[2]
E-mail: lacmatastephen@yahoo.fr
Kuiate Jules-Roger[2]
E-mail: jrkuiate@yahoo.com
[1] Laboratoire de Botanique Appliquée, Faculté des Sciences, Université de Dschang, BP 67, Dschang, Cameroon.
[2] Laboratory of Microbiology and Antimicrobial Substances, Faculty of Science, Department of Biochemistry, University of Dschang, P.O. Box 67, Dschang, Cameroon.
[3] Laboratory of Phytobiochemistry and Medicinal Plants Study, Faculty of Science, Department of Biochemistry, University of Yaoundé 1, P.O. Box 812, Yaoundé, Cameroon.

Cover image
Source: Authors.

References
Afzal, I., Munir, F., Ayub, C., Basra, S., Hameed, A., & Nawaz, A. (2009). Changes in antioxidant enzymes, germination capacity and vigour of tomato seeds in response of priming with polyamines. *Seed Science and Technology, 37*, 765–770. http://dx.doi.org/10.15258/sst

Alagesa Boopathi, C. (2011). Allelopathic effects of *Andrographis paniculata* Nees on germination of *Sesamum indicum* L. *Asian Journal of Experimental Biological Sciences, 2*, 147–150.

Aliloo, A. (2012). Allelopathic potentials of Cupressus arizonica leaves extracts on seed germination and seedling growth of Lolium perenne and Poa pratensis. *International Journal of Agriculture and Crop Sciences, 4*, 1371–1375.

Al-Wakeel, S. A. M., Gabr, M. A., Hamid, A. A., & Abu-El-Soud, W. M. (2007). Allelopathic effects of *Acacia nilotica* leaf residue on *Pisum sativum* L. *Allelopathy Journal, 19*, 411–422.

Anjana, D., & Pramod, K. J. (2010). Seed Germination responses of the medicinal herb *Centella asiatica*. *Brazilian Journal of Plant Physiology, 22*, 143–150.

Badmus, A., & Afolayan, A. (2012). Allelopathic potential of Arctotis arctotoides (L.f.) O. Hoffm aqueous extracts on the germination and seedling growth of some vegetables. *African Journal of Biotechnology, 11*, 10711–10716.

Dayan, F. E., Romagni, J. G., & Duke, S. O. (2000). Investigating the mode of action of natural phytotoxins. *Journal of Chemical Ecology, 26*, 2079–2094. http://dx.doi.org/10.1023/A:1005512331061

FAOSTAT. (2013). *Annuaires statistiques de la FAO – L'alimentation et l'agriculture dans le monde.* Retrieved consulted on the 18 March 2016 from http://www. factfish.com/statistic-country/cameroon/tomatoes,+production+quantity

Fikreyesus, S., Kebebew, Z., Nebiyu, A., Zeleke, N., & Bogale, S. (2011). Allelopathic Effects of *Eucalyptus camaldulensis* dehnh. on germination and growth of tomato. *American-Eurasian Journal of Agricultural & Environmental, 11*, 600–608.

Kiran, B., Lalitha, V., & Raveesha, K. A. (2011). *In vitro* evaluation of aqueous seed extract of *Psoralea corylifolia* L. on seed germination and seedling vigor of maize. *Journal of Applied Pharmaceutical Science, 1*, 128–130.

Kuzyakov, Y., Hill, P. W., & Jones, D. L. (2007). Root exudate components change litter decomposition in a simulated rhizosphere depending on temperature. *Plant Soil, 290*, 293–305. http://dx.doi.org/10.1007/s11104-006-9162-8

Lisanework, N., & Michelsen, A. (1993). Allelopathy in agroforestry systems: The effects of leaf extracts of Cupressus lusitanica and three *Eucalyptus* spp. on four Ethiopian crops. *Agroforestry Systems, 21*, 63–74. http://dx.doi.org/10.1007/BF00704926

Olofsdotter, M. (2001). Rice—A step toward use of allelopathy. *Agronomy Journal, 93*, 3–8. http://dx.doi.org/10.2134/agronj2001.9313

Patel, R. G., & Manhad, A. U. (2014). Effect of Gibberillins on seed germination of *Tithonia rotundifolia* Blake. *International Journal of Innovative Research in Science, Engineering and Technology, 3*, 10680–10684.

Perelló, A., Gruhlke, M., & Slusarenko, A. (2013). Effect of garlic extract on seed germination, seedling health, and vigor of pathogen-infested wheat. *Journal of Pharmacy Practice and Research, 53*, 317–323.

Ruilong, W., Yang, X., Song, Y., Zhang, M., Su, Y., & Rensen, Z. (2011). Allelopathic potential of *Tephrosia vogelii* Hook. f.: Laboratory and field evaluation. *Allelopathy Journal, 28*, 53–62.

Samreen, U., Hussain, F., & Sher, Z. (2009). Allelopathic potential of *Calotropis procera* Ait. *Pakistan Journal of Plant Science, 15*, 7–14.

Singh, N. B., Pandey, B. N., & Singh, A. (2009). Allelopathic effects of *Cyperus rotundus* extract *in vitro* and ex-vitro on banana. *Acta Physiologiae Plantarum, 31*, 633–638. http://dx.doi.org/10.1007/s11738-009-0274-7

Sornchai, A., Saithong, N., Srichompoo, Y., Unartngam, A., & Iamtham, S. (2014). Effect of *Spirulina maxima* aqueous extract on seed germination and seedling growth of mung bean, *Vignaradiate* and rice, *Oryza sativa* var. Japonica.. *Journal of the International Society for Southeast Asian Agricultural Sciences, 20*, 77–84.

Tawaha, A., & Turk, M. (2003). Allelopathic effects of black mustard (Brassica nigra) on germination and growth of wild barley (Hordeum spontaneum). *Journal of Agronomy and Crop Science, 189*, 298–303. http://dx.doi.org/10.1046/j.1439-037X.2003.00047.x

Tonfack, L. B., Bernadac, A., Youmbi, E., Mbouapouognigni, V. P., Ngueguim, M., & Akoa, A. (2009). Impact of organic and inorganic fertilizers on tomato vigor, yield and fruit composition under tropical andosol soil conditions. *Fruits, 64*, 167–177. http://dx.doi.org/10.1051/fruits/2009012

ZhongQun, H., Junan, Z., HaoRu, T., & Zhi, H. (2012). Different vegetables crops in response to Allelopathic of hot pepper root exudates. *World Applied Sciences Journal, 19*, 1289–1294.

3

Comparing cork quality from Hafir-Zarieffet mountain forest (Tlemcen, Algeria) vs. Tagus basin *Montado* (Benavente, Portugal)

Amina Ghalem[1], Inês Barbosa[2], Rachid Tarik Bouhraoua[1] and Augusta Costa[2,3]*

*Corresponding author: Augusta Costa, Center for Environmental and Sustainability Research (CENSE), Nova University of Lisbon, Campus de Caparica, 2829–516 Caparica, Portugal; Instituto Nacional de Investigação Agrária e Veterinária, I.P., Quinta do Marquês, Av. da República, 2780–159, Oeiras, Portugal

E-mails: augusta.costa@iniav.pt, augusta.costa.sousa@gmail.com

Reviewing editior: Raffaele Dello Ioio, Universita degli Studi di Roma La Sapienza, Italy

Additional information is available at the end of the article

Abstract: In the southwestern Mediterranean Basin, cork oaks (*Quercus suber* L.) are periodically harvested for their cork. This natural product is valued by its homogeneity which heightens the importance of characterizing cork tissue discontinuities, or cork pores. Cork porosity profile in natural cork planks has been reported to be affected by forest management practices but, so far, has been scarcely addressed. We characterize the cork porosity profile in two contrasting cork oak woodland; at a mountain forest, in Western Algeria (absence of forest management) and at a peneplain "montado," in southern Portugal (intensively managed toward the optimization of cork production). Image analysis techniques were applied on transverse sections of more than 40 cork samples from both woodland, and a stepwise discriminant analysis was used to discriminate between the cork pore features datasets. Cork porosity profiles were similar between regions but; in the cork samples from Algeria, cork pores were having higher values for linear dimensions of pores (length and perimeter) and contrasting shape values (roundness) which depreciate cork quality, when compared to the cork samples from Portugal. However, improved woodland management strategies at Algeria should ensure adequate cork homogeneity and suitability for more valuable cork products.

ABOUT THE AUTHORS

The authors were motivated by the spatial variability of cork oak woodland across Mediterranean environments, the implications in the cork yield, and cork quality and, consequently, in the economic and ecological sustainability of these sensitive forest ecosystems. Authors used image analysis techniques to study cork porosity in cork planks, in a similar way that cork industry detects cork porosity in cork products such as natural cork stoppers and disks, the most valuable cork product in the cork industrial processing.

This study will be one first original report from a young research group, integrating young MSc and PhD students from the Center for Environmental and Sustainability Research (NOVA University of Lisbon, Portugal) and Tlemcen University (Algeria) (http://www.augustacosta.net/people.html) addressing a comprehensive understanding on cork planks quality and cork oak woodland management, working at INIAV, I.P.

PUBLIC INTEREST STATEMENT

Cork is the outer bark of cork oak trees; one natural product generating US $2 billion annually, and, currently, the sixth most valuable global non-timber forest product. Cork oaks are strictly distributed in the western Mediterranean Basin, between the southwestern Europe and northern Africa. An increasingly important challenge for the later region is to apply adequate forest management practices toward the optimization of high-quality cork production. In their research, authors assessed cork quality of two contrasting cork-producing regions: Portugal, the world leader in cork production; and Algeria, one potential cork-producing region. Results showed strong similarities between cork quality profiles, but higher cork quality heterogeneity at Algeria. Clearly, at this region, adequate forest management practices such as thinning (for selecting the best cork-producing trees) or pruning (for optimizing trees' cork-harvesting surface), and appropriate cork-harvesting cycles should improve the cork yield and quality, at medium term.

Subjects: Environment & Economics; Forestry; Resource Management–Environmental Studies

Keywords: *Quercus suber* **L.; cork oak; Mediterranean Basin; cork porosity; image analysis; lenticular channels; stepwise discriminant analysis**

1. Introduction

Cork oak (*Quercus suber* L.) is a strictly Mediterranean species distributed in the western Mediterranean Basin, between the southwestern Europe and northern Africa, a climatic, ecological, and socioeconomical sensitive region (Bugalho, Caldeira, Pereira, Aronson, & Pausas, 2011; Costa, Madeira, Santos, & Plieninger, 2014; Giorgi, 2006). Cork oak woodland are considered keystone ecosystems (Vicente & Alés, 2006), while enhance other important provisioning/regulating ecosystem services (Plieninger, van der Horst, Schleyer, & Bieling, 2014). Moreover, cork oak woodland are important biodiversity hotspots where cork oak is harvested throughout their lifetime for their bark (Oliveira & Costa, 2012; Ticktin, 2004), the cork, one valuable global non-timber forest product (FAO, 2013).

In Portugal and Algeria, the cork oak woodland ecosystems are under high human pressure; in Portugal, these woodland are iconic examples of domesticated nature featuring savannah-like ecosystems carefully managed for a sustainable cork yield (Costa & Oliveira, 2015; Costa et al., 2014). In Algeria, cork oak woodland are disturbed forest ecosystems with strong, political, economical, and technological underlying factors, leading to the depreciation of cork value in relation to other tree's products such as acorn or (fire)wood (Messaoudène & Merouan, 2009). Despite the differences found between cork oak woodland conditions and management in Portugal and Algeria, a common increasingly important challenge for both regions is maintaining the trees producing high-quality cork. Cork quality depends on the homogeneity of the cork tissue, and on the presence of cork tissue discontinuities, pores or lenticular channels crossing radially the transverse sections of cork planks (Costa & Pereira, 2010).

Given the importance of the quality grading of cork planks in the industrial processing, namely for the production of natural disks and stoppers (Costa & Pereira, 2006; Pereira, 2007), the cork quality has long being determined in raw cork material by visual inspection (Costa & Pereira, 2010). In the surface of the transverse sections of cork planks, the high porosity variability in-between the transverse section of the cork samples have been reported in technical studies in industries, and a strong radial variability from the inner part (with lower porosity) to the outer part (with higher porosity) has been generally accepted. However, it is urgently needed a more comprehensive understanding on the extent of lenticular channels, i.e. on the cork porosity variability profile in the cork plank transverse sections, so far, scarcely addressed.

In this study, the goal is to reinforce the existent knowledge on the cork quality profile of cork planks by a large-scale assessment of cork yield in the Mediterranean area, including those of areas of northern Africa. In fact, given the ever-increasing demand for cork raw material, and with the Portuguese cork industry leading the world market with about a 65% share (APCOR, 2015), all the cork oak woodland areas should constitute potential cork-producing areas, contributing to the high cork-value chain.

We propose to use image vision inspection systems currently used in the industry for stoppers and disks quality classification, to identify and quantify the lenticular channels, in transverse sections of cork planks. The goal is to fill the existent knowledge gaps on the porosity profile of cork by generating comparable data on cork porosity in the cork plank's transverse sections of the two Mediterranean Basin cork-producing regions, Portugal and Algeria. Similar porosity (i.e. cork tissue discontinuities attributed to lenticular channels) features will be addressed and we hypothesized that in both regions the cork porosity variability profile will be similar, with larger and more porosity in outer part of the transverse section of cork planks. Moreover, it is hypothesized that appropriate forest management practices should improve cork quality, given the presence of porosity features, at each study area.

2. Material and methods

2.1. Study areas

Cork sampling was made during the cork-harvesting season, in the summer of 2010, at two study areas in the western Mediterranena Basin: at the cork oak woodland of Hafir-Zarieffet (ZA) (34°50′ N, 1°23′ W), at Tlemcen Mountain, in northwestern Algeria, close to the Mediterranean Sea and; at the cork oak woodland of Benavente (CL) (38°04′ N, 8°40′ W), in the Tagus Basin peneplain, in south-western Portugal, close to the Atlantic Ocean (Supplementary material, Figure S1).

The climate in both study areas is of Mediterranean type, smoothed by the influence of the Mediterranean Sea (at ZA) and the Atlantic Ocean (at CL). Mean annual temperature and annual rainfall are 13.5°C and 654 mm, and 15.3°C and 577 mm at ZA and CL, respectively (Figure 1). The mean temperature of the coldest month (January) is 3.5 and 5.0°C and for the hottest month (July) is 30.7 and 28.7°C, for ZA and CL, respectively. The highest temperatures are in summer, when precipitation is the lowest, and a dry period ($p < 2T$) generally occurs extending from May to September (Figure 1).

The study areas are within the potential vegetation area for cork oak, corresponding to the meso-Mediterranean (ZA) and thermo-Mediterranean (CL) thermotypes and to the sub-humid ombrotype (Capelo et al., 2007; Dehane, 2012).

At ZA, the elevation ranges from 1,000 to 1,220 m a.s.l. and the landscape is steeply undulated (with slopes ranging between 3 and 12%). The dominant Jurassic formations are related to the predominant shallow (less than 30 cm depth) carbonated soils, from sandy loam to sandy clay. At CL, Pliocene formations are dominant, occupying the flat or gently undulating (with slopes lower than 5%) areas adjacent to the Tagus river alluvial area. Mostly related to the nature of these geological formations, the predominant soils are deep sandy soils. In both study areas, ZA and CL, the soils are poor in nutrient and organic matter, and with low water storage capacity (Costa, Madeira, & Oliveira, 2008; Dehane, 2012).

At ZA, the cork oak woodland are mainly open woodland and scattered-tree woodland, with the occurrence of thorny trees due to fire recurrence (Letreuch-Belarouci, Letreuch-Belarouci, Benabdeli, & Medjahdi, 2009) and overgrazing. Exploited tree's density is under 50 trees ha^{-1}, and trees are old and poorly managed, within heterogeneous cork-production cycles, extending largely from typical Iberia's 9 to 12-years. Cork oak management oriented to the cork production management is lacking.

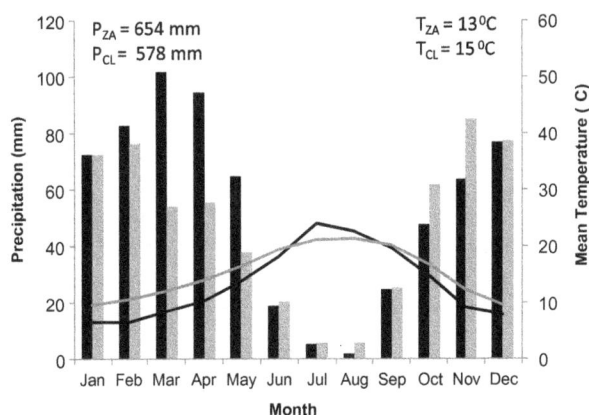

Figure 1. Climatic diagrams for the study areas, Hafir-Zarieffet (in black) and Benavente (in gray).

Notes: Data from the meteorological stations of Mefrouche (Hafir-Zarieffet, ZA) 34°51′ N, 1°16′ W (1961–1990 for the precipitation and 1975–1990 for temperature) and Salvaterra de Magos (Benavente, CL) 39°02′ N, 8°44′ W (1961–1990). Precipitation (in bars) and mean temperature (in lines).

At CL, the cork oak forest management practices, from planting or seeding to pruning, thinning, and cork harvesting, all are oriented toward optimizing cork productivity (Costa, Oliveira, Vidas, & Borges, 2010; Costa et al., 2008). The cork oak woodland are uneven-aged and resulted of assist natural regeneration complemented with local artificial seeding and plantation. Trees are pruned in juvenile ages for maximize harvesting surface, before the beginning of the 9-year cork production cycles, at the 20–25 years of age. Thinning occurs regularly, only to eliminate dead trees. The mean tree density was 86 trees ha^{-1} (with 62 trees ha^{-1} under exploitation) and an average annual cork production of 110 kg ha^{-1} year^{-1} was to be expected (Costa et al., 2010). The cork productivity at CL, Portugal, 110 kg ha^{-1} year^{-1}, is higher than at ZA, Algeria with 33 kg ha^{-1} year^{-1} but mean annual cork growth is similar 2.9 and 3.1 mm year^{-1}, respectively.

2.2. Image acquisition and cork quality assessment

A total of 41 cork samples were randomly selected in the two study areas. The cork samples, with dimensions of 10×10 cm^2, were prepared for image acquisition. The preparation included boiling in water at 100°C for 1 h, and dried in open air until equilibrium (minimum one week), and an optical quality surface finishing: the cutting and the sanding of a plan surfaces in the transversal section in order that cork pores and cork annual rings were clearly visible (Costa, Nunes, Graça, & Spiecker, 2015). The image acquisition of the transversal sections was made through snapshot images scanned at a resolution of 300 dpi and stored in TIF graphic format (Supplementary material, Figure S2). Images were then analyzed using ImageProPlus® image-processing software.

Cork quality assessment was made at the transversal section within a defined rectangular area of interest (AOI). A set of 13 pore variables at cork sample level was selected for cork quality assessment (Table 1). Pore data were filtered out by pore area, only pores with an area equal or superior to 0.5 mm^2 were kept for analysis as small porosity is functionally and esthetically irrelevant and only brings higher variance and variability to the cork sample (Costa & Pereira, 2007).

The exploratory analysis of cork porosity was made in the transverse section of the cork sample at an AOI-level; firstly, considering the AOI and as a whole and then considering the inner part of the AOI (one third of the AOI transect length, from cork belly to cork back); the middle AOI (second third); the external AOI (one third of the AOI length). At these subsections (inner, middle, and external AOI) of the cork sample transverse section, the found cork porosity will be therefore comparable and independent on the cork age, cork thickness, and cork production cycles.

2.3. Statistical analysis

Stepwise discriminant analysis (SDA) was used to investigate differences between the cork-producing regions, ZA and CL, the categorical dependent variable, and the cork porosity features, the independent pore-level metrics variables (Costa & Pereira, 2006; Gonzalez-Adrados, Lopes, & Pereira, 2000). The stepwise discriminant functions were evaluated using the Wilk's λ value: the smaller the Wilk's λ value, the more important is the independent variable or feature to the canonical discriminant function. The purpose of using SDA was to find the pore features that contribute the most to distinguish these regions (SPSS vs. 21.0). This statistical analysis creates a new (reduced) set of variables, called canonical discriminant functions, each one a linear function of the original independent (predictor) variables, in this case the cork porosity at pore level. These discriminant functions would account for most of the variance and would define the maximum possible difference between both regions. The relative importance of the predictor variables is evaluated through their correlation (Pearson coefficients) with each discriminant function.

Table 1. Pore variables measured in the transverse section of the cork samples at both regions

Pore variables	Acronym	Description	Range	P_o[1]	C_o[2]
Dimension					
Area	a	Area of pore (mm²)	≥0	x	–
Perimeter	pe	Perimeter of pore (mm)	≥0	x	–
Diameter	di	Diameter of pore (mm)	≥0	x	–
Length	le	Length of pore (mm)	≥0	x	–
Width	wi	Width of pore (mm)	≥0	x	–
Mean area	Mpa	Average pore area (mm²)	≥0	–	x
Maximum area	Maxpa	Maximum pore area (mm²)	≥0	–	x
Mean perimeter	Mpe	Average pore perimeter (mm)	≥0	–	x
Mean diameter	Mdi	Average pore diameter (mm)	≥0	–	x
Mean length	Mle	Average pore length (mm)	≥0	–	x
Mean width	Mwi	Average pore width (mm)	≥0	–	x
Shape					
Shape factor	sh	Perimeter-area ratio by adjusting for a circle standard	≥1	x	–
Aspect ratio	as	Ratio between width and length of a bouding pore rectangle	≥2	x	–
Fractal dimension	fr	Pore perimeter increase per unit increase in pore area	0–2	x	–
Roundness	ro	Ratio between the largest and smallest equivalent pore ellipse	≥1	x	–
Mean shape factor	Msp	Mean pore shape index	≥1	–	x
Mean aspect ratio	Mas	Mean pore aspect ratio	≥2	–	x
Mean fractal dimension	Mfr	Mean pore fractal dimension	0–2	–	x
Mean roundness	Mro	Mean pore roundness	≥1	–	x
Concentration					
Number	np	Total number of pores per 100 mm² (#)	≥1	–	x
Mean nearest neighbor distance	Menn	Average distance to the nearest neighboring	>0	–	x
Area fraction	af	Area-edge-to-edge squared distance ratio	≥0	–	x

[1]P_o—pore-level variable.
[2]C_o—cork plank-level variable.

3. Results

3.1. Cork porosity
The porosity profile of the transversal sections of the cork planks from Hafir-Zarieffet (ZA) and from Benavente (CL) showed similarities in the range of the values variation for all the selected features (Table 2). Regarding the dimension features, the mean pore area (Mpa) ranged between 4.5 mm² (at ZA) and 5.3 mm² (at CL). The mean values for maximum pore area (Maxpa) ranged between 29.2 mm² (at ZA) and 47.6 mm² (at CL). At the later region, the Maxpa was almost the double, confirming the relatively higher Mpa found in CL.

Lenticular channels are the dominant cork tissue discontinuities, mostly linear shape objects crossing radially the transverse section of the cork planks. Shape features such as aspect ratio, fractal dimension, or roundness showed compatible values in both regions, CL and ZA (Table 2). The mean fractal dimension was 1.1 in average in both regions and the roundness and aspect ratio mean values at ZA, respectively, 8.9 and 6.2, were superior to those found at CL, respectively, 4.6 and

Table 2. Cork porosity of the transverse section of cork samples

Variables	Hafir-Zarieffet (ZA)					Benavente (CL)				
	AOI	Outer cork	Middle cork	Inner cork	Yearly	AOI	Outer cork	Middle cork	Inner cork	Yearly
Dimension										
Mpa	4.5 ± 4.02	3.2 ± 2.55	5.9 ± 7.30	5.6 ± 5.62	3.5 ± 5.17	5.3 ± 3.17	3.7 ± 2.67	17.7 ± 26.12	4.0 ± 1.64	3.1 ± 3.17
Maxpa	29.2 ± 37.70	9.6 ± 10.24	19.9 ± 30.39	20.4 ± 34.16	8.8 ± 16.08	47.6 ± 45.52	16.8 ± 12.57	39.7 ± 48.57	17.2 ± 9.40	11.4 ± 15.94
Mpe	14.6 ± 6.01	10.7 ± 3.39	19.0 ± 14.91	16.7 ± 8.35	11.1 ± 6.46	14.2 ± 4.42	11.1 ± 4.35	32.8 ± 35.47	12.6 ± 3.37	9.2 ± 4.62
Mdi	2.2 ± 0.74	1.8 ± 0.54	2.5 ± 1.43	2.3 ± 1.08	1.9 ± 0.90	2.1 ± 0.36	2.0 ± 0.57	3.3 ± 1.69	2.0 ± 0.46	1.7 ± 0.53
Mle	6.2 ± 2.65	4.5 ± 1.44	8.5 ± 6.75	7.3 ± 3.56	4.4 ± 2.04	4.9 ± 1.01	4.1 ± 1.41	9.9 ± 7.90	4.4 ± 1.45	3.1 ± 1.64
Mwi	1.1 ± 0.47	1.1 ± 0.44	1.0 ± 0.58	0.9 ± 0.56	1.0 ± 0.80	1.5 ± 0.38	1.4 ± 0.47	2.4 ± 1.75	1.5 ± 0.40	1.2 ± 0.21
Shape										
Msp	37.5 ± 29.05	22.0 ± 14.30	33.4 ± 26.08	34.0 ± 25.89	22.9 ± 20.02	21.5 ± 7.76	18.9 ± 13.68	40.9 ± 35.54	16.8 ± 6.49	15.3 ± 12.75
Mas	6.2 ± 2.24	4.7 ± 1.27	7.8 ± 3.37	6.9 ± 4.11	5.8 ± 2.68	4.7 ± 1.15	4.2 ± 1.71	7.4 ± 4.09	4.5 ± 1.22	3.4 ± 1.25
Mfr	1.1 ± 0.02	1.1 ± 0.02	1.1 ± 0.03	1.0 ± 0.36	1.1 ± 0.02	1.1 ± 0.02	1.1 ± 0.02	1.1 ± 0.04	1.1 ± 0.03	1.1 ± 0.02
Mro	8.9 ± 3.60	6.6 ± 2.25	12.5 ± 6.53	9.9 ± 5.62	7.8 ± 4.92	4.6 ± 1.63	4.6 ± 2.27	6.3 ± 2.74	4.1 ± 1.74	3.0 ± 1.27
Concentration										
np	280 ± 116	468 ± 299	165 ± 87	188 ± 135	334 ± 188	250 ± 79	330 ± 149	140 ± 61	230 ± 100	420 ± 143
Menn	3.2 ± 0.78	2.6 ± 0.66	4.3 ± 1.12	3.5 ± 1.69	3.1 ± 0.97	3.3 ± 0.47	2.9 ± 0.58	4.6 ± 1.21	3.7 ± 1.17	2.5 ± 0.42
af	9.2 ± 4.48	11.4 ± 5.84	7.7 ± 8.96	9.2 ± 7.26	7.6 ± 7.45	13.2 ± 8.23	13.9 ± 14.04	16.6 ± 19.69	9.3 ± 5.23	12.7 ± 11.65

Notes: Mean values (mean ± standard deviation) for the study areas, Hafir-Zarieffet (ZA) and Benavente (CL). Range of mean values for all the AOI and along a virtual radial transect across the transverse section, from the outer (older) to the inner (younger) layers of the cork plank.

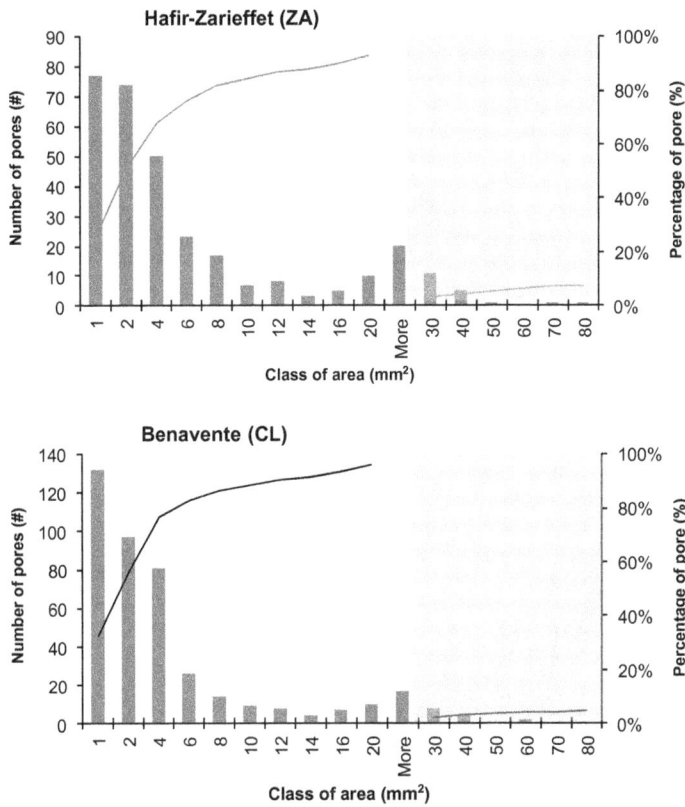

Figure 2. Histogram for the pore area feature in all cork samples from Hafir-Zarieffet (ZA) and Benavente (CL).

Note: In bars is the frequency distribution of the number of pores and in lines is the curve for the cumulative percentage of pores.

4.7. These differences are confirmed by the fact that at ZA, the lenticular channels are longer (6.2 mm mean length) and relatively thinner (1.1 mm mean width) than at CL (4.9 and 1.5 mm, respectively for length and width) (Table 2).

In relation to the concentration features, ZA presents a porosity coefficient of 9.2% against 13.2% at CL. In average, the cork samples from CL showed higher porosity, at CL, the larger Mpa and higher Maxpa directly influences the porosity coefficient as CL presents smaller number of pores, 250 pores per 100 cm^2 against 280 pores per 100 cm^2 at ZA.

The distribution of cork porosity per classes of pore area showed similarities between the two regions of ZA and CL. However, at CL, a larger number of pores with area smaller than 2 mm^2 (# 229 pores) is noticeable when compared to ZA (# 151) (Figure 2) and the correspondent pore area (of pores with area <2 mm^2) was higher (288 mm^2) at CL when compared to the correspondent area of 186 mm^2, at ZA. This difference could significantly increase the porosity coefficient (area fraction) of the cork samples at CL (see Table 2). In addition, at CL, pores of extremely large dimension (>30 mm^2) were noticed in the cork samples, increasing the range of pore area variation (Figure 2).

3.2. Cork quality profile
The porosity from the cork samples of ZA and CL was represented in a pore centroid scatterplot (Figure 3), showing the location of the total number of pores, 205 at ZA and 494 at CL, from cork back (minimum *y*-value) to cork belly (maximum *y*-value) and the correspondent area, represented by a proportional circle.

The spatial distribution of pores in the transverse section of cork samples showed that the porosity at ZA ranged between *y*-values of 0 and 50 mm, while at CL ranged between *y*-values of 0 and 25 mm

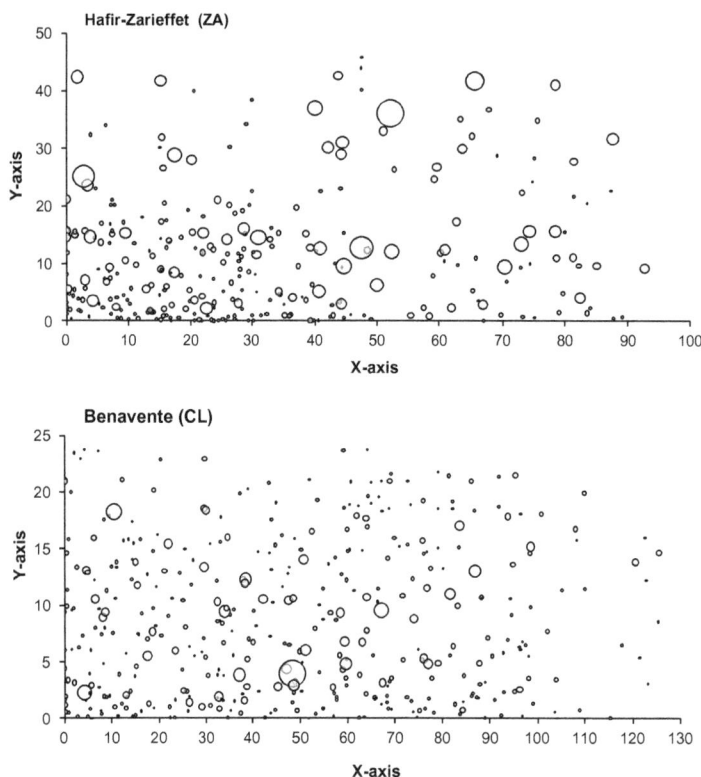

Figure 3. Representation of pore centroids in a *XY*-scatter plot for all cork samplers at Hafir-Zarieffet (ZA) and Benavente (CL).

Note: Circle diameter is proportional to pore area located from cork back (minimum *Y*-value) to cork belly (maximum *Y*-value).

(Figure 3), i.e. cork samples thickness at ZA are almost the double than the ones at CL. At CL, the pore area decreases from the cork back (outer cork) to cork belly (inner cork) (Table 2), while this trend is not as consistent at ZA as there are relatively larger cork pores in the inner cork layers (Figure 3).

The pore centroids of the larger pores at ZA and CL are located in the middle cork layers in result of the linear projection of the lenticular channels in a transverse section, but it is in the yearly cork layers (more than at the inner cork layers) that are located the smallest porosity in both regions (Table 2). Pore area (pa), length (le) and fractal dimension (fr) of the lenticular channels at CL showed a effectively a consistent decreasing trend of from cork back (outer-cork layers) to cork belly (inner- or yearly cork layers). However, at ZA, none of these pore variables allowed such as consistent trend and it is noticeable in the spatial representation of the porosity (Figure 3).

3.3. Cork quality regions
The SDA functions allowed distinguishing the porosity of the two regions (Wilks' Lambda significant at $p < 0.001$). Two SDA functions (*F1* and *F2*) explained about 98% of the variation in porosity of the transverse section of the cork planks: 77.5% variation was explained by *F1* and 20.7% was explained by *F2*.

The strongest correlations between the first SDA function, *F1*, and the independent variables were found for pore length (equal to 0.698) and pore roundness (equal to 0.694). Between the second SDA function, *F2*, which explained less variability, and the independent variables the higher correlation was found with pore perimeter (equal to 0.727) (Table 3).

The results (Table 3) showed that, at the AOI-level, and considering the cork layers, the selected independent pore-level variables, with the highest correlation with the SDA functions were those quantifying pore linear dimension, as pore perimeter and pore length, and related to pore shape, as roundness.

Table 3. Eigenvalues, canonical correlation coefficients, and variance explained by the first two discriminant functions (F1 and F2)		
Variables	**F1**	**F2**
Lenght	0.698	0.497
Perimeter	0.444	0.727
Roundness	0.694	0.677
Eigenvalue	0.212	0.057
Variance (%)	77.5	20.7
Cannonical correlation	0.618	0.331

Notes: Significant variables at pore level for the cork planks porosity by SDA, for the study areas, Hafir-Zarieffet (ZA) and Benavente (CL): pore length, pore perimeter and pore roundness.

The cork layer that was decisive to discriminate between the study regions, ZA and CL was the middle-cork layer. The lenticular channel-driven porosity be the most prominent feature of the transverse section of the cork planks. At ZA, the cork thickness is rather larger when compared to CL, and there are also some cork tissue discontinuities, such as cracks or heavy porosity, that were present in the transverse section, and this abnormal porosity lead to an increasing variability on porosity features, between cork layers when compared to the cork samples at CL (see Figure 3).

4. Discussion

This study show the possibility of using image analysis techniques for quality evaluation of the cork planks, currently made based only on visual appreciation by an experienced operator. Despite the fact that the quality of the cork planks directly conditioned the quality yield profile of the cork products, such as cork stoppers or disks (Costa & Pereira, 2010), the image analysis in the raw material is not done only because as a natural raw material, cork is more heterogeneous and the image analysis is done in the end product.

The porosity values found in the transverse section of the cork samples were in agreement with the values found in cork planks by previous reports for both regions (Dehane, Bouhraoua, González-Adrados, & Belhoucine, 2011; Gonzalez-Adrados et al., 2000; Pereira, Lopes, & Graça, 1996), and porosity was in general mostly related to the radial development of the lenticular channels in the bark of the living tree (Graça & Pereira, 2004).

The intrinsic variability of porosity is higher in the ZA cork samples than in CL cork samples (Table 2) and this could be in result of the difference in the area observed by image analysis, which is higher for the larger cork samples of ZA (Figure 3). The importance of dimension of the observed area for porosity determinations by image analysis was previously addressed for cork planks (Pereira et al., 1996).

According to the results obtained, the high area fraction found at CL, when compared to ZA, was driven by small porosity (mostly inferior to 2 mm²). This *so called* small porosity should not be as detrimental for the cork planks quality as larger cork porosity. However, at CL there are also larger pores as in ZA and this is highly detrimental for cork quality. As previously reported (Costa & Pereira, 2010), small porosity could represent 75% of the total number of the pores and less than 20% of the total pore area.

At ZA, the longer and thinner lenticular channels are mainly related to the larger cork production cycles, larger cork thickness of the cork planks. These longer cork production cycles should be reconsidered in the cork oak woodland management as could also be detrimental to cork quality yield in the sense that the cork quality profile could not be as consistent as in CL. In Algeria regions, where the annual cork growth is relatively high, similar to the one found in the study area where cork samples were collected (annual cork growth rate between 3.1 and 3.6 mm year⁻¹), forest management should include the shortening of the typical cork production cycles (to minimum of 9-year cork production cycle as in Benavente (Portugal). This way, the cork oak woodland would provide a more sustainable productive land use, maintaining their ecological resilience to cork harvest and a fine adaptation to the Mediterranean environments.

The porosity profile found in the transverse section of the cork samples is, however, consistent and emphasis the typical techniques when punching the natural cork stoppers, near the cork belly, in relation to the cork back, as in this region the lenticular channels, and porosity, are a more conspicuous feature. Thus, from the inner (recent) cork layers to the outer (older) cork layers, the porosity increases.

The results on the decision-making of the cork planks porosity discriminated by regions highlighted the importance of dimensional variables at pore level, length, and perimeter, in agreement with previous studies (Pereira et al., 1996). However, in contrast to these previous reports, the importance of the area of the pore was not decisive. This apparent discrepancy could be related to the fact that in this study the cork samples were more heterogeneous and large lenticular channels were found in the cork samples, which could be classified as cracks or heavy porosity, with a relatively poor match for a SDA analysis. Further studies should be made with more cork samples, in order to have a clear understanding on the cork quality profile for both regions, and mainly for Algeria, showing higher heterogeneity in the cork porosity features.

5. Conclusion

The porosity profile of cork samples of the two study areas, Benavente (CL) and Hafir-Zarieffet (ZA), were compared and results showed strong similarities related to lenticular channels presence and their correspondent cork tissue discontinuities. At both regions, CL and ZA, in the transverse section of the cork samples, larger and conspicuous porosity was found in outer part of the transverse section of cork planks, when compared to the one found in the inner part. However, when comparing cork quality of both regions, CL and ZA, clearly ZA had more heterogeneous cork quality, discriminated by porosity linear features, such as length and perimeter. Clearly, at Algeria (ZA), adequate thinning and pruning of cork oak woodland and appropriate cork production cycles should be applied toward the improvement of the cork quality production.

Acknowledgments
Authors acknowledge the collaboration of Beni Mester Cork Industry (Hafir-Zarieffet, Algeria) and of Companhia das Lezírias, S.A. (Benavente, Portugal) for the implementation of the study areas. Authors acknowledge the comments of two anonymous reviewers.

Funding
This research was partially supported by the Foundation for Science and Technology of the Portuguese Ministry of Education and Science (FCT-MEC) through the research projects [grant number UID/AMB/04085/2013], [grant number EXPL/AGR/FOR/1220/2012]. Augusta Costa's contribution was funded by the FCT-MEC under [grant number SFRH/BPD/97166/2013]. Amina Ghalem's internship was supported by University Abou Baker Belkaid of Tlemcen, as part of her PhD studies.

Competing Interests
The authors declare no competing interest.

Author details
Amina Ghalem[1]
E-mail: aminaghalem@ymail.com
Inês Barbosa[2]
E-mail: inesalexandrabarbosa@gmail.com
ORCID ID: http://orcid.org/0000-0003-1827-8660
Rachid Tarik Bouhraoua[1]
E-mail: rtbouhraoua@yahoo.fr
Augusta Costa[2,3]
E-mails: augusta.costa@iniav.pt, augusta.costa.sousa@gmail.com
ORCID ID: http://orcid.org/0000-0002-0400-8523
[1] Faculty of Science, Laboratory n 31, Department of Agro-Forestry, Management Conservatory Water, Soil and Forests, University of Tlemcen, P.O. Box 119, Tlemcen, Algeria.
[2] Center for Environmental and Sustainability Research (CENSE), Nova University of Lisbon, Campus de Caparica, 2829-516, Caparica, Portugal.
[3] Instituto Nacional de Investigação Agrária e Veterinária, I.P., Quinta do Marquês, Av. da República, 2780-159 Oeiras, Portugal.

References
APCOR. (2015). *Anuário Estatístico*. http://www.apcor.pt/wp-content/uploads/2015/12/APCOR-Boletim-Estatistico.pdf
Bugalho, M. N., Caldeira, M. C., Pereira, J. S., Aronson, J., & Pausas, J. G. (2011). Mediterranean cork oak savannas require human use to sustain biodiversity and ecosystem services. *Frontiers in Ecology and the Environment, 9*, 278–286. http://dx.doi.org/10.1890/100084

Capelo, J., Mesquita, S., Costa, J., Ribeiro, S., Arsénio, P., Neto, C., ... Lousã, M. (2007). A methodological approach to potential vegetation modeling using GIS techniques and phytosociological expert-knowledge: Application to mainland Portugal. *Phytocoenologia, 37,* 399–415. http://dx.doi.org/10.1127/0340-269X/2007/0037-0399

Costa, A., & Oliveira, G. (2015). Cork oak (*Quercus suber* L.): A case of sustainable bark harvesting in Southern Europe. In C. M. Shackleton, A. K. Pandey, & T. Ticktin (Eds.), *Ecological sustainability for non-timber forest products dynamics and case studies of harvesting* (pp. 179–198). New York, NY: Earthscan from Routledge.

Costa, A., & Pereira, H. (2006). Decision rules for computer vision quality classification of wine natural cork stoppers. *American Journal of Enology and Viticulture, 57,* 210–219.

Costa, A., & Pereira, H. (2007). Influence of vision systems, black and white, colored and visual digitalization, in natural cork stopper quality estimation. *Journal of the Science of Food and Agriculture, 87,* 2222–2228. http://dx.doi.org/10.1002/(ISSN)1097-0010

Costa, A., & Pereira, H. (2010). Influence of cutting direction of cork planks on the quality and porosity characteristics of natural cork stoppers. *Forest System, 19,* 51–60.

Costa, A., Madeira, M., & Oliveira, Â. C. (2008). The relationship between cork oak growth patterns and soil, slope and drainage in a cork oak woodland in Southern Portugal. *Forest Ecology and Management, 255,* 1525–1535. http://dx.doi.org/10.1016/j.foreco.2007.11.008

Costa, A., Madeira, M., Santos, J. L., & Plieninger, T. (2014). Recent Dynamics of Mediterranean Evergreen Oak Wood Pastures in Southwestern Iberia. In T. Hartel, & T. Plieninger (Eds.), *European wood-pastures in transition—A social-ecological approach* (pp. 70–89). New York, NY: Earthscan from Routledge.

Costa, A., Nunes, L. C., Graça, J., & Spiecker, H. (2015). Insights into the responsiveness of cork oak (*Quercus suber* L.) to bark harvesting. *Economic Botany, 20*(10), 1–14.

Costa, A., Oliveira, A. C., Vidas, F., & Borges, J. G. (2010). An approach to cork oak forest management planning: A case study in southwestern Portugal. *European Journal of Forest Research, 129,* 233–241. http://dx.doi.org/10.1007/s10342-009-0326-y

Dehane, B. (2012). *Incidences de l'etat sanitaire des arbres du chêne-liége sur les accroissements annuels et la qualité du liège de deux suberaies oranaises: M'Sila (W Oran) et Zarieffet (W. Tlemcen)* [Impact of health condition of cork oaks on cork annual increases and quality at cork oak forests in Oran: M'Sila (W Oran) and Zarieffet (W. Tlemcen)] (Dissertation). University of Tlemcen, Algeria.

Dehane, B., Bouhraoua, R. T., González-Adrados, J. R., & Belhoucine, L. (2011). Caractérisation de la qualité du liège selon l'état sanitaire des arbres par la méthode d'analyse d'image Cas des forêts de M'Sila et de Zarieffet (Nord-Ouest Algérien) [Cork quality characterization assessed by image analysis according to the health status of the trees on the forests of M'Sila and Zarieffet (North-West of Algeria)]. *Forêt méditerranéenne, 32,* 39–50.

FAO. (2013). *State of Mediterranean forests.* Rome. Retrieved from http://www.fao.org/docrep/017/i3226e/i3226e.pdf

Giorgi, F. (2006). Climate change hot-spots. *Geophysical Research Letter, 33,* L08707.

Gonzalez-Adrados, J. R., Lopes, F., & Pereira, H. (2000). Quality grading of cork planks with classification models based on defect characterisation. *Holz als Roh- und Werkstoff, 58,* 39–45. http://dx.doi.org/10.1007/s001070050383

Graça, J., & Pereira, H. (2004). The periderm development in *Quercus suber. IAWA Journal, 25,* 325–335. http://dx.doi.org/10.1163/22941932-90000369

Letreuch-Belarouci, A. M., Letreuch-Belarouci, N., Benabdeli, K., & Medjahdi, B. (2009). Impact des incendies sur la structure des peuplements de chêne-liège et sur le liège: le cas de la subéraie de Tlemcen (Algérie) [Impact of wildfires on the structure of cork oak stands and on cork yield: the case of the cork oak forests of Tlemcen (Algeria)]. *Forêt méditerranéenne, 30,* 231–238.

Messaoudène, M., & Merouan, H. (2009). Site profile 1.1: Akfadou, Algeria. In J. Aronson, J. S. Pereira, & J. G. Pausas (Eds.), *Cork oak woodlands on the edge ecology, adaptive management, and restoration* (pp. 22–23). Washington DC: Island Press.

Oliveira, G., & Costa, A. (2012). How resilient is *Quercus suber* L. to cork harvesting? A review and identification of knowledge gaps. *Forest Ecology and Management, 270,* 257–272. http://dx.doi.org/10.1016/j.foreco.2012.01.025

Pereira, H. (2007). *Cork: Biology, production and uses.* Amsterdam: Elsevier.

Pereira, H., Lopes, F., & Graça, J. (1996). The evaluation of the quality of cork planks by image analysis. *Holzforschung, 50,* 111–115. http://dx.doi.org/10.1515/hfsg.1996.50.2.111

Plieninger, T., van der Horst, D., Schleyer, C., & Bieling, C. (2014). Sustaining ecosystem services in cultural landscapes. *Ecology and Society, 19,* 59. doi:10.5751/ES-06159-190259

Ticktin, T. (2004). The ecological implications of harvesting non-timber forest products. *Journal of Applied Ecology, 41,* 11–21. http://dx.doi.org/10.1111/jpe.2004.41.issue-1

Vicente, Á. M., & Alés, R. F. (2006). Long term persistence of dehesas. Evidences from history. *Agroforestry Systems, 67,* 19–28. http://dx.doi.org/10.1007/s10457-005-1110-8

4

Antibacterial and toxicity studies of the ethanol extract of *Musa paradisiaca* leaf

Ememobong Gideon Asuquo[1] and Chinweizu Ejikeme Udobi[1]*

*Corresponding author: Chinweizu Ejikeme Udobi, Faculty of Pharmacy, Department of Pharmaceutics and Pharmaceutical Technology, Pharmaceutical Microbiology Unit, University of Uyo, Uyo, Nigeria
E-mail: ceudobi@yahoo.com
Reviewing editor: Tsai-Ching Hsu, Chung Shan Medical University, Taiwan
Additional information is available at the end of the article

Abstract: The ethanol extract of the leaf of *Musa paradisiaca* and its aqueous fraction were investigated for antibacterial potentials against Gram-positive and Gram-negative organisms of clinical importance. Phytochemical screenings showed the presence of secondary metabolites namely tannin, flavonoids, alkaloids among others. When the extracts were subjected to susceptibility test, results showed that while both the ethanol extract and its aqueous fraction showed activity against the test organisms, the aqueous fraction showed better antibacterial activity. The minimum inhibitory concentration of the aqueous fraction ranged from 3.125 to 25 mg/ml. Acute toxicity test of the aqueous fraction using Swiss Albino mice gave an LD_{50} of 489.9 mg/kg body weight which indicates a relative toxicity of the extract. The time–kill studies demonstrated a reduction in the viable cell count for the extract.

Subjects: Bioscience; Food Science & Technology; Health and Social Care

Keywords: ethanol; *Musa paradisiaca*; phytochemical; toxicity; safety

1. Introduction
For a long period of time, plants have been a valuable source of natural products for maintaining human health (Karadi, Shah, Parekh, & Azmi, 2011). About 80% of individuals from developed countries use traditional medicine, which has compounds derived from medicinal plants (World Health Organisation, 1997). Many plants have been used because of their antimicrobial potentials, which are due to compounds synthesized in the secondary metabolism of the plant such as the phenolic compounds which are part of the essential oils as well as in tannins (Venkanna & Estari, 2012).

ABOUT THE AUTHOR
Chinweizu Ejikeme Udobi, PhD, is a senior lecturer in the pharmacy school of the University of Uyo where he teaches pharmaceutical microbiology and related courses. Among other areas, his research interest includes the development of novel antimicrobial agents from natural sources. Together with members of his research team, Ejikeme has published extensively on the potentials of most plants of tropical Africa for use in the development of new anti-infective agents in these days of continuous development of resistance by organisms. Ememobong Gideon Asuquo is an active member of Ejikeme's research team.

PUBLIC INTEREST STATEMENT
Treatment of infectious diseases has always been a challenge to man. As scientists continue to search for new agents for the treatment of infectious diseases, plants have proved to be a reliable source. This research seeks to provide the scientific basis for the use of *Musa paradisiaca* (plantain) leaf for such purpose and perhaps establish it as a possible source of a new drug against some organisms. The ethanol extract of the leaf and its aqueous fraction were therefore tested against some organisms. The extract showed activity by inhibiting the growth of these organisms and causing a reduction in their viable count with time while showing only a moderate toxicity in mice. These findings provide scientific explanation for the use of this plant leaf in traditional medicines and shows that it can be exploited as a possible source of new antibiotics.

Thousands of plants have been studied and most have been found to possess phytochemicals which are safe and broadly effective alternatives with less adverse effect. Many beneficial biological activities such as antimicrobial, antioxidant, antidiarrhoeal, analgesic, antiulcerative, antihypertensive and wound healing activity have been reported (Agarwal et al., 2009; Alisi, Nwanyanwu, Akujobi, & Ibegbulem, 2008; Goel & Sairam, 2002; Osim & Ibu,1991; Rabbani et al., 2001; Singh & Dryden, 1990; Yin, Quan, & Kanazawa, 2008).

Musa paradisiaca also known as plantain is a familiar tropical fruit and important source of food in the world and is consumed as an energy yielding food and desert (Ighodaro, 2012). It is extensively cultivated in the tropics and is a staple crop for over 70 million people of the Sub-Saharan Africa (Yusoff, 2008). Various parts of the plant such as the leaves, roots and flowers have been reportedly used for medicinal purposes. For example, the leaf juice is used in the treatment of fresh wounds, cuts and insect bites (Onyenekwe, Okereke, & Owolewa, 2013). The sap of the plant has been used as a remedy for epilepsy, hysteria and in dysentery and diarrhoea, the roots as anthelmintic, flowers as astringent and fruits as mild laxatives (Okareh, Adeolu, & Adepoju, 2015). A cold infusion of the root has been used to treat venereal diseases and anaemia. In addition, the fruit has been reportedly used as antiscorbutic, aphrodisiac and diuretic (Gill, 1992). Plantain has been reported to have a high fibre content, and thus is capable of lowering cholesterol and helps to relieve constipation and hence prevention of colon cancer. In Nigeria, traditional medicine practitioners have used the leaf decoction of plantain and banana for treatment of typhoid fever, diarrhoea, malaria, stomach ache or ulcers (Apata, 1979; Okigbo & Omodamiro, 2006). Okorondu, Akujobi, and Nwachukwu (2012) have reported the antifungal properties of the peel and stalk extracts of *M. paradisiacal*. Karadi et al. (2011) and Karuppiah and Mustaffa (2013) have reported the activity of the leaf extract against *Escherichia coli, Staphylococcus aureus* and *Pseudomonas aeruginosa* while Sahaa, Acharyaa, Shovon, and Royd (2011) have reported antibacterial activities of the leaf extract against *Salmonella typhi, Shigella dysenteriae* and *Bacillus cerus*.

A number of new antibiotics have been produced by pharmaceutical industries in the last 30 years. Resistance to these drugs by micro-organisms has been increased day by day (Karuppiah & Mustaffa, 2013). WHO 2014 report on global surveillance of antimicrobial resistance revealed that antibiotic resistance is no longer a prediction for the future; it is happening right now, across the world, and is putting at risk the ability to treat common infections in the community and hospitals. In line with this, new drugs, both synthetic and natural must be sought to treat diseases. This implies that local medicinal plants need to be screened for antimicrobial properties of their extracts against known organisms which rely on the bioactive phytocomponents present in the plants (Karadi et al., 2011; Okorondu, 2011; Okorondu, Sokari, Akujobi, & Braide, 2010; Veeramuthu, Muniappan, & Savarimuthu, 2006).

This research work is aimed at providing scientific basis for the use of this plant traditionally to treat different ailments and also provide evidence for the possible use of this plant to produce new drugs for the treatment of infections caused by micro-organisms used in this study.

2. Materials and methods

2.1. Plant collection, authentication, preparation and extraction

Fresh plantain leaves were collected from the plantain plantation of Chief E.A. Bassey in Use Ndon, Ibiono Ibom Local Government Area, Akwa Ibom state. The botanical authentication was done at the Department of Pharmacognosy, Faculty of Pharmacy, University of Uyo, Uyo. The herbarium number is UUPH 51(a).

The leaves were cut into smaller pieces for easy drying. The dried leaves were powdered using a mortar and pestle. The powdery samples were packed into screwed bottles and labelled appropriately. The ethanol extract of the leaf of *M. paradisiaca* was prepared by soaking 300 g of the dried powdery samples in 2,500 ml of ethanol for 48 h, during which the mixture was intermittently shaken. It was later filtered through Whatman No. 42 filter paper. The extracts were evaporated to dryness at 40°C in a water bath.

2.2. Fractionation of extract

The ethanol extract was fractionated using petroleum ether, chloroform and water according to the method of Udobi, Onaolapo, and Agunu (2008). Twenty grams of the extract was dissolved in 200 ml of water before shaking vigorously in a separating flask. The mixture obtained was filtered using a filter paper to remove debris. Two hundred millilitres of petroleum ether was then added to the mixture in a separating funnel, shaken vigorously and allowed to settle. The petroleum ether layer (on top) was removed and concentrated while a further 200 ml of chloroform was added to the aqueous layer and also vigorously shaken and allowed to settle. The aqueous and the chloroform layers were further separated, the chloroform portion was concentrated to dryness by allowing to stand on the laboratory bench until all the solvent evaporated, while the aqueous layer was concentrated to dryness using the water bath at 40°C.

2.3. Phytochemical screening

The ethanol extract was screened for the presence of alkaloids, saponins, tanins, carbohydrates, flavonoids, anthraquinones, cardiac glycosides, deoxy sugar, steroids and terpenes using standard methods as described by Harborne (1973), Sofowora (1993), Trease and Evans (1986), William, Trease, and Evans (1996).

2.4. Preparation of culture media

The culture media used namely nutrient agar and nutrient broth (Oxiod, England), were prepared according to the manufacturer's instruction, sterilized in an autoclave at 121°C for 20 min and allowed to cool to about 45°C.

2.5. Preparation of inoculum

Test inoculum was prepared according to the method of Chinweizu (2010). Nutrient broth medium was used as the diluent. The Gram-positive organisms were diluted to 1:1000 while the Gram-negative organisms were diluted to 1:5000 to avoid overcrowding.

2.6. Susceptibility testing

The agar well diffusion method was used to carry out this test. 0.1 ml of 1:1000 dilution of test organism for Gram-positive and 1:5000 dilution of test organism for Gram-negative were introduced into labelled sterile Petri dishes. Twenty millilitres of the cooled nutrient agar medium was aseptically poured into each Petri dish and gently swirled to mix. The plates were allowed to set and wells were created using sterile 4-mm cork borer. Different concentrations (200–12.5 mg/ml) of the extract diluted with water was introduced into the different wells and labelled appropriately. The Petri dishes were allowed to stand for 20 min before incubation at 37°C for 24 h. The diameters of the zones of growth inhibition were measured in millimetres (mm).

2.7. Minimum inhibitory concentration

Agar dilution method was used to determine the minimum inhibitory concentration (MIC). The medium used was nutrient agar medium. 50 mg/ml of the extract was prepared in warm water. Ten millilitres of the extract (50 mg/ml) was transferred aseptically into a universal bottle containing 10 ml of warm nutrient agar prepared as double strength. The agar was mixed well and this gave a concentration of 25 mg/ml. A twofold serial dilution was carried out with 10 ml of single strength nutrient agar to have concentrations of 12.5, 6.25, 3.125 and 1.562 mg/ml, respectively. The agar was immediately transferred aseptically into labelled sterile Petri dishes and allowed to set. The plates were inoculated with a 0.1 ml each of 1:1000 dilution of test organism for Gram-positive and 1:5000 dilution of test organism for Gram-negative. Incubation was done at 37°C for 24 h. The plates were examined for the present of microbial growth. The lowest concentration of the extract that showed no visible growth of test organisms was considered as the MIC.

2.8. Acute toxicity testing

Lorke's method (Lorke, 1983) was used to assess the lethal dose (LD_{50}) of the aqueous fraction of the ethanol extract of *M. paradisiaca* leaf that kills 50% of the test animal population. In the first phase, nine healthy mice were divided into three groups of three animals each. The animals were fasted for 24 h. Each group of animals were administered different doses (10, 100 and 1,000 mg/kg body weight) of the plant extract. The animals were placed under observation for 24 h and monitored for mortality. The second phase involved the use of four mice which were distributed into four groups of one animal each. The animals were administered different doses (200, 400, 600, and 800 mg/kg body weight) of the plant extract according to the specifications of the method. They were then observed for 24 h and mortality taken note of. All experimental protocols were in compliance with the Faculty of Pharmacy University of Uyo ethics on research in animals as well as internationally accepted principles for laboratory animal use and care.

2.9. Time–kill test

The time–kill test was carried out according to the method of Coudron and Stratton (1995). This test is used to assess the *in vitro* reduction of a microbial population of test organisms after exposure to a test material. The organisms used were *S. aureus*, *Vibrio cholerae* and *S. dysenteriae*. Three test tubes containing 8 ml of sterile nutrient broth each were labelled according to the organisms; one test tube was used as control. One millilitre of the aqueous fraction of the ethanol extract of *M. paradisiaca* leaf was introduced into the test tubes in concentrations that gave a final concentration corresponding to the MIC of each organism, 3.125 mg/ml for *S. aureus* and *S. dysenteriae*, 12.5 mg/ml for *V. cholerae*. One millilitre of a standardized inoculum of overnight broth culture was introduced into each of the test tube. The test tubes were incubated at 37°C with shaking. At 30 min intervals, 0.1 ml was withdrawn from each test tube and diluted to 10^2 dilution factor. 0.1 ml of this dilution was aseptically transferred into sterile Petri dishes containing nutrient agar medium in triplicates. The plates were incubated at 37°C for 24 h. After the incubation period, the number of colony on each plate was enumerated and the colony forming units calculated, expressed in $\log_{10}$. This was used to generate a time–kill curve for each organism by plotting the $\log_{10}$ colony forming unit against time.

3. Results

3.1. Phytochemical screening

The phytochemical screening showed the presence of some secondary metabolites which includes alkaloids, saponins, tannins, cardiac glycosides, terpenes, deoxy sugar, flavonoids and carbohydrates. The result is presented in Table 1.

Table 1. Result of the phytochemical screening of the ethanol leaf extract of *Musa paradisiaca*

Chemical constituents	Indications
Alkaloids	+
Saponins	+
Tanins	+
Cardiac glycosides	+
Deoxy sugar	+
Flavonoids	+
Carbohydrates	+
Anthraquinones	–
Terpenes	+

Note: Key: + present; – absent.

Table 2a. Growth inhibition zones of different concentrations of the crude ethanol extract of *Musa paradisiaca* leaf

Conc. (mg/ml)	Zones of inhibition (mm)				
	S. aureus	*B. subtilis*	*P. aeruginosa*	*V. cholerae*	*S. dysenteriae*
200	22	16	12	21	18
100	18	14	10	16	15
50	13	12	8	15	13
25	10	10	6	13	12
Control	26	30	–	20	22

Notes: 12.5 mg/ml of cefuroxime was used as control; Zones of inhibition are means of double determination.

Table 2b. Growth inhibition zones of different concentrations of the aqueous fraction of the ethanol extract of *Musa paradisiaca* leaf

Conc (mg/ml)	Zones of inhibition (mm)				
	S. aureus	*B. subtilis*	*P. aeruginosa*	*V. cholerae*	*S. dysenteriae*
200	25	18	14	22	20
100	20	16	12	18	16
50	14	12	8	14	14
25	10	9	6	10	12
12.5	–	–	–	–	10
Control	26	30	–	20	22

Notes: 12.5 mg/ml of cefuroxime was used as control; Zones of inhibition are means of double determination.

3.2. Susceptibility testing

Microbial susceptibility test with the crude ethanol extract and the aqueous fraction of *M. paradisiaca* leaf showed zones of growth inhibitions whose diameters were measured in millimetres (mm) and presented in Tables 2a and 2b, respectively.

3.3. Minimum inhibitory concentration

The results obtained showed the minimum concentration of the aqueous fraction of the ethanol extract of *M. paradisiaca* leaf that inhibits the growth of the different micro-organisms as presented in Table 3.

3.4. Acute toxicity testing

The concentration of the aqueous fraction of the ethanol extract of *M. paradisiaca* leaf that killed 50% of mice, expressed as LD_{50} is presented in Table 4.

3.5. Time–kill test

The mean of the number of colonies obtained per plate in respect to the time interval was used to calculate the colony forming unit, cfu/ml of the isolates and $\log_{10}$ cfu/ml.

4. Discussion

The ability of plants to synthesize a wide variety of chemical compounds that are used to perform important biological functions have been the basis for the use of plants to effectively treat human diseases (Harborne, 1973). Even though pharmaceutical industries have produced a number of new antibiotics in the last four decades, resistance to these drugs by micro-organisms has increased (Karadi et al., 2011) and therefore there is need for research on plant materials with antimicrobial properties.

Table 3. MIC of the aqueous fraction of the ethanol extract of *Musa paradisiaca* leaf

Isolates	Concentration of extract (mg/ml)
S. aureus	3.125
B. subtilis	12.5
P. aeruginosa	25
V. cholera	12.5
S. dysenteriae	3.125

Table 4. Acute toxicity test of the aqueous fraction of the ethanol extract of *Musa paradisiaca* leaf

Phases	No. of mice	Weight of mice (g)	Dose (mg/kg)	Mortality	LD_{50} (mg/kg)
1	3	20	10	0	$\sqrt{D_0 \times D_{100}}$
		19			
		23			
	3	19	100	0	=489.9 (mg/kg)
		23			
		19			
	3	20	1,000	3	
		21			
		23			
2	1	22	200	0	
	1	20	400	0	
	1	21	600	1	
	1	23	800	1	

Notes: D_0 = highest dose that gave no mortality, 400 mg/kg body weight; D_{100} = lowest dose that produced mortality, 600 mg/kg body weight.

In this regard, *M. paradisiaca*, commonly known as plantain was studied to extract the active components of the leaf, carry out phytochemical screening of the crude extract and identify the secondary metabolites present. The antibacterial activities of these metabolites against selected Gram-positive and Gram-negative clinical isolates were further studied.

Bioactive compounds found to be present in this plant include alkaloids, saponins, tannins, cardiac glycosides, terpenes, deoxy sugar, flavonoids and carbohydrates as presented in Table 1. The findings of this study is consistent with reports of the presence of these phytochemicals in various parts of the *M. paradisiaca* plant as documented by Akpuaka and Ezem (2011) and Akpabio, Udiong, and Akpakpan (2012). These phytochemicals have been reported to exert multiple biological and pharmacological effects (antibacterial, antihypertensive, antidiabetic and anti-inflammatory activities) (Ighodaro, 2012). The presence of these bioactive substances in *M. paradisiaca* leaves therefore suggests that the leaves possess valuable medicinal potential that can be explored.

The results of the antimicrobial screening of the crude ethanol extract and the aqueous fraction of *M. paradisiaca* leaf against five human pathogenic microbes showed that both extracts were less effective when compared to a reference antibiotic, cefuroxime. *S. aureus*, *Bacillus subtilis*, *P. aeruginosa*, *V. cholerae* and *S. dysenteriae* were comparatively more inhibited by both extracts (Tables 2a and 2b). The aqueous fraction of the ethanol extract exhibited higher antimicrobial activities compared to the crude ethanol extract (Tables 2a and 2b).This is because the secondary metabolites being polar compounds like water have been pooled into the aqueous fraction during fractionation. *S. aureus* had a wider zone of inhibition than others, indicating a possibility in the use of the plant to treat infections due to the organism which are on the increase. The antimicrobial properties of plant

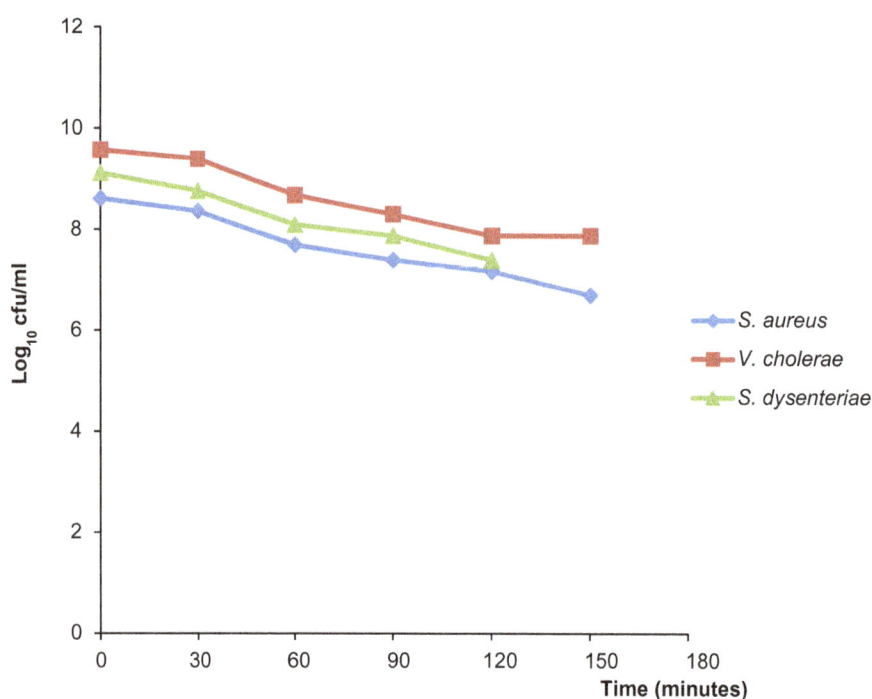

Figure 1. The time–kill test of the aqueous fraction of the ethanol extract of *M. paradisiaca* leaf against *S. aureus, V. cholerae* and *S. dysenteriae.*

Table 5. Results of the time–kill test of the aqueous fraction of the ethanol extract of *M. paradisiaca* leaf against *S. aureus, V. cholerae* and *S. dysenteriae*

Time (min)	S. aureus		V. cholerae		S. dysenteriae	
	cfu/ml	$\log_{10}$ cfu/ml	cfu/ml	$\log_{10}$ cfu/ml	cfu/ml	$\log_{10}$ cfu/ml
0	4.1×10^8	8.613	3.766×10^9	9.576	1.3×10^9	9.114
30	2.3×10^7	8.362	2.475×10^9	9.394	5.75×10^8	8.760
60	5.0×10^7	7.699	4.75×10^8	8.677	1.25×10^8	8.097
90	2.5×10^7	7.398	2.0×10^8	8.301	7.5×10^7	7.875
120	1.5×10^7	7.176	7.5×10^7	7.875	2.5×10^7	7.398
150	5.0×10^6	6.699	7.5×10^7	7.875	–	–

Note: Colony forming units are means of double determination.

extracts had been attributed to the presence of alkaloids and flavonoids (Okorondu et al., 2012) which have been found present in this plant. Alkaloids, flavonoids and tannins have been known to show medicinal activity as well as exhibiting physiological activity (Sofowora, 1993). Tannins in plants have been shown to confer antidiarrhoeic properties on plants (Asquith & Butler, 1986). This is consistent with the traditional use of the leaf of *M. paradisiaca* for the treatment of diarrhoea (Onyenekwe et al., 2013) and is evident in the activity of the extract against *S. dysenteriae* and *V. cholerae* which causes diarrhoea. Flavonoids are known to have antioxidant effects, the presence of these phytochemicals in the leaf of *M. paradisiaca* confers medicinal properties on the plant and this explains the use of this plant for treatment of different ailments (Kim, Kim, Kim, Oh, & Jung, 1994). Okorondu et al. (2012) reported that alkaloids are phytochemical components with bitter taste and possess antimicrobial properties. The microbes against which the extracts were effective are pathogens already implicated in the aetiology and severity of human diseases. Thus, the plant extract may be useful in pharmaceutical and medical formulations. Therefore, further purification of the extract and formulation into antibiotics is suggested.

The lowest concentration of the plant extract required for inhibiting the growth of test organisms (MIC) was considered. The MIC values of the aqueous fraction against the test micro-organisms presented on Table 3, ranged from 3.125 mg/ml for *S. aureus* and *S. dysenteriae* to 25 mg/ml for *P. aeruginosa*.

The safety of the aqueous fraction of the ethanol extract of *M. paradisiaca* leaf was evaluated using Lorke's method of determination of acute toxicity (Lorke, 1983). The median lethal dose that killed 50% of the test animal, LD_{50} was 489.9 mg/kg body weight. Ngulde, Tijjani, Ihopo, and Ya'uba (2013) states that any extract with LD_{50} greater than 5,000 mg/kg is considered as safe. Therefore, the aqueous fraction of the ethanol extract of *M. paradisiaca* leaf can be said to be moderately toxic being less than 5,000 mg/kg.

The *in vitro* time–kill regimes of the aqueous fraction of the ethanol extract of *M. paradisiaca* against *S. aureus*, *V. cholerae* and *S. dysenteriae* was assessed using standard microbiological procedures. The results are presented in Table 5. The extract demonstrated a bacteriostatic activity showing reduction in bacterial count due to growth inhibition since the MIC was used. The log reduction in viable cell count in the time–kill assay within 2 h 30 min ranged from 8.613 to 6.699 log_{10} for *S. aureus*, 9.576 to 7.875 log_{10} for *V. cholerae* and 9.114 to 7.398 log_{10} for *S. dysenteriae*. The downward steep of the time–kill curves (Figure 1) generated shows that the log reduction of microbial population is proportionate with time of exposure. This time–kill studies corroborates the reported efficacies of preliminary antibacterial study of selected plant extracts as reported by Oladosu, Isu, Ibrahim, and Ibrahim (2010) and this supports the folkloric uses of these plants in the treatment of different ailments among the traditional people.

5. Conclusion

The results of this study indicate that the crude and the aqueous fraction of the ethanol extract of *M. paradisiaca* leaf possess antibacterial activity. The aqueous fraction showed greater activity than the crude extract against the test organisms. The Gram-positive organisms responded better than the Gram-negative, this may be due to the difference in the morphology of the cell wall, especially the presence of Gram-negative cell envelope. From the MIC values obtained from this study, it can be concluded that the concentration of *M. paradisiaca* leaf extract required to arrest the growth of test organisms is relatively less. The time–kill curves clearly show the ability of the extract to reduce microbial population exponentially with time. As such, the antibacterial activity of *M. paradisiaca* can be exploited for the treatment of infections caused by the test organism at relatively low doses to avoid toxicity.

Funding
The authors received no direct funding for this research.

Competing Interest
The authors declare no competing interest.

Author details
Ememobong Gideon Asuquo[1]
E-mail: shalomgideon@yahoo.com
Chinweizu Ejikeme Udobi[1]
E-mail: ceudobi@yahoo.com
[1] Faculty of Pharmacy, Department of Pharmaceutics and Pharmaceutical Technology, Pharmaceutical Microbiology Unit, University of Uyo, Uyo, Nigeria.

References
Agarwal, P. K., Singh, A., Gaurav, K., Goel, S., Khanna, H. D., & Goel, R. K. (2009). Evaluation of wound healing activity of extracts of plantain banana (*Musa sapientum* var. *paradisiaca*) in rats. *Indian Journal of Experimental Biology, 47*, 322–340.

Akpabio, U. D., Udiong, D. S., & Akpakpan, A. E. (2012). The physicochemical characteristics of plantain (*Musa paradisiaca*) and banana (*Musa sapientum*) pseudo stem wastes. *Advances in Natural and Applied Sciences, 6*, 167–172.

Akpuaka, M. U., & Ezem, S. N. (2011). Preliminary photochemical screening of some Nigerian dermatological plants. *Journal of Basic Physiological Research, 2*(1), 1–5.

Alisi, C. S., Nwanyanwu, C. E., Akujobi, C. O., & Ibegbulem, C. O. (2008). Inhibition of dehydrogenase activity in pathogenic bacteria isolates by aqueous extracts of *Musa paradisiaca* (var Sapientum). *African Journal of Biotechnology, 7*, 1821–1825.
http://dx.doi.org/10.5897/AJB

Apata, L. (1979). The practice of herbalism in Nigeria. In A. Sofowora (Ed.), *African medicinal plants* (pp. 13–19). Ile Ife: Proceedings of a Conference, University of Ife Press.

Asquith, T. N., & Butler, L. G. (1986). Interactions of condensed tannins with selected proteins. *Phytochemistry, 25*, 1591–1593.
http://dx.doi.org/10.1016/S0031-9422(00)81214-5

Chinweizu, E. U. (2010). *Parkia biglobosa* (African Locust Bean): *The medicinal tree of the future*. Saarbrucken: Lap Lambert Academic publishing/AG and Co. Saarbrucken.

Coudron, P. E., & Stratton, C. W. (1995). Use of time-kill methodology to assess antimicrobial combinations against metronidazole-susceptible and metronidazole-resistant strains of *Helicobacter pylori*. *Antimicrobial Agents and Chemotherapy, 39*, 2641–2644. http://dx.doi.org/10.1128/AAC.39.12.2641

Gill, L. S. (1992). *Ethnomedicinal uses of plants in Nigeria* (pp. 169–170). Benin City: University of Benin Press.

Goel, R. K., & Sairam, K. (2002). Anti-ulcer drugs from indigenous sources with emphasis on *Musa sapientum, Tamrabhasma, Asparagus racemosus* and *Zingiber officinale*. *Indian Journal of Pharmacology, 34*, 100–110.

Harborne, J. B. (1973). *Phytochemical methods: A guide to modern techniques of plant analysis*. London: Chapman and Hall Limited.

Ighodaro, O. M. (2012). Evaluation study on Nigerian species of *Musa paradisiaca* peels: Phytochemical screening, proximate analysis, mineral composition and antimicrobial activities. *Researcher, 4*, 17–20.

Karadi, R. V., Shah, A., Parekh, P., & Azmi, P. (2011). Antimicrobial activities of *Musa paradisiaca* and *Cocos nucifera*. *International Journal of Research in Pharmaceutical and Biomedical Sciences, 2*, 264–267.

Karuppiah, P., & Mustaffa, M. (2013). Antibacterial and antioxidant activities of *Musa* sp. leaf extracts against multidrug resistant clinical pathogens causing nosocomial infection. *Asian Pacific Journal of Tropical Biomedicine, 3*, 737–742. http://dx.doi.org/10.1016/S2221-1691(13)60148-3

Kim, S. Y., Kim, J. H., Kim, S. K., Oh, M. J., & Jung, M. Y. (1994). Antioxidant activities of selected oriental herb extracts. *Journal of the American Oil Chemists' Society, 71*, 633–640. http://dx.doi.org/10.1007/BF02540592

Lorke, D. (1983). A new approach to practical acute toxicity testing. *Archives of Toxicology, 54*, 275–287. http://dx.doi.org/10.1007/BF01234480

Ngulde, S. I., Tijjani, M. B., Ihopo, J. M., & Ya'uba, A. M. (2013). Antitrypanosomal potency of methanol extract of *Cassia arereh* Delile root bark in albino rats. *International Journal of Drug Research and Technology, 3*(1), 1–7.

Okareh, O. T., Adeolu, A. T., & Adepoju, O. T. (2015). Proximate and mineral composition of plantain (*Musa paradisiaca*) wastes flour; a potential nutrient source in the formation of animal feeds. *African Journal of Science and Technology, 6*, 53–57.

Okigbo, R. N., & Omodamiro, O. D. (2006). Antibacterial effects of leaf extracts of pigeon pea (*Cajanuscajan* (L) *Mill* sp.) on some human pathogens. *Journal of Herbs, Spices and Medicinal Plants, 12*, 117–127.

Okorondu, S. I. (2011). Evaluation of the antifungal properties of picralima nitida seed extracts. *International Journal of Natural and Applied Sciences, 7*, 41–46.

Okorondu, S. I., Sokari, T. G., Akujobi, C. O., & Braide, W. (2010). Phytochemical and antibacterial properties of *Musa paradisiaca* stalk plant. *International Journal of Biological Science, 2*, 128–132.

Okorondu, S. I., Akujobi, C. O., & Nwachukwu, I. N. (2012). Antifungal properties of *Musa paradisiaca* (plantain) peel and stalk extracts. *International Journal of Biological Science, 6*, 1527–1534.

Oladosu, P., Isu, N. R., Ibrahim, K., & Ibrahim, J. A. (2010). Ethnobotanical survey and preliminary evaluation of some selected medicinal plants used in the treatment of tuberculosis in some parts of Northern Nigeria. *ZUMA Journal of Applied Sciences, 8*, 1–4.

Onyenekwe, P. C., Okereke, O. E., & Owolewa, S. O. (2013). Phytochemical screening and effect of *Musa paradisiaca* stem extrude on rat haematological parameters. *Current Research Journal of Biologicl Sciences, 5*, 26–29.

Osim, E. E., & Ibu, J. O. (1991). The effect of plantains (*Musa paradisiaca*) on DOCA-induced hypertension in rats. *Pharmaceutical Biology, 29*, 9–13. http://dx.doi.org/10.3109/13880209109082841

Rabbani, G. H., Teka, T., Zaman, B., Majid, N., Khatun, M., & Fuchs, G. J. (2001). Clinical studies in persistent diarrhea: Dietary management with green banana or pectin in Bangladeshi children. *Gastroenterology, 121*, 554–560. http://dx.doi.org/10.1053/gast.2001.27178

Sahaa, K., Acharyaa, S., Shovon, S. S. H., & Royd, P. (2011). *Medicinal activities of leaves of Musa sapientum var. sylvesteris in vitro*. Dhaka: Department of pharmacy, East West University.

Singh, Y. N., & Dryden, W. F. (1990). The augmenting action of banana tree juice on skeletal muscle contraction. *Toxicon, 28*, 1229–1236. http://dx.doi.org/10.1016/0041-0101(90)90122-N

Sofowora, A. (1993). *Medicinal plants and traditional medicine in Africa* (pp. 191–289). Ibadan: Spectrum Books.

Trease, G. E., & Evans, W. C. (1986). *Textbook of Pharmacognosy* (13th ed.) (pp. 343–383). London: Bailliere Tindal.

Udobi, C. E., Onaolapo, J. A., & Agunu, A. (2008). Antibacterial activities and bioactive components of the aqueous fraction of the stem bark of *Parkia biglobosa* (Jacq) (Mimosacea). *Nigerian Journal of Pharmaceutical Science, 7*, 76–82.

Veeramuthu, D., Muniappan, A., & Savarimuthu, I. (2006). Antimicrobial activity of some ethnomedicinal plants used by Paliyar tribe from Tamil Nadu, India. *BMC Complementary and Alternative medicine, 6*, 35.

Venkanna, L., & Estari, M. (2012). In vitro antimicrobial activity of some medicinal plants used by tribes in Warangal district (Andhra Pradesh) India. *Biology and Medicine, 4*, 85–88.

William, C. E., Trease, G. E., & Evans, E. C. (1996). *Pharmacognosy* (14th ed.). WB, Saunders Limited: London.

World Health Organisation. (1997). *Monongraph on selected medicinal plants*. Geneva: WHO VBC/97.1.

Yin, X., Quan, J., & Kanazawa, T. (2008). Banana prevents plasma oxidative stress in healthy individuals. *Plant Foods for Human Nutrition, 63*, 71–76. http://dx.doi.org/10.1007/s11130-008-0072-1

Yusoff, N. A. B. (2008). *Correlation between total phenolics and mineral content with antioxidant activity and determination of bioactive compounds in various local bananas (Musa sp.)* Penang: Universiti Sains Malaysia.

5

Radioprotective effects of diallyl disulphide and Carica papaya (L.) leaf extract in electron beam radiation induced hematopoietic suppression

Yogish Somayaji T[1], Vidya V[1], Vishakh R[1], Jayarama Shetty[2], Alex John Peter [2] and Suchetha Kumari N[3]*

*Corresponding author: Suchetha Kumari N, Department of Biochemistry, K S Hegde Medical Academy, Nitte University, Mangaluru, Karnataka, India
E-mail: kumari.suchetha@gmail.com
Reviewing editor: Tsai-Ching Hsu, Chung Shan Medical University, Taiwan
Additional information is available at the end of the article

Abstract: Numerous herbal and synthetic formulations have been used to determine the protective effects against radiation toxicity. In the present study, the radioprotective effects of aqueous leaf extract of Carica papaya (L.) and diallyl disulfide were studied in irradiated and non-irradiated groups. The optimum dose selection for papaya leaf extract and diallyl disulfide was done using Kaplan-Meier survival analysis. The mice were treated with papaya leaf extract at 5, 50, 100 and 200, 250, 500, 1,000 mg/kg body weight concentration and diallyl disulfide (DADS) at 0.5, 2, 5, 10, 20 mg/kg body weight for five consecutive days and on the 5th day one hour after dose administration they were irradiated with 10 Gy dose of lethal electron beam radiation. The mice were observed for maximum survival within 30 days post irradiation. The survival analysis from Kaplan-Meier curve indicated the optimum radioprotective dose for papaya leaf extract at 100 mg/kg body weight and for diallyl disulfide at 10 mg/kg body weight. The optimum dose was used for further studies with sublethal, lower sublethal, and adaptive split dose responses. The results indicate hematopoietic suppression in irradiated groups with sublethal, lower sublethal, and adaptive split dose groups. An increase in platelet levels and RBC levels were exhibited by papaya leaf extract pretreatment prior to irradiation with sublethal and lower sublethal doses. There was reduction in the antioxidant enzymes, total antioxidant capacity and increased lipid peroxidation seen in sublethal and lower sublethal radiation control groups. The pretreatment groups with papaya leaf extract and DADS prior irradiation have moderately enhanced the total antioxidant capacity, antioxidant enzyme levels, and reduced lipid peroxidation.

ABOUT THE AUTHORS

The authors of Central Research Laboratory, K S Hegde Medical Academy, Nitte University are majorly involved in studies related to radiation protectors, mitigators, and their mode of action in different systems of Invitro and Invivo. Most of the resources that are being studied are of plant origin as they are non-toxic and inexpensive. The present study is an attempt to understand the radioprotective role and radio-adaptive response of papaya leaf extract and diallyl disulfide using electron beam as a source of ionizing radiation.

PUBLIC INTEREST STATEMENT

The study highlights the protective role of papaya leaf extract and diallyl disulfide against ionizing radiation induced tissue damages in Swiss albino mice. To understand the nature of protection offered by them we have studied at different levels like survival, antioxidant levels, changes in DNA integrity, and response at different doses. The results highlight these studies and their significance and relevance in offering the protection. The results indicate a protective role of papaya leaf extract and diallyl disulfide in improving the survival, enhancing the antioxidant levels, reducing the DNA damage, and also a radiation adaptive response in mice.

There was a radio-adaptive response seen with the blood cell DNA damage in mice irradiated with an adaptive dose of 1 Gy 24 h prior to a dose of 5 Gy when compared to a single sublethal radiation dose group. The papaya leaf extract and DADS also contributed this effect and further enhanced the adaptive response.

Subjects: Bioscience; Food Science & Technology; Medicine, Dentistry, Nursing & Allied Health

Keywords: antioxidant; radiation toxicity; dose reduction factor; radioadaptive response

1. Introduction
Carica papaya (L.)—commonly known as papaya, pawpaw has been studied extensively for its invaluable medicinal properties. Its herbal tea has shown to enhance the platelet levels in dengue patients (Kala, 2012). These properties can be attributed to the various bioactive components like alkaloids, flavonoids, tannins, saponins, carbohydrates, proteins, fat, and steroids.

Diallyl disulfide (DADS) is a constituent of garlic which has been studied previously for its anticancer properties (Lawson, Wang, & Hughes, 1991; Yan, Wang, & Barlow, 1992). Garlic and garlic extracts have been reported previously for their antioxidant properties and protection against free radical damage in the body (Block, 1985). Studies on the antioxidant properties of garlic compounds, allyl cysteine, alliin, allicin, and allyl disulfides indicated hydroxyl scavenging potential and prevent microsomal lipid peroxidation (Chung, 2006).

Exposure to ionizing radiations causes deleterious effects on physiology, metabolism, and behavior in almost all organisms. One of the basic mechanisms of radiation damage is the production of free radicals, leading to the formation of peroxides and oxidative reactive species. The free radicals produced may also lead to leakage of lysosomal acid phosphatases, changes in the surface properties of chromosomes leading to stickiness, breakage of double-strands of DNA, and chromosomal aberrations (Kumar et al., 2006). The study of these effects and novel approaches to counter them is of great importance in the field of radiation biology. In the present study, the effects of DADS—A synthetic compound and *C. papaya* (L.) extract—A natural herbal source were compared against different doses of electron beam radiation-induced hematological and cytogenetic changes in mice.

In the field of medicine using radiation treatment against cancers, these effects can be seen in normal tissues also. Thus, various attempts have been made to increase the radiation sensitivity of cancerous cells and or protect the normal cells from radiation. Thus, the use of chemicals along with radiation has proved to be ameliorative in cancer treatment (Upadhyay, Dwarakanath, Ravindranath, & Mathew, 2005). Herbal formulations which are non-toxic and inexpensive have been evaluated for their radioprotective efficacy (Suzen, 1999). WR-2721, also known as amifostine—a phosphorothioate is one of the best known radioprotectors which has undergone extensive preclinical testing and clinical trials for its radioprotective and chemoprotective effects (Citrin et al., 2010). The major reason for its success is its selective concentration in the normal tissues compared to the tumor tissue and its interaction with the DNA (Joseph, 1997; Yuhas, 1980).

2. Materials and methods
The leaves of *C. Papaya* (Linn.) were obtained from the local plantation. The leaves were washed, dried in a hot air oven at 55°C. After drying, the leaves were powdered and stored in airtight containers. The synthetic DADS (>75% pure, HPLC graded) was obtained from TCI chemicals, Japan.

2.1. Preparation of extracts (Vennila et al., 2012; Zohra, Meriem, Samira, & Alsayadi Muneer, 2012)
The aqueous extracts of *C. papaya* (Linn.) were prepared by boiling approximately 5 g of powdered leaves with 100 mL distilled water for 20 min and kept overnight in a refrigerator. The extracts were filtered using Whatmann No.1 filter paper, dried, and stored in airtight containers.

2.2. Ethical clearance

The study has been ethically approved by the Institutional Animal Ethics Committee of the K S Hegde Medical Academy, Nitte University Ref. KSHEMA/IAEC/17/2013 dated 16.12.2013.

2.3. Administration and optimization of doses

The major purpose of a radioprotective drug is to enhance the life span and reduce the toxicities induced because of radiation. The major types of signs and symptoms include the gastrointestinal syndrome that occurs within 10 days after irradiation of mice to a lethal dose and the bone marrow or hematopoietic syndrome that occurs after 10 days post lethal irradiation (Ikpeme, Ekaluo, Kooffreh, & Udensi, 2011; Patil, Shetty, Bhide, & Narayanan, 2013). These syndromes are the primary cause for death and the secondary causes include severe intestinal and lung infections because of the radiation-induced immune suppression. The basis for selecting an optimum dose is to understand the minimum drug load that must be present in the system to ensure maximum survival. Hence, a series of doses ranging from the lowest to the highest non-toxic dose may be used to study the effect. The dose at which maximum survival of mice seen in a 30-day survival study post lethal irradiation is the optimum radioprotective dose.

The aqueous extract of *C. papaya* (L.) was chosen for the study as it was soluble and suitable for oral administration. DADS formed a colloidal suspension in water. The mice were treated with papaya leaf extract at 5, 50, 100, 200, 250, 500, 1,000 mg/kg body weight concentration and DADS at 0.5, 2, 5, 10, 20 mg/kg body weight for five consecutive days and on the 5th day one hour after dose administration they were irradiated with 10 Gy dose of lethal electron beam radiation. The mice were observed for maximum survival within 30 days post irradiation. The survival analysis from Kaplan-Meier curve indicated the optimum radioprotective dose for papaya leaf extract at 100 mg/kg body weight and for DADS at 10 mg/kg body weight.

For the sublethal, low dose and adaptive response studies, a volume of 0.1 mL/10 g of *C. papaya* (L.) aqueous leaf extract and DADS were administered using oral gavages to their respective groups for five consecutive days. On the 5th day, one hour after the administration of drug the mice were placed in well-ventilated perspex boxes with a dimension 3 × 6 cms. They were irradiated with a sublethal dose of 6 Gy electron beam radiation, 1 Gy as low dose, and 1 Gy priming dose with 5 Gy as the challenge dose for adaptive dose were given at a dose rate of 1 Gy/min, with a source to target distance of 100 cms.

The mice were dissected after 24 h of irradiation. They were anesthetized and the whole blood was collected by cardiac puncture for hematological estimations. The organs like liver, kidney, brain, and spleen were dissected and weighed. The bone marrow was removed for cytological studies.

The tissues of liver, kidney, and brain were washed with ice cold phosphate buffered saline and stored in ice cold phosphate buffered saline until processing. A 10% homogenate w/v of tissue was prepared using ice cold PBS with pH-7.4 using Remi (RQ-127A) homogenizer. The homogenized samples were then centrifuged for 20 min at 10,000 rpm at 4°C in Remi cooling centrifuge (C24BL). The supernatant was separated and used for all the estimations.

The whole blood was collected in 2% EDTA tubes and processed for hematological studies within 3 h of collection. The serum/plasma and the tissue homogenates were separated by centrifugation and stored in Panasonic (MDF-U334-PE) biomedical freezer at −30°C until further processing.

2.4. Hematological studies

The hematological studies were done using Erma veterinary blood cell counter (PCE-210VET) using the whole blood collected in 2% EDTA.

2.5. Antioxidant studies
Suitable spectrophotometric methods were used and the measurements recorded in Systronics PC-based double beam UV spectrophotometer 2202.

2.6. Total antioxidant capacity (Prieto, Pineda, & Aguilar, 1999)
The absorbance was read at 695 nm and the values are expressed in terms of ascorbic acid equivalents.

2.7. Superoxide dismutase (Mc Cord & Fridovich, 1969)
The blue-green colored solution was read at 560 nm. The activity of superoxide dismutase (SOD) was expressed in Units/mg of protein for homogenates.

2.8. Glutathione peroxidase (Hafeman, Sundae, & Houestra, 1974)
The absorbance was measured at 412 nm and the activity expressed in units/mg of protein for homogenates.

2.9. Catalase (Aebi, 1974)
The activity was expressed in units/mg of protein for homogenates and Units/mg of protein for homogenates.

2.10. Lipid peroxidation and membrane stabilization

2.10.1. Formation of malondialdehyde (Ohkawa, Ohishi, & Yagi, 1979)
The endpoint was measured at 535 nm and calculated using malondialdehyde standard curve.

2.11. Reduced glutathione (Moron, Depierre, & Mannervik, 1979)
The absorbance was read at 412 nm within 10 min and calculated using a GSH standard curve.

2.12. Genotoxicity and cytotoxicity studies

2.12.1. Comet assay (Singh, McCoy, Tice, & Schneider, 1988; Tice et al., 2000)
The slides were scored using comet score software and the parameters like tail moment, olive moment, and percent DNA in tail were considered to determine level of the genotoxicity and cytotoxicity.

2.12.2. Dose reduction factor (Nageshwar Rao, Satish Rao, Kiran Aithal, & Sunil Kumar, 2009)
The ratio of median lethal doses (LD_{50}) with and without the compound gives the dose reduction factor (DRF). This factor demonstrates the radioprotective efficacy of any compound which in turn has prolonged the life expectancy in irradiated mice.

2.12.3. Statistical analysis of data
The statistical analysis of the data was done using the SPPS software. The dose optimization was done using Kaplan-Meir survival analysis. The survival studies were evaluated using probit analysis. The difference between various groups was analyzed by ANOVA with Tukey's multiple comparison *post hoc* test. The results with p value less than 0.05 was considered significant.

3. Results
Figure 1 shows the survival of mice at different doses of radiation in a 30 day post irradiation observation at different radiation doses. Figures 2 and 3 summarize the Kaplan-Meir survival analysis to extrapolate the optimum survival dose of papaya leaf extract and DADS intervention at a lethal dose EBR. An optimum radioprotective dose of 100 mg/kg body weight and a dose of 10 mg/kg body weight were shown by papaya leaf extract and DADS, respectively.

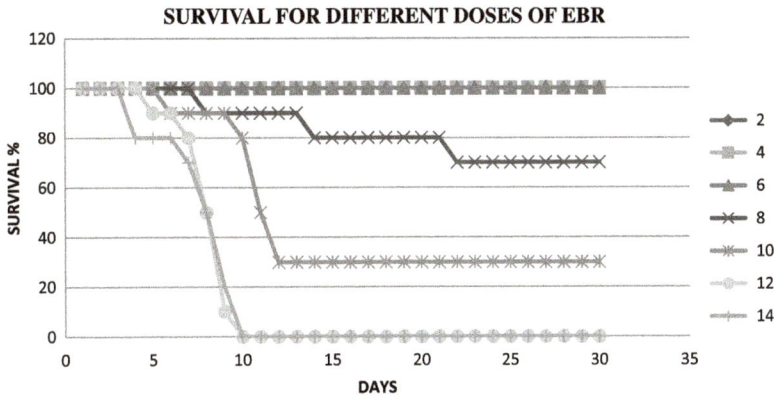

Figure 1. The survival of mice at different doses of electron beam radiation.

The results of hematological parameters like Hb, RBC, hematocrit, WBC, and platelets indicate a significant decrease in the irradiated groups when compared to the normal control groups ($p < 0.05$, $p < 0.05$, $p < 0.05$, $p < 0.001$, $p < 0.05$, respectively). Figure 4 show the hematological changes in control and irradiated groups with respect to papaya extract and DADS intervention. The platelet count is slightly enhanced in irradiated groups pretreated with papaya extract when compared to normal groups, but the results are of no statistical significance ($p > 0.05$). In low dose irradiated groups, there is a decrease in Hb, RBC, and WBC levels ($p < 0.001$, $p < 0.05$, $p < 0.001$, respectively). There is a moderate increase in the RBC levels in papaya leaf extract treatment prior irradiation groups ($p < 0.05$). In adaptive doses there is a decrease in Hb, RBC, and WBC levels in irradiated groups ($p < 0.01$, $p < 0.05$, $p < 0.001$, respectively). But there are no changes seen in papaya leaf extract and DADS pretreatment groups. Figures 5 and 6 summarize the results of the different hematological parameters in response to low dose and adaptive doses of radiation with respect to the effects of papaya leaf extract and DADS intervention.

In the antioxidant studies, there is a significant decrease in catalase ($p < 0.01$), GSH levels ($p < 0.01$), and total antioxidant capacity ($p < 0.05$) in the irradiated groups at sublethal dose. The increase in catalase was only seen in irradiated groups pretreated with DADS ($p < 0.01$). The GSH levels were enhanced in irradiated groups pretreated with papaya extract and DADS ($p < 0.01$). The total antioxidant

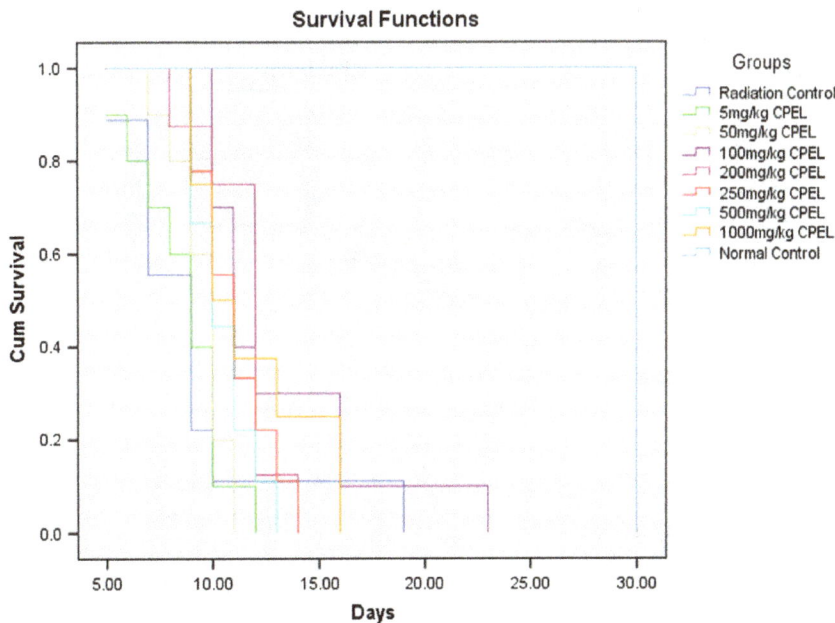

Figure 2. The maximum survival of mice at different doses of *Carica papaya* (L.) aqueous leaf extract.

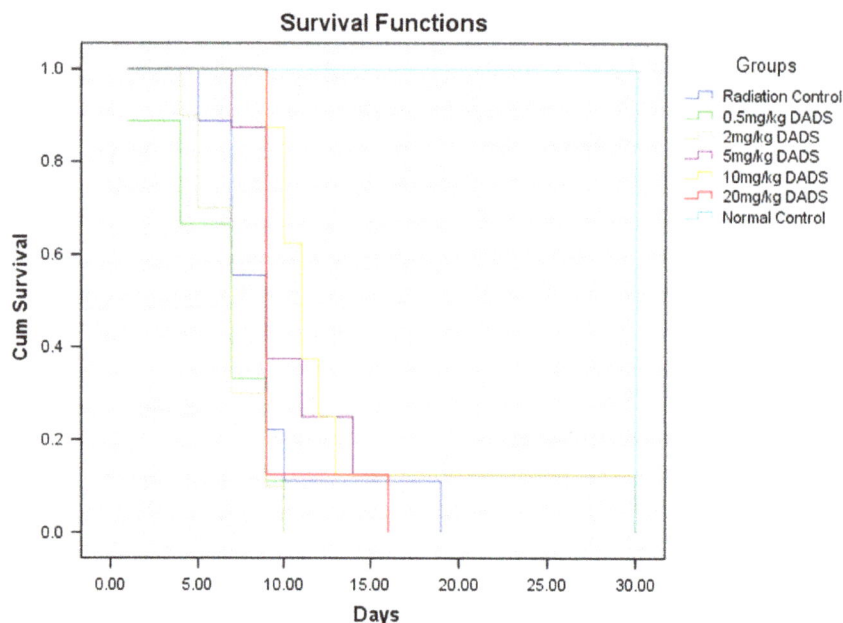

Figure 3. The maximum survival of mice at different doses of diallyl disulfide.

levels were rejuvenated in irradiated groups pretreated with papaya extract ($p < 0.01$), but no significant changes were seen in the irradiated group pretreated with DADS. Figure 7 summarizes the antioxidant and lipid peroxidation parameters of control and irradiated groups with respect to papaya leaf extract and DADS intervention at sublethal doses of EBR. At low and adaptive doses, the total antioxidant capacity also was reduced ($p < 0.05$) and near normal levels were retained in other groups but the difference was not statistically significant. The catalase levels remained unaltered in all the groups with low and adaptive radiation doses. GSH levels remained elevated in the irradiated groups pretreated with papaya extract and DADS at low radiation doses and not significant at adaptive dose. No statistically significant difference was seen in glutathione peroxidase (GPx) and SOD levels in sublethal and low doses whereas the levels of GPx was reduced in radiation control group ($p < 0.01$) but in the other groups the levels remained unchanged. The increased MDA levels in irradiated groups indicate increased lipid peroxidation at sublethal, low, and adaptive radiation doses ($p < 0.01, p < 0.05, p < 0.05$, respectively). The levels were found to be normalized in all the irradiated groups pretreated with

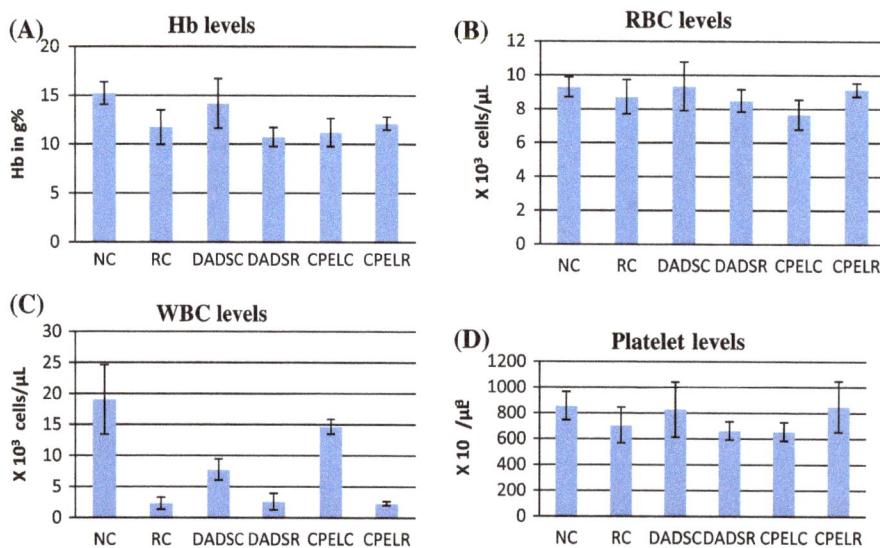

Figure 4. The hematological parameters of control, sublethal dose radiation control, and papaya and DADS pretreatment prior irradiation groups.

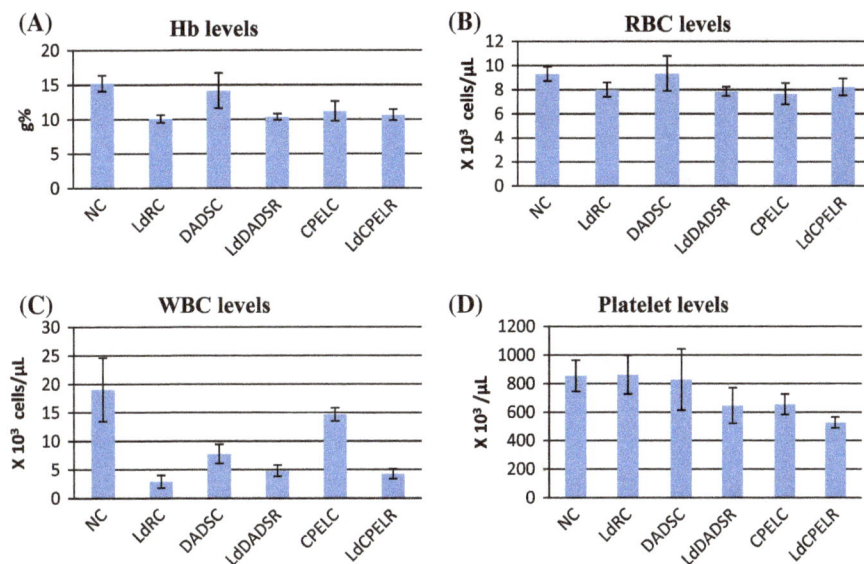

Figure 5. The hematological parameters of control, low dose radiation control, and papaya and DADS pretreatment prior irradiation groups.

papaya extract and DADS. In adaptive doses the MDA levels remained increased in irradiated groups pretreated with DADS and papaya extract. Figures 8 and 9 summarize the antioxidant and lipid peroxidation parameters in control and irradiated groups with respect to papaya leaf extract and DADS intervention at low and adaptive doses of EBR.

There is an overall increase in the olive tail moment (OTM) of the irradiated groups at sublethal and low doses ($p < 0.01$, $p < 0.01$, respectively) of radiation whereas there is decrease in the OTM at adaptive doses of radiation ($p < 0.01$). But the irradiated groups pretreated with papaya extract and DADS did not show any statistically significant results though a trend of reduction in DNA damage was observed in them when compared to the radiation control ($p > 0.05$). The other parameters of comet score like the percentage of DNA in tail and tail moment also showed a trend of decrease in the irradiated groups pretreated with papaya extract and DADS when compared to the radiation control, the results were not statistically significant. Figures 10–12 summarize the olive tail moment

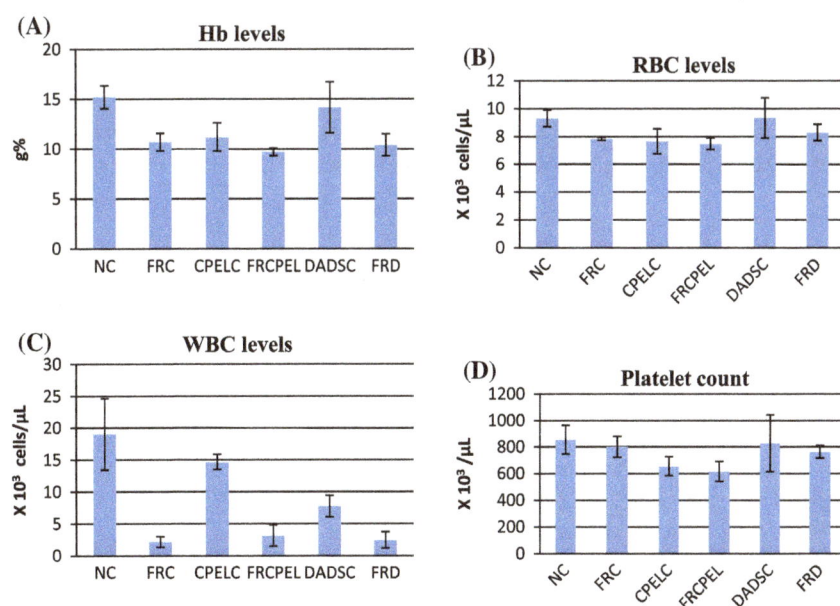

Figure 6. The hematological parameters of control, adaptive dose radiation control, and papaya and DADS pretreatment prior irradiation groups.

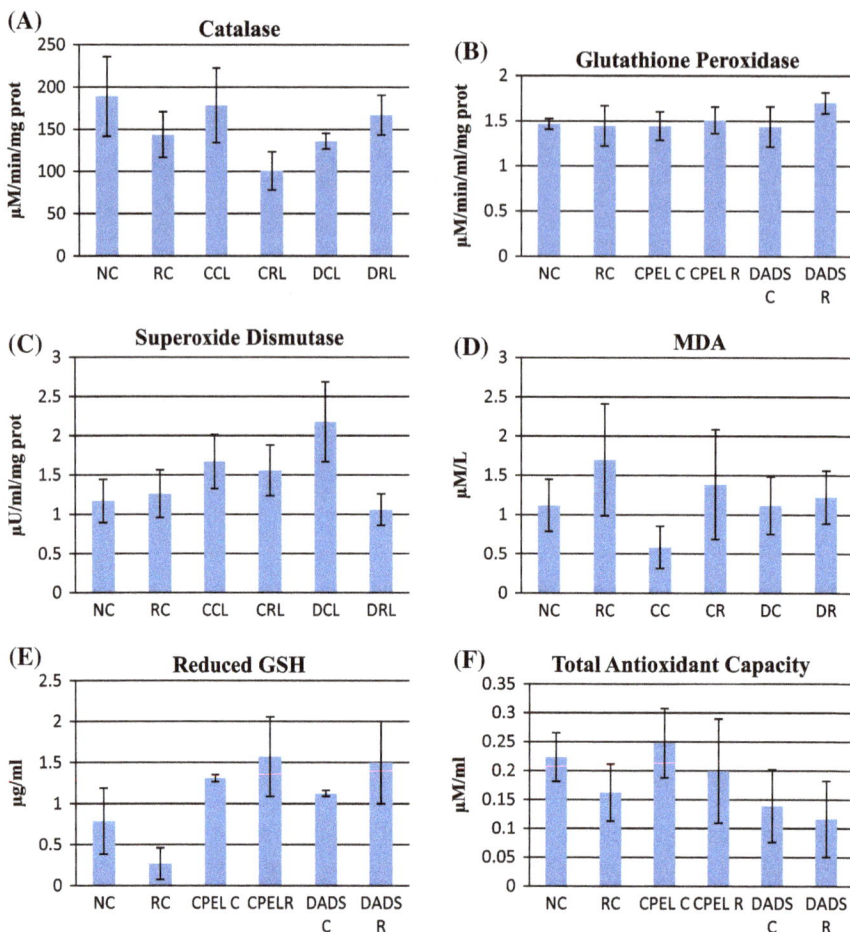

Figure 7. The antioxidant and lipid peroxidation parameters of control, sublethal dose radiation control, and papaya and DADS pretreatment prior irradiation groups.

indicating DNA damage levels in control and irradiated groups with respect to papaya leaf extract and DADS intervention at sublethal, low, and adaptive doses of EBR.

The DRF was calculated using the LD_{50} values obtained with and without the compound of interest. The LD_{50} without any intervention was obtained at 8 Gy and the LD_{50} with the papaya extract intervention shifted the LD_{50} to 9.0 Gy and the LD_{50} with DADS was found to be 10 Gy. Thus on calculating the DRF, a value of 1.12 and 1.25 was obtained for papaya leaf extract and DADS, respectively.

4. Discussion

Previous studies have shown that electron beam radiation at doses of 10 Gy, 12 Gy, and 14 Gy causes severe radiological effects in the gastrointestinal tract and bone marrow (Jagetia & Tiyyagura, 2002). In our study also death occurred within 15 days post radiation and maximum deaths were observed between the 9th and 14th days with 10 Gy, 12 Gy, and 14 Gy radiation. At higher doses of 12 Gy and 14 Gy there was facial edema, hypopigmentation, corneal opacity indicating loss of eyesight in mice. Reduced body weight and ruffled skin are some of the common radiological characteristics seen in mice. There was a drastic reduction in the body weights for initial 15 days post radiation. The present study reports the median lethal dose for electron beam radiation with 15 MeV electrons at a dose rate of 1 Gy/min as 8.933 Gy in Swiss albino mice.

The median lethal dose (LD_{50}) indicating 50% survival was extrapolated as 8.933 Gy using Probit analysis. We are reporting the LD_{50} in the lower range compared to the previous studies for the dose rate of 1 Gy/min at 15 MeV for electron beam radiation.

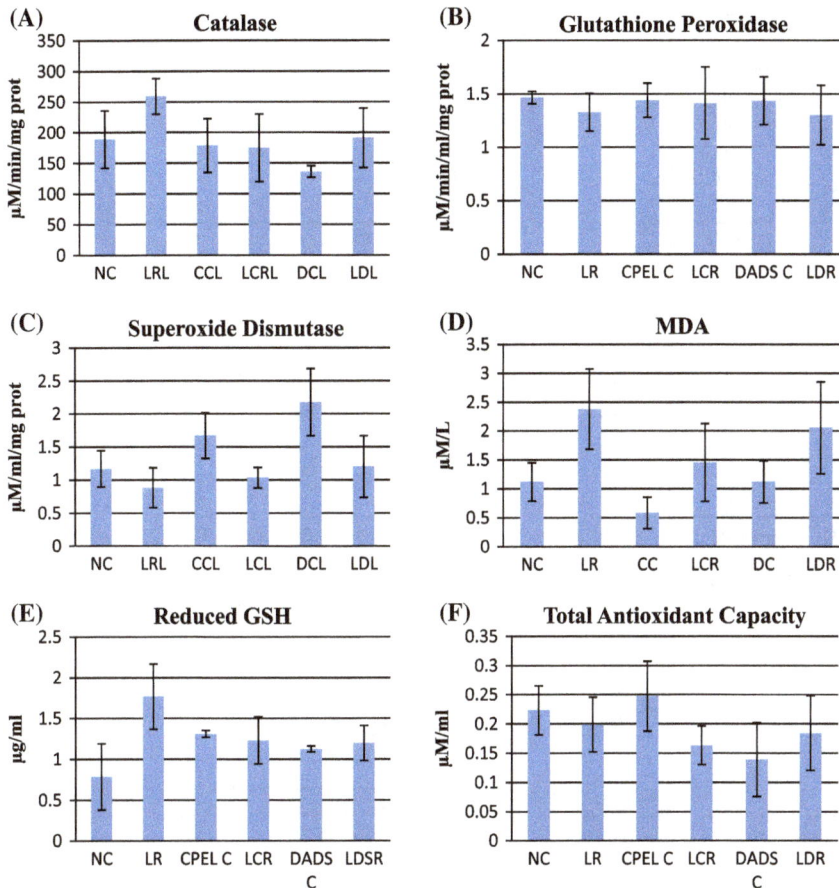

Figure 8. The antioxidant and lipid peroxidation parameters of control, low dose radiation control, and papaya and DADS pretreatment prior irradiation groups.

The median lethal dose for DADS was found to be 75 mg/kg body weight. Previous study done by Amagase (2006) has recorded the median lethal dose at 135 mg/kg body weight in females and 150 mg/kg body weight in males. The results of current study reveal diallyl sulfide to be highly toxic at those doses. From the Kaplan-Meier survival analysis the optimum radioprotective dose for DADS was found to be 10 mg/kg body weight.

Radiation-induced hematological suppression was seen at sublethal doses of 6 Gy irradiated groups. Decreases were seen in hemoglobin, RBC, total WBC count, and platelet counts in irradiated groups. Pretreatment of DADS prior irradiation did not show any ameliorative effects on hemoglobin, RBC, total WBC counts. A decrease in the total antioxidant capacity and catalase levels were seen with irradiated groups when compared to the non-irradiated groups. But no changes were seen in the levels of SOD and glutathione peroxidase levels were seen in the irradiated groups. Reduced levels of GSH and MDA levels were seen in the irradiated groups indicating increased lipid peroxidation. DADS prior treatment to radiation enhanced the levels of catalase, glutathione peroxidase and GSH. It also decreased the levels of MDA, but no changes were seen in the SOD levels and total antioxidant capacity. There was increased olive tail moment in irradiated group indicating radiation induced DNA damage, but no changes were seen with DADS pretreatment prior irradiation groups.

Similar effects were seen in the radiation studies at a lower sublethal dose of 1 Gy, where there was a decrease in the hematological parameters. Papaya leaf extract pretreatment group prior irradiation enhanced the RBC levels. Decrease in total antioxidant capacity, GSH, catalase, SOD, GPx levels, and increase in MDA levels were seen in irradiated groups and ameliorative effects were seen in the groups pretreated with papaya leaf extract and DADS.

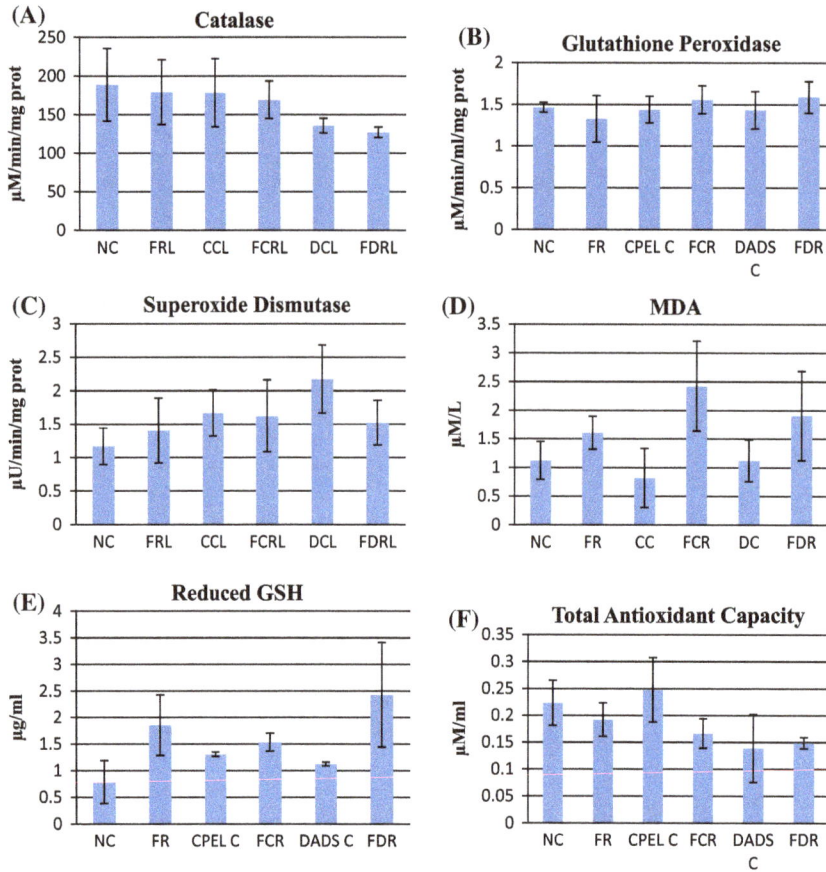

Figure 9. The antioxidant and lipid peroxidation parameters of control, adaptive dose radiation control, and papaya and DADS pretreatment prior irradiation groups.

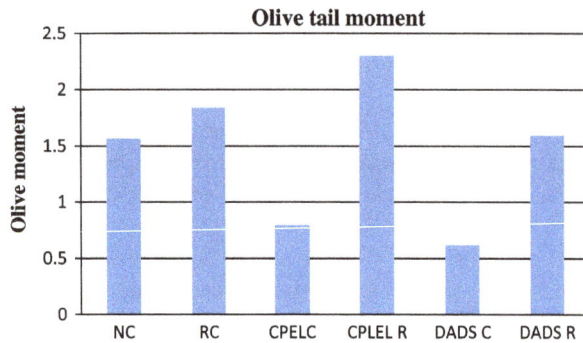

Figure 10. The olive tail moment in control, sublethal dose radiation control, and papaya and DADS pretreatment prior irradiation groups.

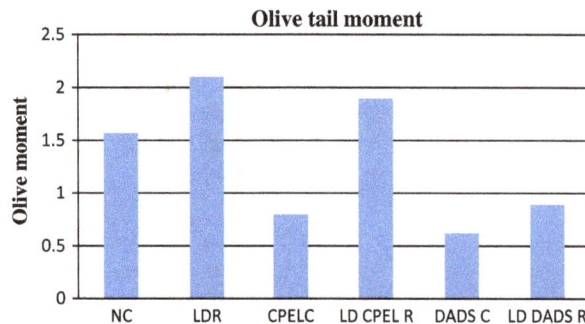

Figure 11. The olive tail moment in control, low dose radiation control, and papaya and DADS pretreatment prior irradiation groups.

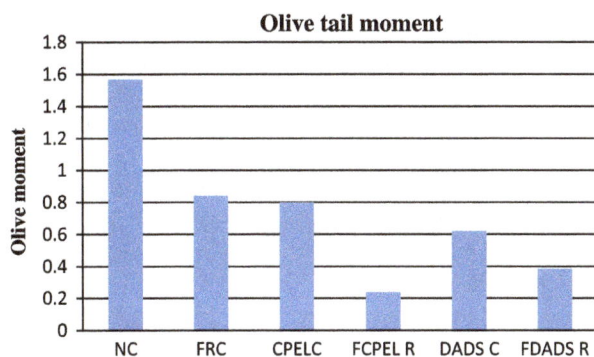

Figure 12. The olive tail moment in control, adaptive dose radiation control, and papaya and DADS pretreatment prior irradiation groups.

In order to study the adaptive response, mice were irradiated initially with 1 Gy of priming dose and after 24 h a challenge dose of 5 Gy was given. There was no adaptive response seen in the hematological parameters, as there was decrease in the Hb, RBC, and WBC counts seen in irradiated groups with single sublethal dose and adaptive split dose groups. The unaltered levels of catalase, GPx, and slight increase in SOD and GSH levels provide a circumstantial evidence of dose adaptation, but decrease in the total antioxidant capacity and increase in MDA indicate deteriorating antioxidant levels. But a distinct decrease in the olive tail moment of the group irradiated with adaptive split dose on comparison with single sublethal dose indicated an adaptive response in the DNA damage levels. The DADS pretreatment prior adaptive split dose irradiation showed an enhanced adaptive response to DNA damage when compared to its respective adaptive split dose radiation control group.

Studies have shown that radiation reduces the hematological parameters like hemoglobin, red blood cell count, white blood cell count, platelet count, and inhibits hematopoiesis (Begum, Prasad, & Thayalan, 2012). The present study also shows similar effects in sublethal, lower sublethal, and adaptive split dose radiation groups. The deleterious effects of radiation include formation of free radicals and development of oxidative stress. Radiation has been shown to induce oxidative stress by increasing membrane peroxidation and reducing antioxidants in the blood. There is an overall reduction in the blood antioxidants, reduced glutathione, and the antioxidant enzymes like SOD, catalase, glutathione S transferase, and glutathione peroxidase (Verma et al., 2011). The current study reveals a similar response with sublethal, lower sublethal, and adaptive split dose irradiation. Radiation also inflicts damage to the DNA and leads to loss of cell viability and increased apoptosis (Madhu, Suchetha Kumari, Naveen, & Sanjeev, 2012). From the present study, we report an increased DNA damage in sublethal and lower irradiation groups and a reduced DNA damage in adaptive split dose groups.

Sulfhydryl compounds like amifostine (Kouvaris, Kouloulias, & Vlahos, 2007), cysteine, cysteamine, amino ethyl thiourea, glutathione (Nair, Parida, & Nomura, 2001) have been shown to reduce the radiation induced tissue damage by scavenging the free radicals or by repairing the radioactivated polymers or activating the repair enzymes (Suzen, 1999). The naturally occurring organosulfur compounds diallyl sulfide (DAS), diallyl disulfide (DADS), allyl methyl sulfide (AMS), allyl isothio cyanate (AITC), and phenyl isothio cyanate (PITC) have also been shown for their radioprotective activity by enhancing the glutathione content (Chittezhath & Kuttan, 2006). The studies have also revealed that these compounds must be present in the system before radiation (Montoro et al., 2011). Hence, sulfhydryl compounds have only prophylactic effect and no therapeutic effect. Intra cellular non-protein sulfydryl groups are mainly involved in enhancing the membrane stability and antioxidant capacity and DADS in the present study has shown enhanced GSH levels, total antioxidant capacity, and reduced the malondialdehyde levels in sublethal and lower sublethal dose radiation groups. DADS has shown reduced DNA damage in adaptive split dose groups and also has contributed to the radio-adaptive response.

Recent studies have concluded that *C. papaya* (L.) leaf extract significantly increased the platelet and RBC counts in murine models (Dharmarathna, Wickramasinghe, Waduge, Rajapakse, & Kularatne, 2013). Studies done by Joseph (1997) have demonstrated the hematological potential of *C. papaya* (Linn.) leaf extracts. Our present study also indicates a thrombopoietic and erythropoietic potential of *C. papaya* (L.) leaf extract in radiation induced-hematopoietic suppression. According to a study done by Patil et al. (2013) *C. papaya* (L.) leaves contain various phytoconstituents like saponins, tannins, cardiac glycosides, and alkaloids. The alkaloids present include carpaine, pseudocarpaine, and dehydrocarpaine I and II. These constituents may act on the bone marrow, prevent its destruction and enhance its ability to produce platelets. Moreover, it can also prevent platelet destruction in the blood and thereby increase the life of the platelet in circulation. In the present study, papaya leaf extract has shown moderate increase in antioxidant levels and reduced lipid peroxidation at sublethal and lower sublethal doses. At adaptive split doses, it reduced the DNA damage levels thus contributing to the radio-adaptive response in mice.

The overall survival of mice against a lethal dose of radiation is dependent on the resistance acquired by mice or the antioxidant load given to it (Jagetia, 2007). This demonstrates the amount of radiation dose the animal can withstand with minimal damages. The median lethal dose with and without the compound and extract was calculated. Without any intervention the LD_{50} was found to be 8 Gy and with intervention the LD_{50} was found to be 9 Gy and 10 Gy for *C. papaya* (L.) and DADS, respectively. Enhancement of the median lethal dose for the normal cells will increase the efficacy of the radiotherapy. A similar DRF of 1.12 and 1.25 for *C. papaya* (L.) and DADS, respectively, obtained have increased the percentage survival of pretreated mice by approximately 0.1 Gy.

5. Conclusions

The results indicate a potential hematoprotective property of *C. papaya* (L.) and DADS against radiation-induced hematological and biochemical alterations in Swiss albino mice. An adaptive response was also seen in *C. papaya* (L.) and DADS pretreatment. Further studies may be carried out to understand the dose and time dependant modifications that take place post irradiation and development of specific target molecules to counter the radiation-induced hematological and biochemical damages.

Acknowledgements
The authors are grateful to the board of research in nuclear sciences (BRNS) sanction No. 2012/34/26/BRNS for funding the study and Nitte University for providing the laboratory facility. The authors acknowledge the support of Dr Ganesh Sanjeev, Professor, Microtron center, Mangalore University, Dr Damodhara Gowda, Department of Physiology and Dr Chandrika Rao, Department of Pathology, K.S. Hegde Medical Academy, Nitte University. The authors also thank Dr Sanal T.S., Associate professor, Department of Biostatistics, K.S. Hegde Medical Academy, Nitte University.

Funding
This work was supported by Board of Research in Nuclear Sciences [grant number 2012/34/26/BRNS].

Competing Interest
The authors declare no competing interest.

Author details
Yogish Somayaji T[1]
E-mail: Somayaji387@gmail.com
Vidya V[1]
E-mail: vidyavr27@gmail.com
Vishakh R[1]
E-mail: vishakh.kedilaya@yahoo.co.in
Jayarama Shetty[2]
E-mail: jairamshettyk@gmail.com
Alex John Peter[2]
E-mail: alexau2008@gmail.com
Suchetha Kumari N[3]
E-mail: kumari.suchetha@gmail.com
[1] Central Research Laboratory, K S Hegde Medical Academy, Nitte University, Mangaluru, Karnataka, India.
[2] Department of Oncology, Nitte Leela Narayan Shetty Memorial Institute of Oncology, Mangaluru, Karnataka, India.
[3] Department of Biochemistry, K S Hegde Medical Academy, Nitte University, Mangaluru, Karnataka, India.

References
Aebi, H. (1974). Catalase estimation. In H. V. Berg Meyer (Ed.), *Methods of enzymatic analysis* (pp. 673–684). New York, NY: Verlag Chemie.
http://dx.doi.org/10.1016/B978-0-12-091302-2.50032-3
Amagase, H. (2006). Clarifying the real bioactive constituents of garlic. *Journal of Nutrition, 136*, 716S–725S.
Begum, N., Prasad, N. R., & Thayalan, K. (2012). Apigenin protects gamma-radiation induced oxidative stress, hematological changes and animal survival in whole body

irradiated Swiss albino mice. *International Journal of Nutrition, Pharmacology, Neurological Diseases, 2*, 45–52.

Block, E. (1985). The chemistry of garlic and onions. *Scientific American, 252*, 94–99.

Chittezhath, M., & Kuttan, G. (2006). Radioprotective effects of naturally occurring organosulfur compounds. *Tumori, 92*, 162–169.

Chung, L. Y. (2006). The antioxidant properties of garlic compounds: Allyl cysteine, alliin, allicin and allyl disulphide. *Journal of Medicinal Food, 9*, 205–213.

Citrin, D., Cotrim, A. P., Hyodo, Fuminori, Baum, B. J., Krishna, M. C., & Mitchell, J. B. (2010). Radioprotectors and mitigators of radiation-induced normal tissue injury. *The Oncologist, 15*, 360–371. http://dx.doi.org/10.1634/theoncologist.2009-S104

Dharmarathna, S. L. C. A., Wickramasinghe, S., Waduge, R. N., Rajapakse, R. P. V. J., & Kularatne, S. A. M. (2013). Does Carica papaya leaf-extract increase the platelet count? An experimental study in a murine model. *Asian Pacific Journal of Tropical Biomedicine, 3*, 720–724. http://dx.doi.org/10.1016/S2221-1691(13)60145-8

Hafeman, D. G., Sundae, R. A., & Houestra, W. G. (1974). Effect of dietary selenium on erythrocyte and liver glutathione peroxidase in the rat. *Journal of Nutrition, 104*, 580–587.

Ikpeme, E. V., Ekaluo, U. B., Kooffreh, M. E., & Udensi, O. (2011). Phytochemistry and heamatological potential of ethanol seed leaf and pulp extracts of carica papaya (Linn.). *Pakistan Journal of Biological Sciences, 14*, 408–411. doi:10.3923/pjbs.2011.408.411

Jagetia, G. C. (2007). Recent advances in Indian herbal drug research guest editor: Thomas Paul Asir Devasagayam radioprotective potential of plants and herbs against the effects of ionizing radiation. *Journal of Clinical Biochemistry and Nutrition, 40*, 74–81. http://dx.doi.org/10.3164/jcbn.40.74

Jagetia, G. C. R., & Tiyyagura, K. (2002). The grapefruit flavanone naringin protects against the radiation-induced genomic instability in the mice bone marrow: A micronucleus study. *Mutation Research/Genetic Toxicology and Environmental Mutagenesis, 519*, 37–48. http://dx.doi.org/10.1016/S1383-5718(02)00111-0

Joseph, W. F. (1997). Pharmacologic approaches to protection against radiation- induced lethality and other damage. *Environmental Health Perspectives, 105*, 1473–1478.

Kala, C. P. (2012). Leaf juice of *Carica papaya* (L) A remedy of dengue fever. *Medicinal Aromatic Plants, 1*, 1–2.

Kouvaris, J. R., Kouloulias, V. E., & Vlahos, L. J. (2007). Amifostine: The first selective-target and broad-spectrum radioprotector. *The Oncologist, 12*, 738–747. doi:10.1634/theoncologist.12-6-738 http://dx.doi.org/10.1634/theoncologist.12-6-738

Kumar, M., Samarth, R., Kumar, M., Senthamil, S. R., Saharan, B., Kumar, A. (2006). Protective effect of *Adhatoda vascia* nees against radiation-induced damage at cellular, biochemical and chromosomal levels in Swiss albino mice. *eCAM*, 1–8.

Lawson, L. D., Wang, Z. J., & Hughes, B. G. (1991). Identification and HPLC quantitation of the sulfides and dialk(en)yl thiosulfinates in commercial garlic products. *Planta Medica, 57*, 363–370. http://dx.doi.org/10.1055/s-2006-960119

Madhu, L. N., Suchetha Kumari, N., Naveen, P., & Sanjeev, G. (2012). Protective effect of *Nardostachys jatamansi* against radiation-induced damage at biochemical and chromosomal levels in swiss albino mice. *Indian Journal of Pharmaceutical Sciences, 74*, 460–465.

Mc Cord, J. M., & Fridovich, I. (1969). Superoxide dismutase enzyme function for erythrocraprein. *Journal of Biochemistry, 244*, 6049–6056.

Montoro, A., Barquinero, J. F., Almonacid, M., Montoro, A., Sebastià, N., Verdù, G., ... Soriano, J. M. (2011).

Concentration-dependent protection by ethanol extract of propolis against γ-ray-induced chromosome damage in human blood lymphocytes. *Evidence-Based Complementary and Alternative Medicine*, Article ID 174853, 7 doi:10.1155/2011/174853

Moron, M. A., Depierre, J. W., & Mannervik, B. (1979). Levels of glutathione, glutathione reductase and glutathione S-transferase activities in rat lung and liver. *Biochimica et Biophysica Acta (BBA)—General Subjects, 582*, 67–78. http://dx.doi.org/10.1016/0304-4165(79)90289-7

Nageshwar Rao, B., Satish Rao, B. S., Kiran Aithal, B., & Sunil Kumar, M. R. (2009). Radiomodifying and anticlastogenic effect of Zingerone on Swiss albino mice exposed to whole body gamma radiation. *Mutation Research, 677*, 33–41.

Nair, C. K. K., Parida, D. K., & Nomura, T. (2001). Radioprotectors in radiotherapy. *Journal of Radiation Research, 42*, 21–37. http://dx.doi.org/10.1269/jrr.42.21

Ohkawa, H., Ohishi, N., & Yagi, K. (1979). Assay for lipid peroxides in animal tissues by thiobarbituric acid reaction. *Analytical Biochemistry, 95*, 351–358. http://dx.doi.org/10.1016/0003-2697(79)90738-3

Patil, S., Shetty, S., Bhide, R., & Narayanan, S. (2013). Evaluation of platelet augmentation activity of carica papaya leaf aqueous extract in rats. *Journal of Pharmacognosy and Phytochemistry, 1*, 58–61.

Prieto, P., Pineda, M., & Aguilar, M. (1999). Spectrophotometric quantitation of antioxidant capacity through the formation of a phosphomolybdenum complex: Specific application to the determination of Vitamin E. *Analytical Biochemistry, 269*, 337–341. http://dx.doi.org/10.1006/abio.1999.4019

Singh, N. P., McCoy, M. T., Tice, R. R., & Schneider, E. L. (1988). A simple technique for quantitation of low levels of DNA damage in individual cells. *Experimental Cell Research, 175*, 184–191. http://dx.doi.org/10.1016/0014-4827(88)90265-0

Suzen, S. (1999). Radioprotective agents in the treatment of cancer and the role of amifostine. *Journal of the Faculty of Pharmacy of Ankara University, 28*, 117–128.

Tice, R. R., Agurell, E., Anderson, D., Burlinson, B., Hartmann, A., Kobayashi, H., ... Sasaki, Y. F. (2000). Single cell gel/comet assay: Guidelines for in vitro and in vivo genetic toxicology testing. *Environmental and Molecular Mutagenesis, 35*, 206–221. http://dx.doi.org/10.1002/(SICI)1098-2280(2000)35:3<>1.0.CO;2-I

Upadhyay, S. N., Dwarakanath, B. S., Ravindranath, T., & Mathew, T. L. (2005). Chemical radioprotectors. *Defence Science Journal, 55*, 403–425. http://dx.doi.org/10.14429/dsj.55.2003

Vennila, S., Mohana, S., Bupesh, G., Mathiyazhagan, K., Dhanagaran, D., Baskar, M., ... Leeba, B. (2012). Qualitative phytochemical screening and invitro antioxidant activity of *Helicteres isora* L. *Herbal Tech Industry*, 14–18.

Verma, P., Sharma, P., Parmar, J., Sharma, P., Agrawal, A., & Goyal, P. K. (2011). Amelioration of radiation-induced hematological and biochemical alterations in Swiss albino mice by panax ginseng extract. *Integrative Cancer Therapies, 10*, 77–84. http://dx.doi.org/10.1177/1534735410375098

Yan, X., Wang, Z., & Barlow, P. (1992). Quantitative estimation of garlic oil content in garlic oil based health products. *Food Chemistry, 45*, 135–139. http://dx.doi.org/10.1016/0308-8146(92)90024-V

Yuhas, J. M. (1980). Active versus passive absorption kinetics as the basis for selective protection of normal tissues by S-2-(3- aminopropylamino)-ethylphosphorothioic acid. *Cancer Research, 40*, 1519–1524.

Zohra, S. F., Meriem, B., Samira, S., & Alsayadi Muneer, M. S. (2012). Phytochemical screening and identification of some compounds from Mallow. *Journal of Natural Product and Plant Resources, 2*, 512–516.

6

Effect of crystallization of caffeic acid enhanced stability and dual biological efficacy

Md Shamsuddin Sultan Khan[1]*, Muhammad Adnan Iqbal[2] and Amin Malik Shah Abdul Majid[1]

*Corresponding author: Md Shamsuddin Sultan Khan, School of Pharmaceutical Sciences, University of Science Malaysia, Penang, Malaysia
E-mail: jupitex@gmail.com
Reviewing editor: Tsai-Ching Hsu, Chung Shan Medical University, Taiwan
Additional information is available at the end of the article

Abstract: Caffeic acid (CA) is a well-known natural compound used for both health supplement and metabolic diseases. In the present study, single, easy, and quick isolation procedure of caffeic acid was described from rosmarinic acid. A quick crystal formation also was described for highly hygroscopic CA which was spontaneous and stable with lower hygroscopic nature. In addition, the produced crystal CA was highly stable and showed neuroprotective activity with <6 µg/ml in R28 cell line and cytotoxic activity with >100 µg/ml in U87 MG which revealed its dual biological effect for the treatment of Alzheimer's disease and cancer, respectively.

Subjects: Analysis & Pharmaceutical Quality; Analytical Chemistry; Cancer; Natural Products; Organic Chemistry; Pharmacology

Keywords: caffeic acid; rosmarinic acid; neuroprotective activity; cytotoxicity; crystallography; cancer; Alzheimer's diseases

1. Introduction

Caffeic acid (CA) is an interesting small molecule due to its chemical stability and antioxidant power (Chen & Ho, 1997; Gulcin, 2006; Hong, Yang, Nam, Koo, & Lee, 2015). It is a potent molecule which started a rational-based drug discovery for wide range of therapeutic application (Rezg et al., 2015; Rosendahl et al., 2015). Chemical structure of caffeic acid is delicate to drive any kind of desired scaffold hope. In our lab, we routinely check the phytochemicals and computationally design their newer derivatives which ultimately may become suitable candidates for the treatment of Alzheimer's disease and metabolic disorder like cancer. Earlier studies on caffeic acid showed its mild stimulant effect (Liszt, Eder, Wendelin, & Somoza, 2015; Shin et al., 2015; Toth et al., 2015). The past studies also showed its cytoprotective effect (Lee et al., 2015; Nunes et al., 2015), antiangiogenic effect, and moderate toxicity (Akyol et al., 2015; Kosova, Kurt, Olmez, Tuğlu, & Arı, 2015) that can be affected due to the stability of the compounds. This earlier observation postulates that caffeic acid can show

ABOUT THE AUTHORS

The authors are working to find out the differential efficacy of a drug and that solution. This kind of problem is seen during the early development of the drug in clinical case and thereby their failure of approval and marketing. Here, authors are trying to discuss the physical development of a candidate drug which might at least reduce risk of potency of a drug in different clinical cases.

PUBLIC INTEREST STATEMENT

The emerging growing interest to design the formulation in improving drug's shelf life is becoming viable to eradicate their differential efficacy. In the current study, a kind of viable natural solution of increasing shelf-life of the chemical compound and/or drug was described that showed worth even the smallest risk of differential efficacy in clinical set. The formulation scientist's genuine interest should look after the drug's behavior in different states of them under clinical use.

its dual biological efficacy due to half-life, solubility, stability, and physicochemical properties of the compound. Caffeic acid is highly hygroscopic and degrade at room temperature and in the presence of light. To determine the stability impact of the CA in biological activity is essential to assess the therapeutic efficacy. There is no evidence yet to explain the dual biological efficacy of the CA yet. The hygroscopic nature has greater impact for the CA bioactivity which might lead to the dual efficacy of this compound. In the present study, our objective was to stabilize the CA using natural ways and to find the reason of the dual efficacy of the same compound.

2. Experimental

2.1. Crystallography

All reagents were purchased from Fluka™, Aldrich™, and JT Baker™, USA. All solvents were used as received without further purification. The isolated caffeic acid was dissolved by adding methanol dropwise and kept under vacuum hood for two weeks at room temperature. After slow evaporation, faint yellow microcrystals were obtained. The crystals were filtered off and washed by cold methanol at room temperature. The microcrystals solution of Dimethylformamide (DMF) was slowly evaporated to determine the crystal's chemical structure suitable for X-ray diffraction studies. Crystal data were obtained on a Bruker SMART APEXII CCD area-detector diffractometer using graphite monochromated MoKα radiation ($\lambda = 0.71073$ Å) at 100 K. A crystal with dimension 9 × 8 × 31 mm was used. The data were collected and reduced using Cambridge online crystallography program and Crystallo Chemcraft win version 1.8. The structure of all compounds was solved using the Crystallo Chemcraft win version 1.8 program package, and refined using the Crystallo Chemcraft win version 1.8 program package. The molecular graphics were created using Crystallo Chemcraft win version 1.8.

2.2. Cell culture and reagents

The rat retinal cell line R28 was purchased from Kerafast, USA, and human glioblastoma cell line U87-MG was bought from the American Type Culture Collection (ATCC, Manassas, VA, USA) and cultured in Dulbecco's modified eagle medium (DMEM) (GIBCO, USA) supplemented with 10% heat inactivated fetal bovine serum (HIFBS) (GIBCO), 100 units/mL penicillin–streptomycin (GIBCO), and 2 mM glutamine (GIBCO) in a humidified incubator with 5% CO_2 at 37°C. Rosmarinic acid (RA) was purchased from Sigma (Germany). MTT, (3-(4,5-Dimethylthiazol-2-yl)-2,5-diphenyltetrazolium bromide), trypsin, and phosphate-buffered saline (PBS) were purchased from Sigma. Matrigel matrix (10 mg/mL) was obtained from BD Bioscience (USA). About 10 mg of powdered RA (96% purity) was dissolved in sterile, cell-cultured tested DMSO (Sigma, USA) to prepare a 10 mg/mL stock solution, which was stored in a moisture controlled environment at RT. Other chemicals used were of analytical grade.

2.3. In vitro cytotoxicity and cytoprotection

The effects of CA on R28 cell and U87MG toxicity were determined using the MTT at 0.004 µg/ml (i.e. 20 µl from 5 mg/ml stock solution) cell proliferation assay. Briefly, R28 and U87MG cells (70% confluent) were seeded in 96-well plates and cultured overnight followed by treatment with different concentrations of CA (6.25, 12.5, 25, 50, 100, and 200 µg/mL). Cells were incubated for 48 h at 37°C and 5% CO_2, and cell proliferation and cytotoxicity were calculated using a regression equation (below). After 48 h of treatment, MTT solution (20 µl) was added to each well and the mixture was re-incubated for 4 h. Absorbance of the samples then was read at 490–570 nm, and percentage proliferation and cytotoxicity (percent inhibition (i.e. IC_{50})) were calculated as follows:

$$\% \text{ Inhibition} = 1 - \frac{\text{Absorbance of treated}}{\text{Absorbance of untreated}} \times 100$$

The % inhibition was plotted against the concentration tested using Microsoft Excel, and the IC_{50} was calculated using the non-linear regression equation. Effect of cytoprotective activity of the crystal CA was determined using serum deprivation assay. Cells were treated with and without serum and after 48 h of treatment and MTT assay was conducted as discussed above. Percent proliferation was determined according to earlier study of Khan et al. (2016).

2.4. Stability study

The thermal, chemical, biological, and environmental stability of the CA was determined according to earlier described study by Khan et al. (2016). In brief, thermal stability was determined by HPLC after treating the caffeic acid (50 mg of powder and crystal) at 60°C in the oven for 48 h. For chemical stability study, caffeic acid was treated with LiOH and quantity of caffeic acid was determined by mass balance and HPLC. Conversely, HCl and Krebs-Heneseleit bicarbonate buffer (pH 7.4) treated caffeic acid at 37°C for 2 h was quantified by HPLC peak and time. To determine the biological stability, caffeic acid (10 mM of powder and crystal) was treated by human serum which was then deproteinized using 500 µL of acetonitrile containing 0.1% trifluoroacetic acid and filtered off to determine the quantity in HPLC. The HPLC method was validated according to earlier described method by Khan et al. (2016).

3. Results and discussion

Keeping in view the literature information of CA, we successfully hydrolyzed the rosmarinic acid (Scheme 1) in the presence of acid as catalysis (polyphosphoric acid) and obtained crystals of CA (He, Li, & You, 2015). The reaction was carried out by mixing 1 equivalent of RA with 15 equivalents of H_3PO_4 to get a thick paste. The paste was heated to 150°C and poured into a large volume of stirring cold water. The solution was then neutralized by sodium hydroxide and the white precipitates were collected. Recrystallization by methanol provided colorless crystals of CA. The melting point of the crystals was found to be more than 300°C. The expected component β-aryllactic acid (β-AA) could not be isolated from the reaction mixture (Scheme 1). The crystals of CA so obtained were exposed to X-ray crystallographic machine. The molecular structure of CA along with the atomic numbering scheme and its packing in the crystal lattice are given in Figure 1(A) and (B).

The summary of the crystal data of CA is provided in the Table 1. Table S1 shows the selected bond lengths (Å) and angles (°), see supplementary file. The compound crystallizes into a monoclinic crystal system with space group P21/n. According to the molecular structure obtained from X-ray crystallographic technique, the CA contains a carboxylic group connected to a benzene ring containing two hydroxy groups through an alkyl chain (C6–C7–C8), see Figure 1(A). The molecular structure of the CA can be divided into two planar moieties that are connected by a single C–C bond. The first plane, the pyrocatechol plane, includes OH, OH, and benzene ring, whereas the second plane, the (E)-but-2-enoic acid plane, includes CO, OH, and C7–C8. Dihedral angle between these two planes is

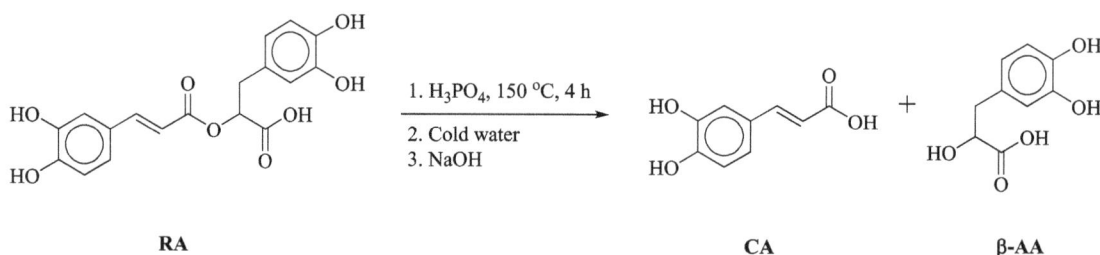

Scheme 1. Isolation of caffeic acid (CA) from rosmarinic acid (RA).

(A)

(B)

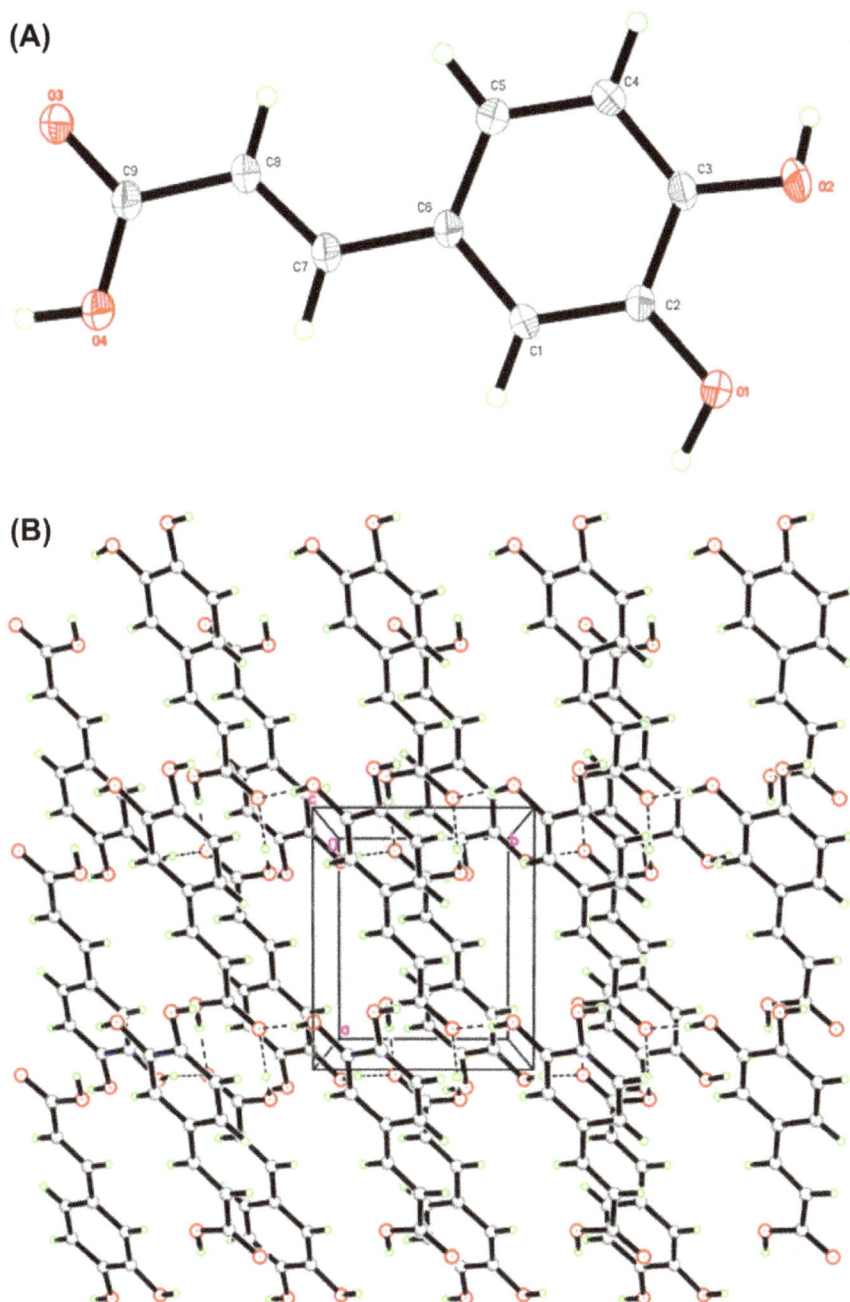

Figure 1. (A) The molecular structure of caffeic acid (CA) draw at 50% probability (B) shows the molecular packing.

Note: Crystal CCDC Depository CRM: 0001000261615, CCDC 1433127.

122.55(3)°. In CA, the C7–C8–benzene torsion angle is 178.94(15)°. In the crystal lattice, the packing of the molecules is stabilized by intermolecular hydrogen bondings. Further stabilization is provided by C-H–π and π–π interactions.

The ring and (E)-but-2-enoic acid are not in the same plane. The reason behind this is the packing effects where the molecules have low π–π stacking interactions. Figure 1(B) illustrates the crystal packing pattern, which shows that the molecules are connected through the C–O–H bonding in a three-dimensional network. The details of bond angles, lengths and torsions were presented in the Figure 3.

Table 1. Crystal data and structure refinement parameters for CA	
Compound	CA
Empirical formula	C9 H8 O4
Formula weight	180.15
Temperature (K)	294 K
Wavelength (Å)	0.71073
Crystal system	Monoclinic
Space group	P 21/n
Unit cell dimensions	
a (Å)	6.7154(5)
b (Å)	5.7725(4)
c (Å)	21.1336(16)
α (°)	90
β (°)	93.9107(11)
γ (°)	90
Volume (Å³)	817.33(10)
Z	4
Calculated density (Dx, g cm⁻³)	1.464
Radiation type λ (Å)	$M_0 K/\alpha$
F(0 0 0)	376.0
Crystal size (mm) (h, k, l_{max})	9 × 8 × 31
Crystal color	Faint yellow
Scan range θ(°)(max)	31.622
Absorption coefficient (μ) (mm⁻¹)	0.117
Maximum and minimum transmission	0.986 and 0.950
Goodness-of-fit on F^2	0.995
Data/restrains/parameters	2736/1.119/130
Final R indices [$I > 2\sigma(I)$]	$R_1 = 0.0629$, $wR_2 = 0.1861$

The crystals of CA were evaluated for the safety in rat neurosensory retinal (R28), and human glioblastoma U87MG cells by inducing serum insult. The R28 retinal precursor cell line is useful to determine the retinal cell behavior including differentiation, neuroprotection, cytotoxicity, light stimulation, retinal gene expression, and neuronal function. R28 cell line is nontumorigenic and the toxic effect on this cell line refers to the injury and inflammation of retina. R28 cells and U87MG cells were treated with 200 µg/mL to 6.25 µg/mL of CA for 48 h. Cell viability was measured using MTT assay. The cell viabilities of R28 cells, and U87MG cells treated with CA were significantly different ($p > 0.05$) from that of untreated controls. The IC_{50} values were found in R28 as 109.53 µg/ml (Figure 2(A), (B), and (F)) and in U87MG as 532.84 µg/ml (Figure 2(C), (D), and (G)). There was significant difference ($p > 0.05$) between viabilities of R28 and U87 MG cells treated with CA. This report suggests that CA, at the given concentrations or above the dose normally used in clinical practice, is toxic to rat neurosensory retinal, and human glioblastoma cells *in vitro*. The reason for this dual effect of crystal CA may be due to stability and solubility parameters. The effect of crystal CA with 6.25 µg/mL had obvious cytoprotective effect which was slightly different with the CA contained moisture (Figure 2(E) and (H)). In each

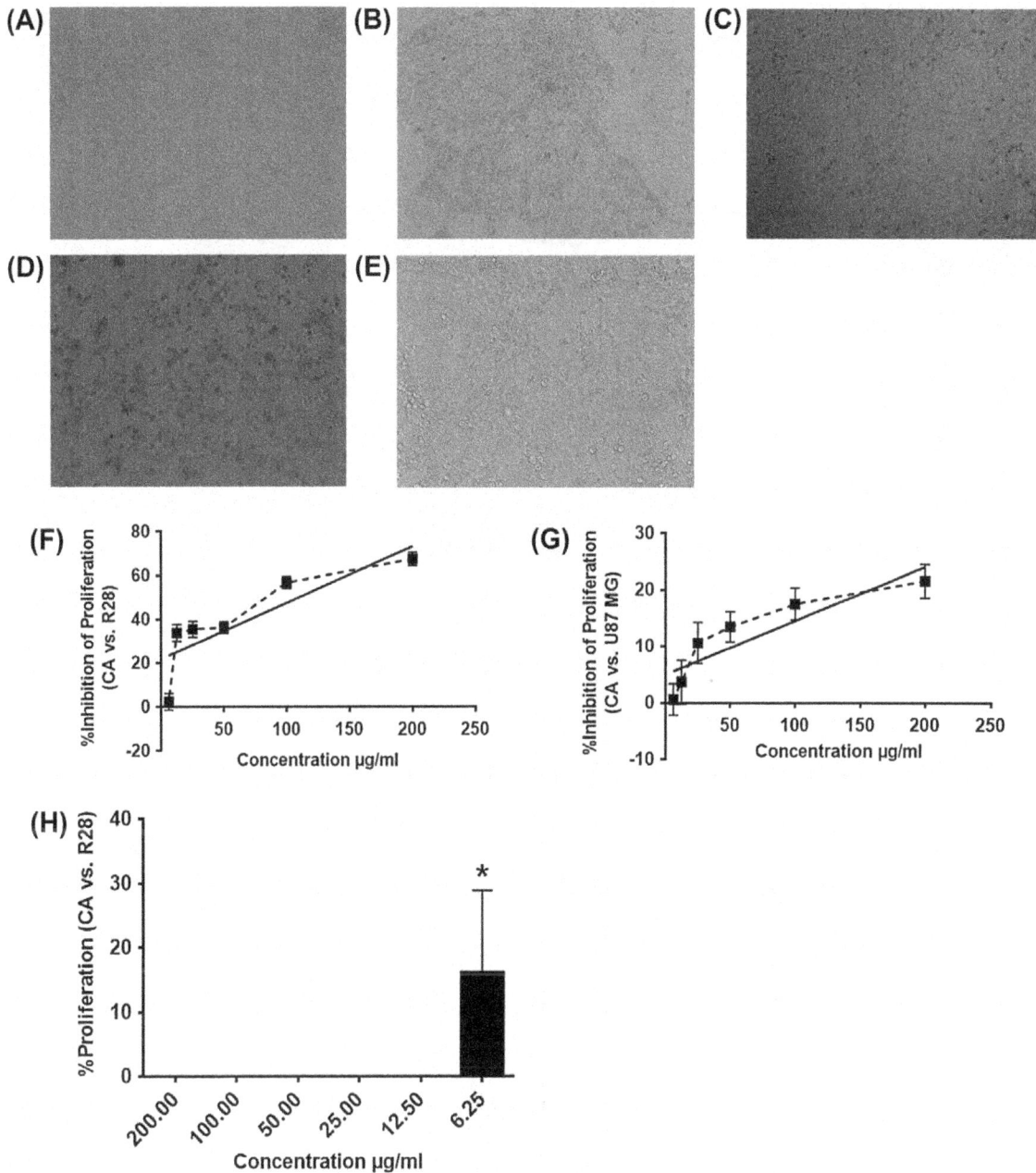

Figure 2. *In vitro* toxic effects of CA in the R28 cell lines (A, B) and U87 MG cells (C, D). The IC_{50} of CA was 109.52 µg/ml (F) and 532.83 µg/ml (G) for R28 and U87 MG cells, respectively. The R28 cells were not affected and proliferated at lowest dose (6.25 µg/ml) of caffeic acid (E, H) in serum insult assay.

Note: The dose dependent effects of CA was plotted and IC_{50} was determined using non-linear regression analysis.

case of stability study, caffeic acid was determined by HPLC before and after treatment of chemical, thermal, biological, and environmental factors. The elution time of caffeic acid was 3–5 min that revealed the affected stability for its solid composition of powder and crystal form. The crystal form of the caffeic acid was more stable than powdered form that the elution time was constant for crystal but it was deviated with multiple peaks at different elution time for powdered form which might be the reason of moisture content (Figures S1 and S2).

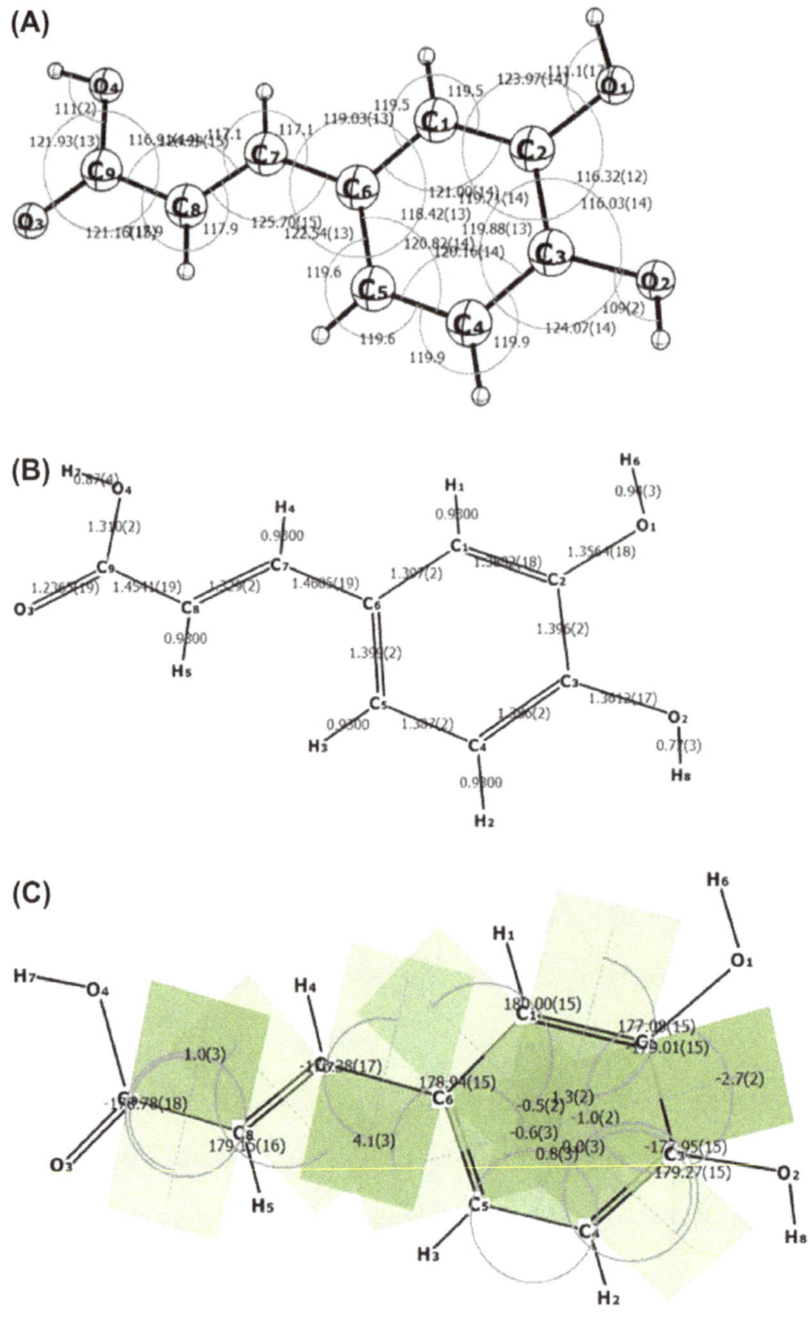

Figure 3. The molecular structure of CA labeled with: (A) bond angles, (B) bond lengths, and (C) torsion angles based on the crystallographic details.

4. Conclusion

The biological activity of caffeic acid is dependent on its delicate dose and stability which determines its efficacy to be nutrient or toxic. The lower dose of the stable CA showed cytostatic properties while the dose above 100 µg/ml could be cytotoxic based on the cell origin. The activity of CA is dependent on the stability and kind of cells origin that less sensitive cell such as highly survival capacity of cells might not be affected by the CA in compared to delicate cells such as U87 MG. CA crystallography provides more information for the resolution of atomic structure, stability impact and their study in the cellular environment. The further impact of the stability of CA will be explored more in future studies.

Acknowledgments
We are grateful to the EMAN LAB researcher who helped us during the analysis and experimentation. We are thankful to the USM doctoral fellowship scheme 2015–2016.

Funding
The entire study is supported by Universiti Sains Malaysia (USM) under the Research University (RU) [grant number 1001/PKIMIA/811217].

Author details
Md Shamsuddin Sultan Khan[1]
E-mail: jupitex@gmail.com
Muhammad Adnan Iqbal[2]
E-mail: adnan_chem38@yahoo.com
ORCID ID: http://orcid.org/0000-0001-6241-7547
Amin Malik Shah Abdul Majid[1]
E-mail: aminmalikshah@gmail.com
[1] School of Pharmaceutical Sciences, University of Science Malaysia, Penang, Malaysia.
[2] School of Chemical Sciences, University of Science Malaysia, Penang, Malaysia.

References
Akyol, S., Akbas, A., Butun, I., Toktas, M., Ozyurt, H., Sahin, S., & Akyol, O. (2015). Caffeic acid phenethyl ester (CAPE) as a remedial agent for reproductive functions and oxidative stress-based pathologies of gonads. *Journal of Intercultural Ethnopharmacology, 4*, 187–191. http://dx.doi.org/10.5455/jice.

Chen, J. H., & Ho, C. T. (1997). Antioxidant activities of caffeic acid and its related hydroxycinnamic acid compounds. *Journal of Agricultural and Food Chemistry, 45*, 2374–2378. http://dx.doi.org/10.1021/jf970055t

Gulcin, I. (2006). Antioxidant activity of caffeic acid (3, 4-dihydroxycinnamic acid). *Toxicology, 217*, 213–220. http://dx.doi.org/10.1016/j.tox.2005.09.011

He, X., Li, Z., & You, Q. (2015). Use of polyphosphoric acid in the hydrolysis of chromone esters. *Synthetic Communications, 32*, 709–714.

Hong, Y. J., Yang, S. Y., Nam, M. H., Koo, Y. C., & Lee, K. W. (2015). Caffeic acid inhibits the uptake of 2-amino-1-methyl-6-phenylimidazo[4,5-b]pyridine (PhIP) by inducing the efflux transporters expression in caco-2 cells. *Biological & Pharmaceutical Bulletin, 38*, 201–207. http://dx.doi.org/10.1248/bpb.b14-00495

Khan, M. S. S., Salam, M. A., Haque, R. S. M. A., Abdul Majid, A. M. S., Abdul Majid, A. S. B., Asif, M., ... Tabana, Y. M. (2016). Synthesis, cytotoxicity, and long-term single dose anti-cancer pharmacological evaluation of dimethyltin(IV) complex of N (4)-methylthiosemicarbazone (having ONS donor ligand). *Cogent Biology, 2*, 1154282. http://dx.doi.org/10.1080/23312025.2016.1154282

Kosova, F., Kurt, F., Olmez, E., Tuğlu, I., & Arı, Z. (2015). Effects of caffeic acid phenethyl ester on matrix molecules and angiogenetic and anti-angiogenetic factors in gastric cancer cells cultured on different substrates. *Biotechnic & Histochemistry, 91*, 38–47. doi:10.3109/10520295.2015.1072769

Lee, S.-Y., Ku, H.-C., Kuo, Y.-H., Yang, K.-C., Tu, P.-C., Chiu, H.-L., Su, M.-J. (2015). Caffeic acid ethanolamide prevents cardiac dysfunction through sirtuin dependent cardiac bioenergetics preservation. *Journal of Biomedical Science, 22*(1), 1–11.

Liszt, K. I., Eder, R., Wendelin, S., & Somoza, V. (2015). Identification of catechin, syringic acid, and procyanidin B2 in wine as stimulants of gastric acid secretion. *Journal of Agricultural and Food Chemistry, 63*, 7775–7783. http://dx.doi.org/10.1021/acs.jafc.5b02879

Nunes, S., Madureira, R., Campos, D., Sarmento, B., Gomes, A. M., Pintado, M., & Reis, F. (2015). Therapeutic and nutraceutical potential of rosmarinic acid-cytoprotective properties and pharmacokinetic profile. *Critical Reviews in Food Science and Nutrition.* doi:10.1080/10408398.2015.1006768

Rezg, R., Mornagui, B., Santos, J. S. O., Dulin, F., El-Fazaa, S., El-Haj, N. B., ... Gharbi, N. (2015). Protective effects of caffeic acid against hypothalamic neuropeptides alterations induced by malathion in rat. *Environmental Science and Pollution Research, 22*, 6198–6207. http://dx.doi.org/10.1007/s11356-014-3824-5

Rosendahl, A. H., Perks, C. M., Zeng, L., Markkula, A., Simonsson, M., Rose, C. I., ... Jernstrom, H. (2015). Caffeine and caffeic acid inhibit growth and modify estrogen receptor and insulin-like growth factor I receptor levels in human breast cancer. *Clinical Cancer Research, 21*, 1877–1887. http://dx.doi.org/10.1158/1078-0432.CCR-14-1748

Shin, H. S., Satsu, H., Bae, M.-J., Zhao, Z., Ogiwara, H., Totsuka, M., & Shimizu, M. (2015). Anti-inflammatory effect of chlorogenic acid on the IL-8 production in Caco-2 cells and the dextran sulphate sodium-induced colitis symptoms in C57BL/6 mice. *Food Chemistry, 168*, 167–175. http://dx.doi.org/10.1016/j.foodchem.2014.06.100

Toth, B., Bartho, L., Vasas, A., Sandor, Z., Jedlinszki, N., Pinke, G., Hohmann, J. (2015). Dual excitatory and smooth muscle-relaxing effect of *Sideritis montana* extract on guinea-pig ileum. *Natural Product Communications, 10*, 487–490.

7

An efficient strategy for expression and purification of spider neurotoxic peptide YC1 in *E. coli*

Hui Wu[1†], Wen-Ying Li[1†], Lei Wu[1], Ling-Yun Zhu[1], Er Meng[1]* and Dong-Yi Zhang[1]*

*Corresponding authors: Dong-Yi Zhang and Er Meng, Research Center of Biological Information, College of Science, National University of Defense Technology, Changsha, Hunan 410073, China
E-mails: dongyizhang@nudt.edu.cn (D.-Y. Zhang), er_meng@qq.com (E. Meng)
Reviewing editor: Yusuf Akhter, Central University of Himachal Pradesh, India
Additional information is available at the end of the article

Abstract: The peptide toxin GsAF II (Kappa-theraphotoxin-Gr2c) is a 31-amino acid peptide, recently isolated from the venom of the tarantula spider *Grammostola rosea*. The peptide toxin ProTX II (β/ω-theraphotoxin-Tp2a), is a 30-amino acid peptide toxin recently isolated from the venom of the tarantula spider *Thrixopelma pruriens*. The GsAF II and ProTX II have similar sequence but have an impact on different activities. To find a method of obtaining toxin and to explore whether amino acid sequences affect activities or not, an amino acid mutant was constructed that the first two amino acids are tyr (Y) and cys (C). The YC1 sequence is as follows: YCQKWMWTCDSERKCCEGLVCRLWCKKKIEW. Then, we constructed the YC1 vector (pET-GST-YC1), which was transformed into the *Escherichia coli* strain SHuffle™. rYC1 was expressed using auto-induction medium. After using a GST column, the expressed fusion protein was digested using SUMO protease (ULP1) to remove the GST-SUMO tag, and then RP-HPLC and ultrafiltration were used for further purification. rYC1 was further analyzed using SDS-PAGE. Then, the purified rYC1 was verified by MALDI-TOF/TOF mass spectrometry. Finally, the IC_{50} of rYC1 was determined to be 2.94 µM for the rabbit $Na_v1.3$ ($rNa_v1.3$) and the activity is between the ProTX II and GsAF II. Finally, the described method is economical and convenient, and toxins obtained using this method can be used for the study of in channels, neurobiology, pharmacology, or other fields.

Subjects: Biomedical Engineering; Drug Targeting; Technology

Keywords: YC1; auto-induction; mass spectrometry; whole-cell patch-clamp experiments

1. Introduction

The peptide toxin GsAF II is a 31-amino acid peptide, recently isolated from the venom of the tarantula spider *Grammostola rosea*. GsAF II modifies gating in voltage-gated Na^+ channels, and it has the following amino acid sequence: YCQKWMWTCDEERKCCEGLVCRLWCKKKIEW (Ono, Kimura, & Kubo, 2011).

ABOUT THE AUTHOR

Hui Wu is presently perusing PhD in biomedical engineering from School of Research Center of Biological Information, College of Science, National University of Defense. Wu is presently working on protein expression and purification. Protein expression and purification are the key progress in the study and can obtain many purified and active peptide. It is more convenient and cost-effective alternative than traditional methods. Wu et al. use this method to obtain the purified and large-scale preparation of toxins.

PUBLIC INTEREST STATEMENT

The spider toxin is isolated from the venom of the spider. Some researchers found that the spider toxin may have therapeutic implications. But the extraction of the native toxin is too difficult to make it to be a common drug. In this paper, our group used gene engineering to obtain many purified and active toxins and it is convenient and cost-effective alternative method. This method provides a new idea for large-scale preparation of toxins or drugs.

The GsAF II is reported to block the following voltage-gated Na⁺ channels: $Na_v1.3$, $Na_v1.4$, $Na_v1.5$, and $Na_v1.7$, which have IC_{50} values of 24, 4, 44, and 1 μM, respectively. ProTX II, is a 30-amino acid peptide toxin recently isolated from the venom of the tarantula spider *Thrixopelma pruriens* and modifies gating in voltage-gated Na⁺ and Ca²⁺ channels (Bladen, Hamid, Souza, & Zamponi, 2014; Edgerton, Blumenthal, & Hanck, 2010; Richard et al., 2002; Salari, Vega, Milescu, & Milescu, 2016). The amino acid sequence of ProTX II is as follows: YCQKWMWTCDSERKCCEGMVCRLWCKKKLW (Smith, Alphy, Seibert, Seibert, & Blumenthal, 2005). ProTX II is reported to block the following voltage-gated Na⁺ channels: $Na_v1.3$, $Na_v1.4$, $Na_v1.5$, and $Na_v1.7$, with IC_{50} values of 0.102, 0.039, 0.079, and 0.0003 μM, respectively (Deuis et al., 2016; Edgerton, Blumenthal, & Hanck, 2011; Schmalhofer et al., 2008; Smith, Cummins, Alphy, & Blumenthal, 2007; Xiao, Blumenthal, Jackson, Liang, & Cummins, 2010). Thus, both GsAF II and ProTX II are potent inhibitors of the sodium channel subtypes tested ($Na_v1.3$, $Na_v1.4$, $Na_v1.5$, and $Na_v1.7$). GsAF II and ProTX II both have the six cysteine residues with spacing consistent with the ICK motif and also share some sequence homology (Kolmar, 2010).

The YC1 sequence YCQKWMWTCDSERKCCEGLVCRLWCKKKIEW which is a mutation of the 11th amino acid in which the glutamic acid (E) of GsAF II is replaced with serine (S) in ProTX II. In the present study, we used single site of 11th amino acid to research this site' affect. We used *Escherichia coli* to produce recombinant GsAF II mutant (YC1). The toxin has three pairs of disulfide bonds. A pET-GS-YC1 vector was constructed and transformed into the *E. coli* strain SHuffle™. rYC1 was expressed using auto-induction medium, and purified with a GST column (Wu et al., 2016; Zhang et al., 2015). After the expressed fusion protein was digested using a SUMO protease (ULP1) to remove the GST-SUMO tag, reversed phase HPLC (RP-HPLC) and ultrafiltration were used for further purification (Eshaghi et al., 2015). The rYC1 protein was further analyzed using SDS-PAGE, then the purified rYC1 was verified by MALDI-TOF/TOF mass spectrometry. According to previous research the IC_{50} of rYC1 is 2.94 μM for $rNa_v1.3$, which is between ProTX II and GsAF II' activity. The mutation of the 11th amino acid had an effect on the activity of $rNa_v1.3$. It developed a new method for toxin production and the method described herein is economical and convenient. The obtained toxins can be used in the study of ion channels, neurobiology, pharmacology, or other fields of study. In addition, this methodology may provide an effective method for searching for other mutants that affect other Voltage-gated sodium channels (VGSCs).

2. Materials and methods

2.1. Materials
E. coli strain DH5α competent cells were purchased from *TaKaRa* (Otsu, Shiga, Japan). The pET-43a (+) vector was purchased from *Novagen* (Madison, WI, USA). The SHuffle™ strain was obtained from *NEB* (Beverly, MA, USA). The cDNA for GsAF II and the GST tag were kept in Prof S. Liang's laboratory. The KOD DNA polymerase was purchased from *TOYOBO* (Osaka, Japan). DNA sequencing and primer synthesis were performed by *Sangon* (Shanghai, China). All of the chemicals and reagents were purchased from *Sigma* (St. Louis, MO, USA). Glycine-SDS-PAGE gels (NuPAGE) and the Seeblue Plus2 pre-stained molecular weight marker were purchased from *Invitrogen* (Carlsbad, CA, USA).

2.2. Construction of the pET-GST-GsAF II vector
Restriction free (RF) cloning technology was used to construct the expression vector (Kozlov, 2007; Meng et al., 2011; Van & Löwe, 2006). The GST tag and SUMO sequences were previously cloned into pET-43a. The GsAF II gene was generated with a stop codon behind the SUMO sequence using the following primer pair: 5′-GCTCACCGTGAACAGATCGGTGGTTACTGCCAGAAATGGATGTG-3′(GsAF II-Forward) and 5′-GATGGTACCGTCGACGTCCTGCAGTTACCATTCGATTTTTTTTTTGC-3′(GsAF II-Reverse). Briefly, the underlined sequences of the forward primer and the reverse primer were complementary to the 5′ end and the 3′ end of the insertion or replacement point of the recipient vector pET-43a (+) with a T m value of 80°C, respectively. The italicized portions were complementary to the 5′ end and the 3′ end of GsAF II with T m value of 60°C, respectively. The resultant construct was verified by sequencing and named pET-GST. The resulting vector was named pET-GST-GsAF II.

2.3. Construction of the pET-GST-YC1 vector

The pET-GST-YC1 vector was constructed by combining the techniques of point mutagenesis and RF cloning. pET-GST-GsAF II was used as a template with the following primer pair: 5′-GGACCTGCGACT CTGAACGTAAATGCTGCGAAGGTCTGGTTTGCCGTCTGTGGTG-3′(YC1-Forward) and 5′-AGCATTTACGTTCA GAGTCGCAGGTCCACATCCATTTCTGGCA-GTAACCACC-3′(YC1-Reverse). The resulting vector was named pET-GST-YC1.

2.4. Expression of the recombinant YC1

The pET-GST-YC1 vector was transformed into *E. coli* SHuffle™ cells. Using the auto-induction method established by Studier FW, the GST-YC1 protein was expressed as described below (Studier, 2005). A single colony from a plate incubated overnight was picked and inoculated into 3 mL ZYM-505 medium containing 100 µg/mL ampicillin inoculated at 37°C in a shaker at 220 rpm. After eight hours, 1 mL of the medium was transferred into 500 mL of fresh ZYM-5051 medium (auto-induction medium and containing 100 µg/mL ampicillin). The culture was then incubated at 30°C in a shaker at 220 rpm. After 16 hours, the cells were harvested by centrifugation at 4,000 rpm at 4°C for 10 min, and the supernatant was removed and discarded.

2.5. Purification of GST-SUMO-YC1 fusion protein by GST affinity chromatography

The cell pellet was resuspended in 3 mL ice-cold Phosphate-buffered saline (PBS) buffer per 50 mL culture and centrifuged at 4,000 rpm at 4°C for 10 min. The supernatant was discarded and the pellet was suspended in 3 mL of ice-cold PBS buffer per 50 mL culture. The cells were lysed on ice by sonication for 10 min at 40% amplitude using a Branson SONIFIER S-250D (Emerson, USA). After centrifugation at 12,000 g at 4°C for 10 min, the supernatant was carefully transferred (soluble fraction) to a clean, pre-chilled tube. Aliquots of 10 µl from both soluble and insoluble fractions were used for SDS-PAGE analysis by adding an equal volume of 2 × SDS sample loading buffer. The samples were boiled for 5 min and used for SDS-PAGE to determine the yield and solubility of the GST-fusion protein. The supernatant was applied to a gravity GST column, and the eluent was desalinated using a 30 kDa ultrafiltration tube with distilled water.

2.6. Preparation of recombinant SUMO protease and removal of the GST-SUMO tag by ULP1

The preparation of the recombinant SUMO protease was performed as previously described and was stored at −20°C (Zhang et al., 2015). The purificated GST-SUMO-YC1 was digested with the recombinant SUMO protease (ULP1). SUMO protease digestion was performed for 16 h at 4°C to remove the GST-SUMO part of the fusion protein.

2.7. Purification of the recombinant YC1 (rYC1)

The digested mixtures were centrifuged at 12,000 rpm for 5 min to remove insoluble particles and were then filtered through a 0.45 µm filter before loading. The supernatant was then subjected to RP-HPLC purification using a C18 column (4.6 × 250 mm, 5 µm) with a linear acetonitrile gradient.

2.8. Identification of rYC1 by MALDI-TOF/TOF mass spectrometry

The molecular mass identification of rYC1 was performed on a Voyager-DE TM STR MALDI-TOF mass spectrometer (Applied Biosystems, Voyager-DE STR Biospectometry work station). Ionization was achieved by irradiation with a nitrogen laser (337 nm), with a 20 kV acceleration voltage. We used α-cyano-4-hydroxy-cinnamic acid as the matrix.

2.9. Cell culture and whole-cell patch-clamp experiments

HEK293T cells (ATCC) were grown under standard cell culture conditions (5% CO_2 and 37°C) in Dulbecco's Modified Eagle Medium (DMEM, Life Technology) supplemented with 10% fetal bovine serum. The Na_v constructs were co-transfected into the HEK293T cells with a plasmid containing the β1 subunit and PEGFP-N1 using Lipofectamine 2000 (Life Technology) according to the manufacturer's instructions.

(A)

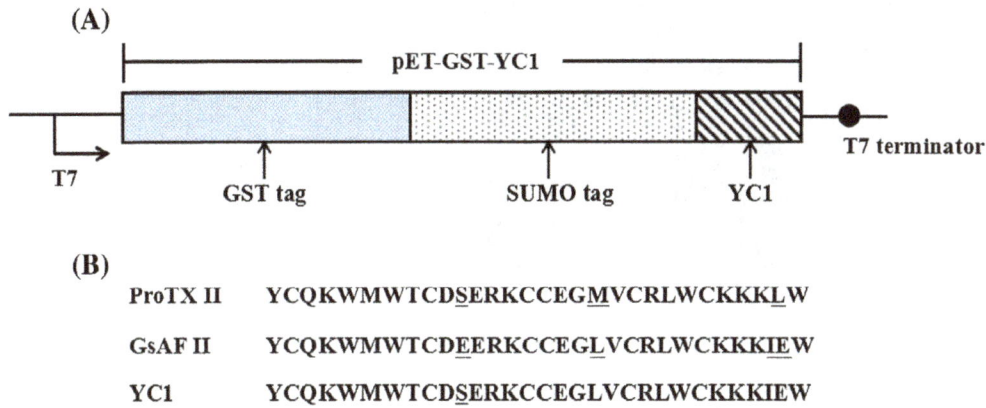

(B)

ProTX II	YCQKWMWTCDSERKCCEGMVCRLWCKKKLW
GsAF II	YCQKWMWTCDEERKCCEGLVCRLWCKKKIEW
YC1	YCQKWMWTCDSERKCCEGLVCRLWCKKKIEW

Figure 1. Construction of pET-GST-YC1 expression vector. (A) The composition of the pET-GST-YC1 vector. (B) The amino acid sequences of ProTX II, GsAF II, and YC1.

The whole-cell-patch-clamp experiments were performed as described below (Wu et al., 2016). Briefly, purified rYC1 was dissolved in distilled water to a concentration of 1 mmole/L and stored at −20°C. Dilute the stock solution to the working concentrations with fresh bathing solution before use. Whole-cell patch-clamp assay recordings were performed at room temperature using an EPC-9 amplifier (HEKA, Lambrecht, Germany). Sodium currents were elicited at 210 mV from a holding potential of 280 mM. The recording and analysis of data were carried out using the Pulse + Pulsefit 8.0 (HEKA, Lambrecht, Germany) and the Sigmaplot 9.0 (Systat Software Inc.) programs, respectively.

3. Results

3.1. Construction of the pET-GST-YC1 vector
YC1 was cloned into the pET-GST vector, and the new vector was named pET-GST-YC1 (Figure 1).

3.2. Expression of GST-YC1 in E. coli SHuffle™
YC1 is similar to ProTX II and GsAF II in that they all have three pairs of disulfide bonds. Therefore, we utilized a SHuffle™ strain for protein expression as it is more conducive to the formation of disulfide bonds compared to other expression strains. The SUMO-YC1 fusion protein was expressed in SHuffle™ (Figure 2, lane 2), and no inclusion bodies were observed after sonication (Figure 2, lanes 3 and 4). Lastly, we observed a very bright fusion protein band after elution from the GST column (Figure 2, lane 6).

3.3. Purification of GST-YC1 by affinity chromatography and ultrafiltration
The glycine-SDS-PAGE data are shown in Figure 2. After elution, the YC1 fusion protein was purified using a 30 kDa ultrafiltration tube (Figure 2, lane 7). YC1 was then subjected to ULP1 cleavage and purified by RP-HPLC (Figure 2, lane 8). YC1 was eluted in 17 min (Figure 3, asterisks). The obtained sample was ultrafiltrated with a 10 kDa ultrafiltration tube, and we collected the filtered liquid in the bottom of the ultrafiltration tube. The liquid was freeze dried to obtain purified rYC1 (Figure 4(A), lane 2). Obviously, the samples in the bottom of the ultrafiltration tube (Figure 4(A), lane 2) were richer and more pure than those in the ultrafiltration tube (Figure 4(A), lane 1).

3.4. Identification of rYC1 by MALDI-TOF/TOF mass spectrometry
The molecular weight of the eluted fraction was further verified by mass spectrometry and determined to be 3,939.9 Da, which is identical to the theoretical value of YC1 (Figure 4(B)).

3.5. Identification and characterization of rYC1 in whole-cell patch-clamp experiments
We observed that 1 µM rYC1 inhibited $rNa_v1.3$ current by approximately 20%, while 10 µM rYC1 inhibited $rNa_v1.3$ current by approximately 70% (Figure 5(A)). Figure 5(B) showed that the IC_{50} value for rYC1 was 2.94 µM for $rNa_v1.3$.

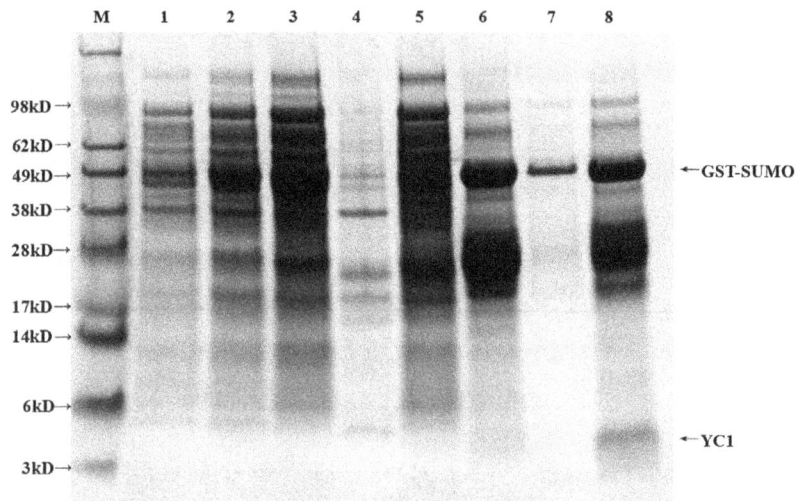

Figure 2. Expression of GST-YC1 in SHuffle™.

Notes: The protein was isolated by 10% glycine-SDS-PAGE and used for experiments as described in the Materials and Methods. Lane M, marker; lane 1, uninduced bacteria total cells protein using the ZYM- 505 medium; lane 2, induced bacterial total cells protein using the ZYM-5051 medium; lane 3, the centrifuged supernatant after sonication (soluble fractions); lane 4, the centrifuged precipitate after sonication (insoluble fractions); lane 5, the flow through from the GST column; lane 6, the flow through from the GST column; lane 7, the fusion protein after ultrafiltration desalination; lane 8, the fusion protein after ULP1 cleavage. The arrow indicates GST-SUMO-YC1 (~49 kDa) and YC1 (~4 kDa).

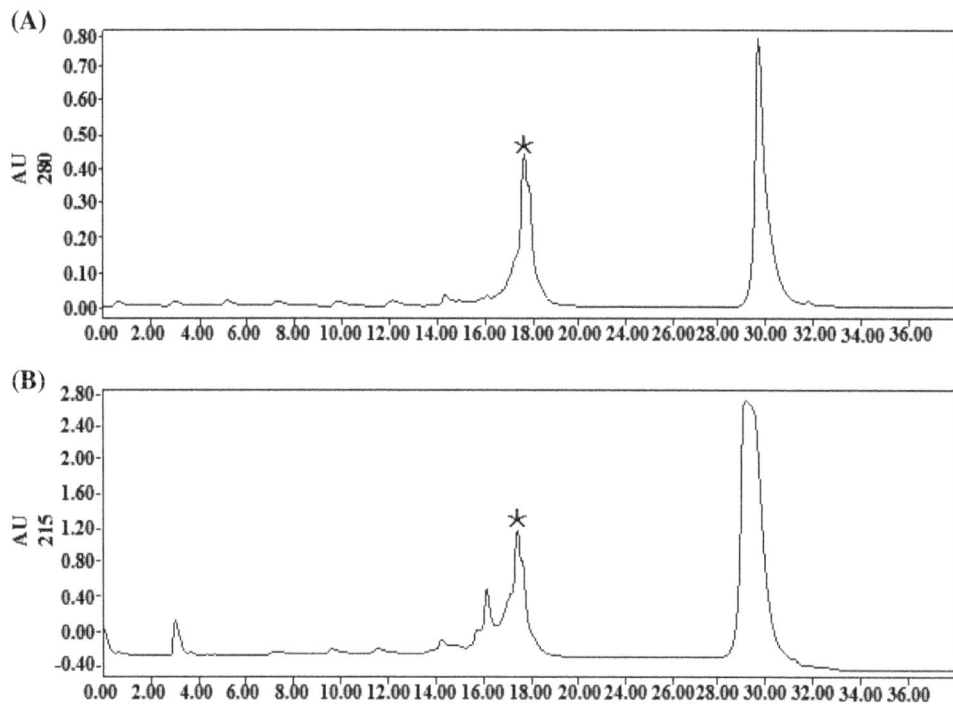

Figure 3. RP-HPLC chromatography of rYC1. (A) Absorption of the eluted proteins was measured at 280 nm, and fractions corresponding to detected peaks were collected. (B) Absorption of the eluting solution was measured at 215 nm, and fractions corresponding to detected peaks were collected.

4. Discussion

Since ancient times polypeptide compounds and small organic molecules isolated from natural venoms have been used in folk medicine worldwide. Numerous extracts, ointments, and decoctions have been prepared from toxic spiders, poisonous plants, venomous serpents, lizards, arthropods,

(A)

(B)   4700 linear spec #1 MC[BP=3939.9,1414.0]

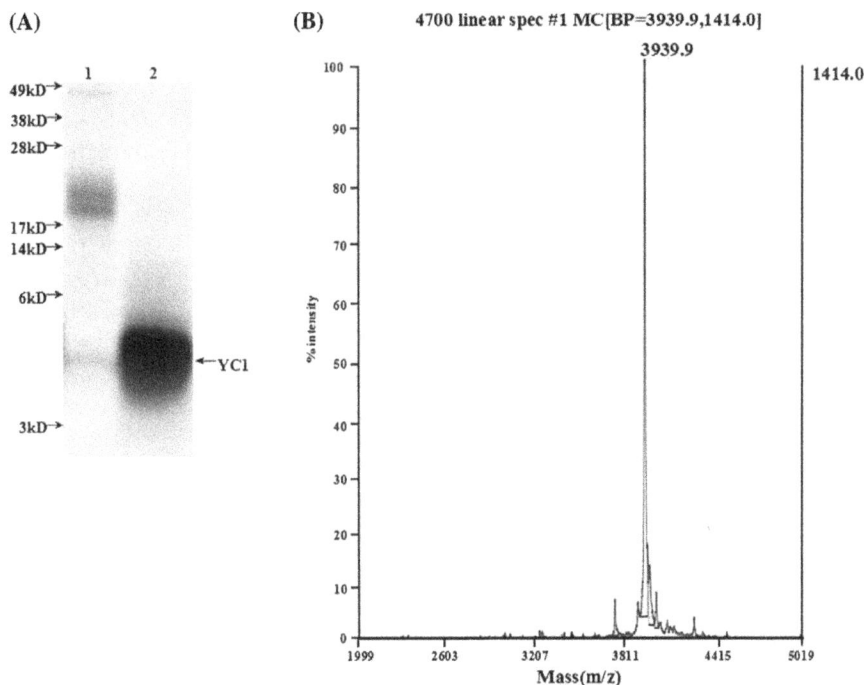

Figure 4. Identification of rYC1. (A) The results of the glycine-SDS-PAGE analysis. Lane 1, protein from eluent marked with asterisk in Figure 3; lane 2, the protein after the ultrafiltration purification of eluent marked with asterisk (Figure 3). (B) Mass spectra of the rYC1. The calculated theoretical molecular weight of rYC1 was 3,937.7 Da, and the measured molecular weight was 3,939.9 Da,

(A)

(B)

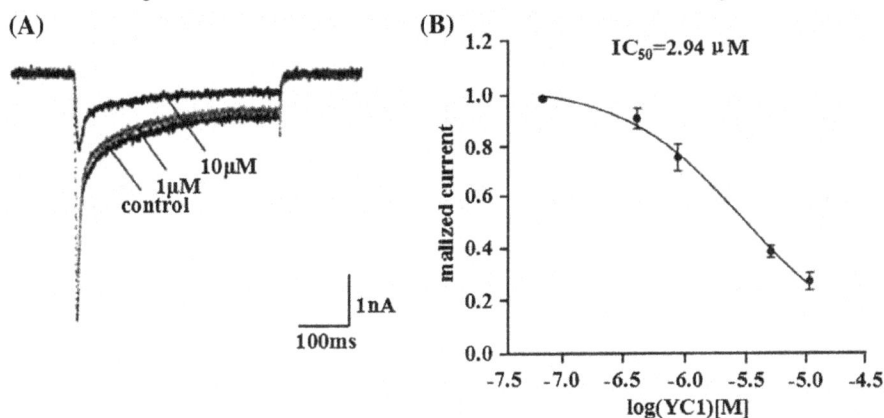

Figure 5. Inhibitory effects of rYC1 on the $rNa_v1.3$ currents. (A) The inhibitory effects of 1 μM or 10 μM rYC1 on $rNa_v1.3$ current. (B) The effects of rYC1 on $rNa_v1.3$ channels expressed in HEK293 cells. The IC_{50} value of rYC1 for $rNa_v1.3$ was 2.94 μmol/L.

and marine residents. The most interesting compounds that show excellent stability and specificity of action are being patented locally or worldwide. Spider venom is a source of many biologically active peptides that may have therapeutic implications. Researchers have studied spider toxins for their potential to target ion channels and receptors (Klint et al., 2012; Saez et al., 2010).

But the extraction of the native toxin is too difficult to make it to be a common drug. In general, there are three methods for researchers to obtain spider toxins. The first is to obtain toxins directly from the spider; however, this method produces relatively little toxin. The second way is to purchase spider toxin from a company, and this method is too costly. Spider toxin can cost as much as $432.30 per milligram when purchased from Sigma (United States). A third way is chemical synthesis, although this method may require complex refolding steps and result in a low refolding rate. In this paper, prokaryotic expression as a means to obtain biological polypeptides was found to be a convenient and cost-effective alternative to these other techniques.

The GsAF II mutant rYC1 was expressed using a prokaryotic expression system. GST column, RP-HPLC, and ultrafiltration were used to further purify rYC1. The resulting protein was assayed using glycine-SDS-PAGE and mass spectrometry to identify the isolated peptide as rYC1. In addition, the mass spectrometry assays showed that rYC1 has three disulfide bonds. The purified rYC1 has an IC_{50} value of 2.94 μM for $rNa_v1.3$, an activity that is between that of ProTX II and GsAF II. Thus, the 11th amino acid, which was altered in the mutant, was observed to have an effect on the activity of $rNa_v1.3$.

Compared with other reported procedures for the expression of disulfide bond-rich peptides in the periplasm of *E. coli*, four significant improvements are presented in this work. First, the SHuffle™ expression system was found to be good for the formation of disulfide bonds and GST, SUMO was used in this system to lead the periplasmic expression of rYC1. The tags are essential for the functional expression of disulfide bond-rich toxins and useful for fusion protein solubility. . Second, instead of IPTG induction, an auto-induction medium was used in these procedures. Auto-induction with lactose is more convenient and simpler than IPTG induction. Third, the use of RP-HPLC and ultrafiltration steps can save time and obtain highly purified rYC1. Fourth, we provide an effective method for searching other mutants that had an effect on other VGSCs. It developed a simple method for active toxin production. We suppose that other disulfide bond-rich peptides can also be produced by this efficient prokaryotic expression strategy.

Author contributors
HW, EM, and DYZ conceived and designed the experiments. HW and WYL performed the experiments. HW, LW, and LYZ analyzed the data. EM and DYZ contributed reagents/materials/analysis tools. HW and EM wrote the paper.

Funding
This work was supported by the Research Program of the National University of Defense Technology [grant number JC2006-02-01] to Prof D.Y. Zhang and Lateral Research Funds [grant number 1501020414029 to Dr. Meng].

Competing Interest
The authors declare no competing interest.

Author details
Hui Wu[1]
E-mail: wu2hui218@sina.com
Wen-Ying Li[1]
E-mail: justliwenying@qq.com
Lei Wu[1]
E-mail: wulei@nudt.edu.cn
Ling-Yun Zhu[1]
E-mail: lyzhucn@gmail.com
Er Meng[1]
E-mail: er_meng@qq.com
Dong-Yi Zhang[1]
E-mail: dongyizhang@nudt.edu.cn
[1] Research Center of Biological Information, College of Science, National University of Defense Technology, Changsha, Hunan 410073, China.
[†] These authors contributed equally to this work.

References
Bladen, C., Hamid, J., Souza, I., & Zamponi, G. W. (2014). Block of T-type calcium channels by protoxins I and II. *Molecular Brain, 7*, 1–8.

Deuis, J., Wingerd, J., Winter, Z., Durek, T., Sousa, S., Zimmermann, K., ... Vetter, I. (2016). Analgesic effects of GpTx-1, PF-04856264 and CNV1014802 in a mouse model of $Na_v1.7$-mediated pain. *Toxins, 8*, 1–19.

Edgerton, G., Blumenthal, K., & Hanck, D. (2010). Inhibition of the activation pathway of the T-type calcium channel $Ca_v3.1$ by ProTxII. *Toxicon Official Journal of the International Society on Toxinology, 56*, 624–636. http://dx.doi.org/10.1016/j.toxicon.2010.06.009

Edgerton, G., Blumenthal, K., & Hanck, D. (2011). Evidence for multiple effects of ProTxII on activation gating in $Na_v1.5$. *Toxicon Official Journal of the International Society on Toxinology, 52*, 489–500.

Eshaghi, S., Hedrén, M., Nasser, M., Hammarberg, T., Thornell, A., & Nordlund, P. (2005). An efficient strategy for high-throughput expression screening of recombinant integral membrane proteins. *Protein Science, 14*, 676–683. http://dx.doi.org/10.1110/ps.041127005

Klint, J., Senff, S., Rupasinghe, D., Er, S., Nicholson, G., & King, G. (2012). Spider-venom peptides that target voltage-gated sodium channels: Pharmacological tools and potential therapeutic leads. *Toxicon, 60*, 478–491. http://dx.doi.org/10.1016/j.toxicon.2012.04.337

Kolmar, H. (2010). Biological diversity and therapeutic potential of natural and engineered cystine knot miniproteins. *Current Opinion in Pharmacology, 41*, 608–614.

Kozlov, S. (2007). Polypeptide toxins from animal venoms. *Recent Patents on DNA & Gene Sequences, 1*, 200–206. http://dx.doi.org/10.2174/187221507782360254

Meng, E., Cai, T., Li, W., Zhang, H., Liu, Y., Peng, K., ... Zhang, D. (2011). Functional expression of spider neurotoxic peptide huwentoxin-I in *E. coli*. *PLoS ONE, 6*, 1–6.

Ono, S., Kimura, T., & Kubo, T. (2011). Characterization of voltage-dependent calcium channel blocking peptides from the venom of the tarantula *Grammostola rosea*. *Toxicon Official Journal of the International Society on Toxinology, 58*, 265–276. http://dx.doi.org/10.1016/j.toxicon.2011.06.006

Richard, E., Vivien, A., Richard, L., Jeremy, C., Chou, J., Ge, D., ... Mchardy, M. (2002). Two tarantula peptides inhibit activation of multiple sodium channels. *Biochemistry, 41*, 14734–14747.

Saez, N., Senff, S., Jensen, J., Sing, Er., Herzig, V., Rash, L., & King, G. (2010). Spider-venom peptides as therapeutics. *Toxins, 2*, 2851–2871.

Salari, A., Vega, B., Milescu, L., & Milescu, L. (2016). Molecular interactions between tarantula toxins and low-voltage-activated calcium channels. *Scientific Reports, 6*, 1–12.

Schmalhofer, W., Calhoun, J., Burrows, R., Bailey, T., Kohler, M., Kaczorowski, G., ... Priest, B. (2008). ProTx-II, a selective inhibitor of Na$_v$1.7 sodium channels, blocks action potential propagation in nociceptors. *Molecular Pharmacology, 74*, 1476–1484. http://dx.doi.org/10.1124/mol.108.047670

Smith, J., Alphy, S., Seibert, A., Seibert, A., & Blumenthal, M. (2005). Differential phospholipid binding by site 3 and site 4 toxins. *Journal of Biological Chemistry, 280*, 11127–11133. http://dx.doi.org/10.1074/jbc.M412552200

Smith, J., Cummins, T., Alphy, S., & Blumenthal, K. (2007). Molecular interactions of the gating modifier toxin ProTx-II with Na$_v$1.5: Implied existence of a novel toxin binding site coupled to activation. *Journal of Biological Chemistry, 282*, 12687–12697. http://dx.doi.org/10.1074/jbc.M610462200

Studier, F. (2005). Protein production by auto-induction in high density shaking cultures. *Protein Expression & Purification, 41*, 207–234. http://dx.doi.org/10.1016/j.pep.2005.01.016

Van, D., & Löwe, J. (2006). RF cloning: A restriction-free method for inserting target genes into plasmids. *Journal of Biochemical & Biophysical Methods, 67*, 67–74.

Wu, H., Chen, B., Jiang, H., Wu, L., Zhu, L., Meng, E., & Zhang, D. (2016). Heterologous expression and purification of neurotoxic hainantoxin-III in *E. coli. Preparative Biochemistry & Biotechnology, 47*, 158–162.

Xiao, Y., Blumenthal, K., Jackson, J., Liang, S., & Cummins, T. (2010). The tarantula toxins ProTx-II and Huwentoxin-IV differentially interact with human Na$_v$1.7 voltage sensors to inhibit channel activation and inactivation. *Molecular Pharmacology, 78*, 1124–1134. http://dx.doi.org/10.1124/mol.110.066332

Zhang, H., Huang, P., Meng, E., Li, W., Zhou, L., Zhu, L., ... Zhang, D. (2015). An efficient strategy for heterologous expression and purification of active peptide hainantoxin-IV. *PLoS ONE, 10*, 1–13.

8

Anti-inflammatory and anticancer activities of extract and fractions of *Rhipsalis neves-armondii* (Cactaceae) aerial parts

Theophine Chinwuba Akunne[1]*, Peter A. Akah[1], Ifeoma A. Nwabunike[1], Chukwuemeka S. Nworu[1], Emeka K. Okereke[1], Nelson C. Okereke[1] and Francis C. Okeke[1]

*Corresponding author: Theophine Chinwuba Akunne, Faculty of Pharmaceutical Sciences, Department of Pharmacology and Toxicology, University of Nigeria, Nsukka 410001, Enugu State, Nigeria

E-mail: theophine.okoye@unn.edu.ng

Reviewing editor: Tsai-Ching Hsu, Chung Shan Medical University, Taiwan

Additional information is available at the end of the article

Abstract: *Rhipsalis neves-armondii* K. Schum. (Cactaceae) aerial parts preparation is used traditionally in southern Nigerian for the treatment of rheumatic disorders and cancer and so the aim of this study was to evaluate these folkloric claims. Carragenaan-induced pedal edema and cotton pellet-induced granuloma were the anti-inflammatory models employed, while the antiploliferative study was performed using the MTT [3-(4,5-dimethylthiazol-2-yl)-2,5-diphenyl-tetrazolium-bromide] assay method on Pancreatic tumor (PANC-1) and Henrietta Lacks' cervical (HeLa) cell lines. The dried pulverized Rhipsalis aerial parts were extracted with methanol to obtain *Rhipsalis neves-armondii* extract (RNE). The RNE was fractionated using chromatographic techniques to obtain the hexane fraction (HF) and ethylacetate fraction (EF). Also acute toxicity and phytochemical studies were performed using standard procedures. Results showed that the extract and fractions significantly ($p < 0.05$) inhibited the development of rat pedal edema. The EF (400 mg/kg) exhibited the highest percentage edema inhibition of 53.33% after 3 h, compared to 36.67% of indomethacin. Also the extract and fractions significantly ($p < 0.05$) suppressed granuloma formation in rats, with HF showing the highest percentage

ABOUT THE AUTHOR

Theophine Chinwuba Akunne, PhD, is a senior lecturer and researcher in the Department of Pharmacology and Toxicology, Faculty of Pharmaceutical Sciences, University of Nigeria, Nsukka. Akunne obtained doctoral degree in Pharmacology and Toxicology, with bias in Natural Products Pharmacology. Akunne's research interest lies in the isolation of phytoconstituents or bioactive compounds with potent anticonvulsant, anticancer, and anti-diabetic effects. Medicinal plants have been the mainstay of Complementary and Alternative Medicine (CAM) therapy in both developed and developing nations. Reduced side effects and possibility of isolation and characterization of novel compounds with clinical potentials have made natural product pharmacology very interesting, especially in the area of new drug development and research. In this study, the aerial parts of *Rhipsalis nerves-armondii* showed anticancer and anti-inflammatory properties, thus buttressing its use in folkloric medicine.

PUBLIC INTEREST STATEMENT

The use of medicinal plants in disease management is as old as man. In both developed and developing nations, extracts and preparations from medicinal plants are being used in the folkloric treatment of cancer and other diseases. Medicinal plants often serve as a reservoir for plants secondary metabolites with potent pharmacological effects from which novel compounds have been and could still be isolated and characterized for drug development and research. Several standard anti-inflammatory and anticancer agents in clinical use have their origin from plants. In chemotherapy, some naturally isolated phytoconstituents used in cancer treatment such as the taxanes, vinca alkaloids, and podophyllotoxins have their origin from plants. In Nigerian ethnomedicine, the aerial preparation of *Rhipsalis neves-armondii* (Cactaceae) is used in the treatment of cancer and rheumatic diseases. This study has scientifically proven the folkloric use through *in vivo* and *in vitro* models of inflammation and cancer.

suppression of 59.4% compared to the control. The RNE exhibited potent cytotoxic effect against PANC-1 and HeLa cells with an estimated IC_{50} value of 30.12 and 45.1 µg/ml, respectively, and showed an LD_{50} greater than 5,000 mg/kg. The findings showed that the extract and fractions of *Rhipsalis neves-armondii* (Cactaceae) possess anti-inflammatory and antiproliferative effects.

Subjects: Clinical Pharmacology and Therapeutics; Medicine; Pharmacology

Keywords: *Rhipsalis neves-armondii*; anti-inflammatory; anticancer; cotton pellet granuloma; carragenaan

1. Introduction

Inflammation is an integral part of the body's defense mechanisms. It has various components such as edema formation, leukocyte infiltration, and granuloma formation, all of which contribute to the associated symptoms and tissue injury typical of an inflammatory reaction (Mitchell, Kumar, Abbas, & Fausto, 2006). Acute inflammatory response is characterized by vasodilatation, exudation of plasma, release of various inflammatory mediators such as cytokines, prostaglandins, and leukocytes, while the typical features of chronic inflammation include infiltration of mononuclear cells, proliferation of fibroblasts, blood vessels, and increased connective tissue formation (Benni, Suresha, & Jayanthi, 2011). Studies have shown that inflammation is one of the major physiological events in cancer development (Aggarwal, Shishodia, Sandur, Pandey, & Sethi, 2006; Lu, Ouyang, & Huang, 2006). Chronic inflammatory diseases are frequently associated with increased risk of cancers and chronic inflammation has emerged as an important risk factor in the pathogenesis of cancer (Aggarwal et al., 2006). Globally, cancer is a leading cause of death worldwide, accounting for about 8.2 million deaths and approximately 14 million new cases in 2012 (World Health Organization, 2014). It is a serious health burden in both developing and civilized nations. Non-steroidal anti-inflammatory drugs (NSAIDs) are the mainstay in the management of inflammatory and other painful conditions. However, their use is often limited by associated unwanted gastro-intestinal (GI) side effects, which have been minimized since the introduction of specific cyclo-oxygenase-2 (COX-2) inhibitors (Rang, Dale, Ritter, & Flower, 2008). Documented experimental, epidemiologic, and clinical studies have shown that non-steroidal anti-inflammatory drugs (NSAIDs), particularly the highly selective cyclooxygenase (COX)-2 inhibitors, have potent anticancer effects (Rayburn, Ezell, & Zhang, 2009; Thun, Henley, & Patrono, 2002). Therefore, eliminating inflammation may be a valid strategy for cancer prevention and therapy (Rayburn et al., 2009). Moreover, side effects associated with standard agents used in cancer chemotherapy have made the need for new and safer agents imperative. The use of herbal medicines is often sought as one of the vital means to obviate the unwanted side effects associated with standard agents. In ethnomedicine, treatment of various forms of cancer with herbal or medicinal plants preparations has been a documented regular practice (Harlev, Nevo, Solowey, & Bishayee, 2013). There is an increase in the use of herbal medicines and other natural products in the management of various ailments such as inflammatory disorders and cancer, both in the developed and the developing nations. It has been reported that apart from their food value, medicinal plants in the cactus genera including *Rhipsalis neves-armondii* K. Schum. (Cactaceae) have been used in ethnomedicine for the treatment of whooping cough, diabetes, cancer, nose bleeding, and rheumatic pains (Anderson, 2001). However, in Nigerian traditional medicine, preparation from the juicy aerial parts of *Rhipsalis neves-armondii* (Cactaceae) is used in the management of inflammatory conditions and cancer. *R. neves-armondii*, an epiphytic cactus herb of Brazilian origin (Anderson, 2001; Barthlott & Taylor, 1995), is commonly distributed in the tropical rainforest zone of Nigeria. It is often used in horticulture as an ornamental plant (Korotkova et al., 2011). This study, therefore, was aimed to investigate the anti-inflammatory and anticancer potentials of the methanol extract and fractions of the aerial parts of *Rhipsalis neves-armondii* in rodents and cell lines, respectively, to scientifically ascertain the folkloric use.

2. Materials and methods

2.1. Cell lines
Human embryonic kidney cells expressing SV40 Large T-antigen (293 T), Pancreatic tumor (PANC-1) cell line, and Henrietta Lacks' cervical (HeLa) cancer cell line were propagated in D-10 medium, consisting of Dulbecco's modified Eagle's medium (DMEM) with high glucose, 2 mM L-glutamine and supplemented with 10% heat-inactivated fetal bovine serum (FBS), 100 U/ml penicillin and 100 µg/ml streptomycin. Tissue culture medium and supplements were purchased from Invitrogen (Karlsruhe, Germany). The cell cultures were maintained in a humidified 5% CO_2 atmosphere at 37°C.

2.2. Animals
Adult Sprague-Dawley rats (150–250 g) and mice of either sex, obtained from the Animal House Facility of the Department of Pharmacology and Toxicology, University of Nigeria, Nsukka, were used. They were housed under standard conditions to acclimatize and fed on standard pellet and water *ad libitum* for 7 days. They were fasted overnight before their use for each experiment. All animal experiments were conducted in compliance with the National Institute of Health Guide for Care and Use of Laboratory Animals (Pub No. 85–23, revised 1985) and in accordance with the University of Nigeria Ethics Committee on the use of laboratory animals, registered by the National Health Research Ethics Committee (NHREC) of Nigeria, with the number; NHREC/05/01/2008B.

2.3. Preparation of plant material
Fresh succulent aerial parts of *Rhipsalis neves-armondii* were collected in the month of August from Nsukka, Enugu state, Nigeria. The plant was identified and authenticated by a taxonomist, Mr Alfred Ozioko of the International Centre for Ethno-medicine and Drug Development (InterCEDD), Aku Road, Nsukka, Enugu State, Nigeria. There the voucher specimen was preserved with the number; INTERCEDD 08 11. The plant material was cut, cleaned, air-dried, and pulverized using a milling machine.

2.4. Extraction and solvent-guided fractionation of RNE
About 2.1 kg of the powdered plant material was extracted by cold maceration with methanol at room temperature for 48 h. The plant material was repeatedly washed with fresh solvent until the filtrate became clear. The filtrate was concentrated using a rotary vacuum evaporator under reduced pressure (40–50°C) to obtain 120 g (5.71% w/w) of *Rhipsalis neves-armondii* extract (RNE). The RNE (50 g) was subjected to solvent-guided fractionation in a silica gel (70–120 mesh size) column (60 cm in length and 7.5 cm in diameter) successively eluted with n-hexane and ethyl acetate in order of increasing polarity. The collected solvent fractions were concentrated under reduced pressure in a rotary evaporator (40–50°C) to obtain the n-hexane (HF; 29 g, 58% w/w), and the ethyl acetate (EF; 8.5 g, 17% w/w) fractions, respectively.

2.5. Phytochemical analysis
The methanol extract (RNE), n-hexane (HF), and ethyl acetate (EF) fractions were subjected to standard phytochemical analysis for identification of plants phytoconstituents using standard methods (Trease & Evans, 1989). Briefly, frothing test for saponins, Salkowski test for terpenoids, Liebermann–Burchard tests for steroids, ferric chloride test for tannins, Keller–Killiani test for cardiac glycosides, Dragendorff's and Mayer's test for alkaloids, Fehling's test for reducing sugars, xanthoproteic test for proteins, iodine test for carbohydrates or starch, and ammonia test for detection of flavonoids were performed for qualitative identification of the phytoconstituents present.

2.6. Acute toxicity study
The acute lethal dose (LD_{50}) of RNE was evaluated using the method described by Lorke (1983). Briefly, the study was performed in two phases. In the first phase, nine mice were divided into three groups of three mice per group, and treated with the RNE at the doses of 10, 100, and 1,000 mg/kg (p.o.), respectively. The animals were observed for 24 h for signs of toxicity. In the second phase, out of the four mice used, three were treated with RNE doses of 1,600, 2,900, and 5,000 mg/kg, while the fourth mouse served as the control (5 ml/kg of distilled water). The animals were observed for 24 h.

2.7. Carragenaan-induced rat paw edema

Adult rats (120–200 g) of both sexes were distributed into five treatment groups containing six animals per group ($n = 6$). Group I (control) received the vehicle (5 ml/kg, 20% Tween 80 + propylene glycol (1:1) solution, p.o.), Group II received indomethacin (10 mg/kg, p.o.), while Groups III to V received 100, 200, and 400 mg/kg of the extract (RNE, p.o.), respectively. Thirty minutes later, acute inflammation was induced by sub plantar injection of 0.1 ml of 1% w/v freshly prepared carrageenan solution into the right hind paw of the rats according to the method described by Winter, Risley, and Nuss (1962) and Vogel (2002) with some modifications. Edema was quantified in terms of volume of the inflamed paw and measured by water displacement using an improvised plethysmometer before and at 0.5, 1, 2, 3, 4, 5, and 6 h after carrageenan injection. The same procedure was performed for fractions, HF and EF (100, 200, 400 mg/kg, p.o.). Inflammation was assessed as the difference between the volume at zero time of the treated paw and the volume at the various times after the administration of the phlogistic agent. Inhibition of edema (%) was calculated using the relation; Inhibition of edema (%) = $100[1-(a - x/b - y)]$; where a = mean paw volume of treated rats at various times after carrageenan injection; x = mean paw volume of treated rats before carrageenan injection; b = mean paw volume of control rats at various times after carrageenan injection; y = mean paw volume of control rats before carrageenan injection. The area under the curve (AUC) was calculated for each of the dose levels of the extract, fractions, and indomethacin using the trapezoid rule as described by Aguwa, Ejiekpe, Okoli, and Ezike (2009). The percentage levels of inhibition was obtained with the formula; Percentage Inhibition (PI) = $1-(AUC_t/AUC_c) \times 100$. where AUC_t = Area under the curve of treatment groups and AUC_c = Area under the curve of control group (Aguwa et al., 2009).

2.8. Cotton pellet granuloma test

The effect of the extract on chronic inflammation was evaluated using cotton-pellet granuloma test in rats (Swingle & Shideman, 1972). Sprague-Dawley rats (120–200 g) of either sex were randomly grouped into four ($n = 5$). The animals were anesthetized by intraperitoneal (IP) administration of ketamine (0.2 ml). Sterilized cotton pellet (100 mg) was aseptically implanted subcutaneously in the depilated neck region of each rat, and the incision sutured. Groups I and II of the animals were then treated by the oral administrations of extract (RNE) 200 and 400 mg/kg respectively, group III received indomethacin (5 mg/kg) while group IV received the vehicle (5 ml/kg), for 7 days. Animals were sacrificed by excess anesthesia on the 8th day and the cotton pellets removed surgically. Pellets were separated from extraneous tissues and dried at 60°C to a constant weight, and the weight of the granuloma tissue was recorded. Granuloma formation was evaluated by determining the average weights of the dry pellets and the percentage inhibition of granuloma formation calculated. Inhibition (%) of granuloma tissue development was calculated using the relation: Inhibition (%) = $[WC-WT/WC] \times 100$; where WC = weight of granuloma tissue of control group; WT = weight of granuloma tissue of treated group (Mukherjee, 2007).

2.9. Cytotoxicity studies

The cytotoxicity assay was performed using the MTT [3-(4,5-dimethylthiazol-2-yl)-2,5-diphenyl-tetrazoliumbromide] assay method as previously described (Romijn, Verkoelen, & Schroeder, 1988) on PANC-1 and HeLa cancer cell lines, while 293T cell act as control. In the MTT assay, cells were seeded onto a 96-well plate at a concentration of 104 cells/well and a volume of 100 µl per well. Different concentrations of the test extracts (31.5–1,000 µg/ml) were added to culture wells in triplicate. Culture medium without any drug was used as the "no-drug" control. After incubation at 37°C under 5% CO_2 for 2 days, a solution of MTT (3 mg/ml, 50 µl per well) was added to each well and further incubated at 37°C + 5% CO_2 for 4 h to allow formazan formation. Subsequently, the medium was removed and 150 µl of DMSO was used to dissolve the resulting blue formazan crystals in living cells. The optical density was determined at 550 nm using a multi-well microtiter plate reader (Tecan, Austria). Each single value of the triplicates was expressed as percent of the mean of triplicates of the "no-drug" control cultures and the mean and standard deviation of the percent values were calculated for each triplicate. The concentration of 50% cellular toxicity (IC_{50}) of the test extracts was calculated by non-linear regression.

2.10. Statistical analysis

Data obtained were analyzed using a one-way analysis of variance (ANOVA) subjected to Dunnett multiple comparison *post hoc* test. The Graph Pad Prism version 5 was also used. Differences between means were accepted to be significant at $p < 0.05$ and the results expressed as mean ± SEM.

3. Results

3.1. Phytochemical analysis

The extract, RNE, showed the presence of alkaloids, flavonoids, terpenoids, and steroids. HF also tested positive for carbohydrates, flavonoids, resins, steroids, and terpenoids, while EF contains alkaloids, flavonoids, resins, steroids, and terpenoids (Table 1).

3.2. Acute toxicity study

The RNE exhibited an estimated LD_{50} greater than 5,000 mg/kg (p. o.) and did not cause any lethality or show any signs of acute intoxication after a 48-h observation period.

3.3. Carragenaan-induced rat paw edema

The RNE, EF, and indomethacin exhibited significant ($p < 0.05$) inhibitions that were non-dose related but were sustained within 2–5-h post induction of edema at all doses treated compared to the control. Also the HF exhibited significant ($p < 0.05$) inhibitions of edema at 400 mg/kg dose (Table 2). However, EF (400 mg/kg) showed the highest magnitude of anti-inflammatory activity considering its lowest AUC quantification of the global edematous response of 2.28 ml/h, while indomethacin gave 2.92 ml/h (Table 2).

3.4. Cotton pellet-induced granuloma test

The RNE, HF, EF, and indomethacin exhibited appreciable dose-related inhibition of granuloma tissue formation. HF (400 mg/kg) gave a significant ($p < 0.05$) effect with a magnitude suppression of 59.4% which was equal and comparable to effect shown by the standard agent, indomethacin. Also EF (400 mg/kg) showed 18.8% inhibition of granuloma tissue formation (Table 3).

3.5. Cytotoxicity studies

The RNE exhibited cytotoxic effect against PANC-1, HeLa, and 293 T cells with estimated IC_{50} values of 30.12, 45.1, and 83.5 µg/ml, respectively (Figures 1 and 2). Hence the potency of antiproliferative activity of RNE against the cells in the order of increasing activity can be shown as; PANC-1 > HeLa > 293T cells.

Table 1. Phytochemical constituents of extract and fractions			
Constituent	RNE	HF	EF
Alkaloids	+	−	+
Carbohydrates	+	+	−
Flavonoids	+	+	+
Fats and oil	−	−	−
Glycosides	−	−	−
Reducing sugar	+	+	−
Resins	+	+	+
Saponins	−	−	−
Steroids	+	+	+
Tannins	−	−	−
Terpenoids	+	+	+

Notes: + = present; − = absent; HF = n-hexane fraction; EF = Ethyl acetate fraction; RNE = *Rhipsalis neves-armondii* extract.

Table 2. Effect of extract and fractions on carrageenan-induced rat paw edema

Treatment	Dose (mg/kg)	Edema (ml)						AUC (ml/h)
		0.5 h	1 h	2 h	3 h	4 h	5 h	
Control	–	0.54 ± 0.02	0.58 ± 0.03	0.78 ± 0.04	0.90 ± 0.03	0.85 ± 0.04	0.83 ± 0.03	3.80
RNE	100	0.66 ± 0.04 (−22.22)	0.62 ± 0.03 (−6.90)	0.57 ± 0.04* (26.92)	0.63 ± 0.06* (30.00)	0.61 ± 0.04* (28.24)	0.62 ± 0.08* (25.30)	3.07
	200	0.81 ± 0.02 (−50.00)	0.69 ± 0.05 (−18.97)	0.60 ± 0.02* (23.08)	0.65 ± 0.03* (27.78)	0.58 ± 0.03* (31.76)	0.58 ± 0.04* (30.12)	3.30
	400	0.72 ± 0.03 (−33.33)	0.60 ± 0.03 (−3.45)	0.58 ± 0.02* (25.64)	0.54 ± 0.02* (40.00)	0.57 ± 0.05* (44.19)	0.63 ± 0.10* (24.10)	2.94
HF	100	0.75 ± 0.03 (−38.89)	0.69 ± 0.02 (−18.97)	0.72 ± 0.03 (7.69)	0.81 ± 0.03 (10.00)	0.85 ± 0.05 (0.00)	0.89 ± 0.06 (−7.23)	3.89
	200	0.76 ± 0.04 (−40.74)	0.72 ± 0.04 (−24.14)	0.68 ± 0.03 (12.82)	0.73 ± 0.07 (18.89)	0.83 ± 0.13 (2.35)	0.87 ± 0.09 (−4.82)	3.78
	400	0.74 ± 0.01 (−37.04)	0.68 ± 0.03 (−17.24)	0.63 ± 0.03* (19.23)	0.63 ± 0.02* (30.00)	0.65 ± 0.02* (23.53)	0.72 ± 0.04 (13.25)	3.32
EF	100	0.62 ± 0.04 (−14.81)	0.52 ± 0.04* (10.34)	0.50 ± 0.03* (35.90)	0.55 ± 0.00* (38.89)	0.64 ± 0.06* (24.71)	0.72 ± 0.07 (13.25)	2.88
	200	0.74 ± 0.05 (−37.04)	0.63 ± 0.04* (−32.61)	0.57 ± 0.04* (26.92)	0.53 ± 0.02* (41.11)	0.57 ± 0.07* (32.94)	0.63 ± 0.05* (24.10)	2.99
	400	0.60 ± 0.05 (−11.11)	0.43 ± 0.00* (25.86)	0.45 ± 0.02* (42.30)	0.42 ± 0.02* (53.33)	0.43 ± 0.02* (49.41)	0.50 ± 0.03* (39.76)	2.28
INDO	10	0.58 ± 0.02 (−7.40)	0.60 ± 0.04 (−3.45)	0.63 ± 0.02* (19.23)	0.57 ± 0.04* (36.67)	0.55 ± 0.02* (35.29)	0.56 ± 0.03* (32.53)	2.92

Notes: Values are mean ± SEM; $n = 6$.

*$p < 0.05$ compared to control (One Way ANOVA; Dunnet post hoc test); INDO = Indomethacin; HF = n-hexane fraction; EF = Ethyl acetate fraction; RNE = Rhipsalis neves-armondii extract. Values in parenthesis represent percentage inhibition (PI) of edema compared to control.

Table 3. Effect of extract and fractions on cotton pellet-induced granuloma in rat

Treatment	Dose (mg/kg)	Granuloma tissue weight (g)	Inhibition %
Control	–	0.64 ± 0.17	–
RNE	200	0.62 ± 0.14	3.1
	400	0.53 ± 0.21	18.0
HF	200	0.58 ± 0.12	9.4
	400	0.26 ± 0.12*	59.4
EF	200	0.54 ± 0.09	15.6
	400	0.54 ± 0.09	18.8
INDO	10	0.26 ± 0.68*	59.4

Notes: Values of edema shown as Mean ± SEM; $n = 6$;

*$p < 0.05$, compared to control (One Way ANOVA; Dunnet *post hoc* test); INDO = Indomethacin; HF = n-hexane fraction; EF = Ethyl acetate fraction; RNE = *Rhipsalis neves-armondii* extract.

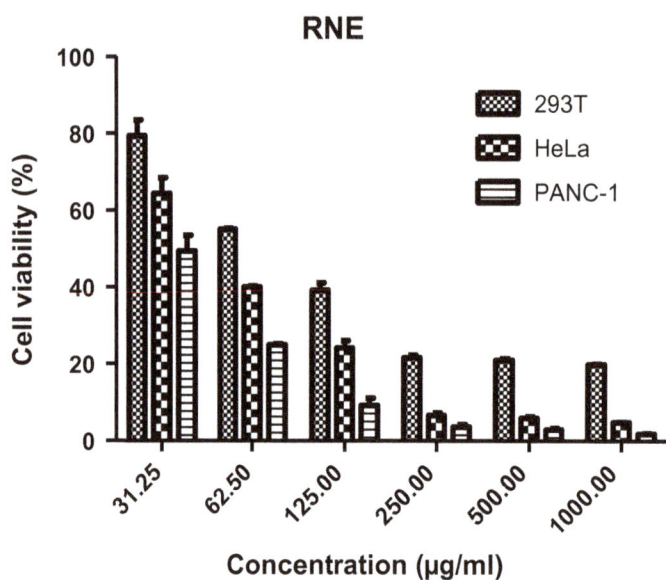

Figure 1. Cytotoxic effect of RNE on cancer cell lines.

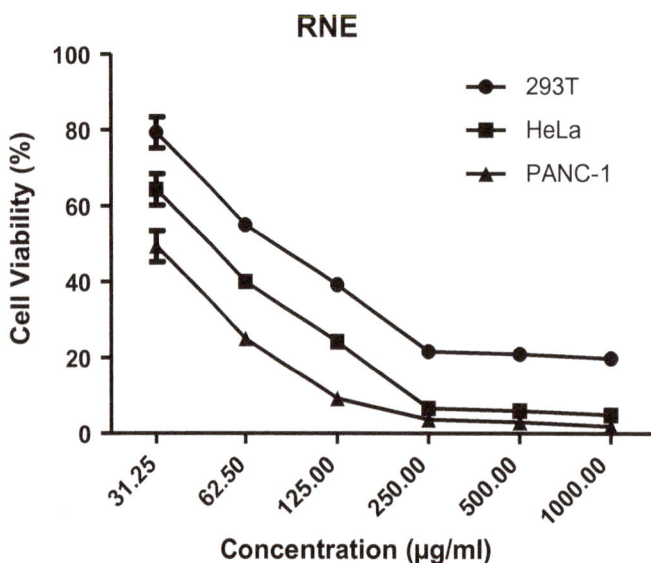

Figure 2. Graph of percentage viability of RNE against tested cancer cell lines.

4. Discussion

Evaluation of the anti-inflammatory potentials of the methanol extract (RNE) and fractions of *Rhipsalis neves-armondii* using acute and chronic inflammation models in rats revealed that the plant possesses potent anti-inflammatory activities on both acute and chronic phases of inflammation. The RNE also exhibited potent anticancer effects against PANC-1 and HeLa cancer cells. Similarly, medicinal plants with both anti-inflammatory and anticancer activities have been reported (Siriwatanametanon, Fiebich, Efferth, Prieto, & Heinrich, 2010). The RNE and its fractions showed a potent suppression of systemic acute inflammation of the rat paw which was evident from 2–5-h post administration of the phlogistic agent. Carrageenan-induced paw edema is widely used for determining the acute phase of inflammation (Prakash, Prasad, Nitin, & Vijay Kumar, 2011). Edema formation in the paw is as a result of synergism between various inflammatory mediators that increase vascular permeability and/or the mediators that increase blood flow (Harriot, Marion, Martha, Wellford, & William, 2004; Lalenti, Ianaro, Moncada, & Di Rosa, 1995). Acute inflammation following subcutaneous injection of carrageenan into the rat paw is usually a biphasic event with the initial phase beginning at about 1 h after irritant administration and associated with massive release of mediators such as histamine and serotonin causing vasodilatation and increased permeability of capillaries, whereas the second phase occurring about 3–5 h later is being mediated by the release of bradykinin, prostaglandins, protease, and lysosomal enzymes which regulate the process of adhesion of molecules (Aruna, Vinod, & Ajudhia, 2010; Brooks & Day, 1991). Administration of carrageenan also produces accumulation of plasma fluid, at same time plasma protein exudation also takes place along with neutrophil extravasations (Chatpaliwar, Johrapurkar, Wanjari, Chakraborty, & Kharkar, 2002). This implies that the anti-inflammatory activity within 1-h post induction of edema may be considered to be more of antihistaminic activity, whereas suppression of inflammation after 4–6 h represents inhibition of arachidonic acid pathway. Hence, the observed anti-inflammatory effect of *Rhipsalis neves-armondii* suggests a possible combination of antihistaminic and arachidonate pathway inhibitory activities considering the time-course of abolishing edema events *vis a vis* the significant suppression of inflammation from 2- to 5-h post administration of the phlogistic agent (Vinegar, Schreibar, & Hugo, 1969). This has correlated with the mechanisms of action of NSAIDs via cyclooxygenase inhibition, which have also been reported to possess antiproliferative and anticancer effects (Rayburn et al., 2009; Thun et al., 2002). Epidemiological studies have shown a 40–50% reduction in mortality from colorectal cancer in individuals who take NSAIDs on regular basis compared with those not taking these agents (Aggarwal et al., 2006).

Granuloma tissue formation in the cotton pellet granuloma model is usually a prolonged process developing during period of several days and involving infiltration and proliferation of mononuclear inflammatory cells such as macrophages, lymphocytes, and fibroblasts which are basic sources of tissue granuloma formation. There is often the appearance of nodules of epithelioid macrophages surrounded by a collar of lymphocyte elaborating factors like IFN-γ hence granuloma tissue formation (Mitchell et al., 2006), with vascular proliferation attempted at healing. The cotton pellet-induced granuloma test as a model of chronic inflammation is thus indicative of the proliferative phase of inflammation since the dry weight of the cotton pellets gives a good correlation with the amount of granulomatous tissue formed (Thangam & Dhananjayan, 2008). Therefore, the observed anti-inflammatory effect of *Rhipsalis neves-armondii* extract and fractions against acute and chronic inflammatory responses may possibly be the mechanism of its antiproliferative action observed against PANC-1 and Hela cancer cells. If left untreated, chronic inflammation may often progress to pathological inflammatory conditions, leading to cancer and other disease. Key molecular players linking inflammation to cancer are cytokines (such as chemokines, IL, TNF-α), nuclear factor-kB (NF-kB), inducible nitric oxide synthase (iNOS), and cyclooxygenase-2 (COX-2). It has been reported that cytokine signaling could contribute to the progression of tumors by stimulation of cell growth and differentiation as well as by the inhibition of apoptosis of altered cells at the inflammatory site (Lu et al., 2006). However, NF-kB has been shown to be activated by inflammatory stimuli and its constitutive activation is found in cancer, hence it has long been suspected to be a critical promoter facilitating the development from inflammation into cancer (Lu et al., 2006). Therefore, inhibition of the effects of these compounds by NSAIDS and anticancer agents contributes to the suppression of tumor. In this study, RNE and fractions showed inhibition of granuloma tissue formation, a marker of chronic inflammation, also a possible antiproliferative mechanism of the extract and fractions, although the specific mechanism of anti-inflammatory and anticancer actions of RNE may not be ascertained at this stage of the work. Furthermore, the RNE's IC_{50} value of 83.50 µg/ml, against the human 293-T cell is an indication of relative safety on human cells when compared to those of 30.12 and 45.1 µg/ml for PANC-1 and HeLa cells, respectively. In addition, EF when compared with HF, showed better anti-inflammatory effect, indicating that the active phytoconstituent(s) responsible for these activities may be polar or non lipid in nature. The EF also gave a better control by exhibiting the lowest AUC quantification of the global edematous response comparable to indomethacin. Lower AUC values indicate more concentration of the agent at the given time thereby exhibiting better anti-inflammatory effect. This is in correlation with the work reported by Aguwa et al. (2009).

Qualitative phytochemical tests revealed abundant presence of alkaloids, flavonoids, terpenoids, and steroids in RNE and its fractions. However, alkaloids being a polar constituent may be among the suspected constituents responsible for the observed anti-inflammatory and anticancer effects of RNE, although at this stage of the study one will not categorically attribute the effects to any specific constituent.

5. Conclusion

Findings from this study clearly suggest that the extract and fractions of the aerial parts of *Rhipsalis neves-armondii* K. Schum. (Cactaceae) possess significant anti-inflammatory properties in both acute and chronic inflammation with anticancer effects. Further studies to isolate and identify the precise constituent(s) responsible for the activities are recommended.

Acknowledgment
The authors appreciate the Department of Molecular & Medical Virology, Ruhr University, Bochum, Germany, for granting us the access to perform the cytotoxicity studies in their laboratory. Gratitude also goes to Mr Alfred Ozioko of the International Centre for Ethno-medicine and Drug Development (InterCEDD), Nsukka, Nigeria for identification of the plant.

Funding
The authors received no direct funding for this research.

Competing Interests
The authors declare no competing interest.

Author details
Theophine Chinwuba Akunne[1]
E-mail: theophine.okoye@unn.edu.ng
Peter A. Akah[1]
E-mail: peter.akah@unn.edu.ng
Ifeoma A. Nwabunike[1]
E-mail: ifeoma.nwabunike2@unn.edu.ng
Chukwuemeka S. Nworu[1]
E-mail: chukwuemeka.nworu@unn.edu.ng
Emeka K. Okereke[1]
E-mail: kings_e2000@yahoo.com
Nelson C. Okereke[1]
E-mail: nelson.okereke@yahoo.com
Francis C. Okeke[1]
E-mail: lasky_bond@yahoo.com
[1] Faculty of Pharmaceutical Sciences, Department of Pharmacology and Toxicology, University of Nigeria, Nsukka 410001, Enugu State, Nigeria.

Authors contributions
Theophine Chinwuba Akunne, Peter A. Akah and Ifeoma A. Nwabunike designed the work, participated in the experiment and write up; Chukwuemeka S. Nworu performed the cytotoxic experiment during his stay in Ruhr-University Bochum, Germany, while Emeka K. Okereke, Nelson C. Okereke and Francis C. Okeke contributed in performing the experiment.

Cover image
Source: Authors.

References
Aggarwal, B. B., Shishodia, S., Sandur, S. K., Pandey, M. K., & Sethi, G. (2006). Inflammation and cancer: How hot is the link? *Biochemical Pharmacology, 72,* 1605–1621. http://dx.doi.org/10.1016/j.bcp.2006.06.029

Aguwa, C. N., Ejiekpe, F. N., Okoli, C. O., & Ezike, A. C. (2009). Anti-inflammatory effects of leaf extracts of *Buchholzia coriacea* Engl. (Capparidaceae). *NPAIJ, 5,* 98–103.

Anderson, E. F. (2001). *The Cactus family* (pp. 51–54). Portland, OR: Timber Press.

Aruna, D., Vinod, G., & Ajudhia, N. K. (2010). Evaluation of anti-inflammatory activity of methanolic extract of *Cassia obtusifolia* seeds in Wistar rats. *Journal of Chemical and Pharmaceutical Research, 2,* 696–700.

Barthlott, W., & Taylor, N. P. (1995). Notes towards a monograph of Rhipsalidaea (Cactaceae). *Bradleya, 13,* 43–79.

Benni, J. M., Suresha, R. N., & Jayanthi, M. K. (2011). Evaluation of the anti-inflammatory activity of *Aegle marmelos* (Bilwa) root. *Indian Journal of Pharmacology, 43,* 393–397. http://dx.doi.org/10.4103/0253-7613.83108

Brooks, P. M., & Day, R. O. (1991). Nonsteroidal anti-inflammatory drugs: Differences and similarities. *The New England Journal of Medicine, 324,* 1716–1725.

Chatpaliwar, V. A., Johrapurkar, A. A., Wanjari, M. M., Chakraborty, R. R., & Kharkar, V. T. (2002). *Indian Drugs, 39,* 543–545.

Harlev, E., Nevo, E., Solowey, E., & Bishayee, A. (2013). Cancer preventive and curative attributes of plants of the cactaceae family: A review. *Planta Medica, 79,* 713–722.

Harriot, M., Marion, E., Martha, A., Wellford, S., & William, A. (2004). Inflammation induced by histamine, serotonine, bradykinin and compound 48/480 in the rat. Antagonists and mechanisms of action. *Journal of Pharmacology and Experimental Therapeutics, 191*, 300–302.

Korotkova, N., Borsch, T., Quandt, D., Taylor, N. P., Muller, K. F., & Barthlott, W. (2011). What does it take to resolve relationships and to identify species with molecular markers? An example from the epiphytic Rhipsalideae (Cactaceae). *American Journal of Botany, 98*, 1549–1572. http://dx.doi.org/10.3732/ajb.1000502

Lalenti, A., Ianaro, A., Moncada, S., & Di Rosa, M. (1995). Modulation of acute inflammation by endogenous nitric oxide. *European Journal of Pharmacology, 211*, 177–184.

Lorke, D. (1983). A new approach to practical acute toxicity testing. *Archives of toxicology, 54*, 272–289.

Lu, H., Ouyang, W., & Huang, C. (2006). Inflammation, a key event in cancer development. *Molecular Cancer Research, 4*, 221–233. http://dx.doi.org/10.1158/1541-7786.MCR-05-0261

Mitchell, N. R., Kumar, V., Abbas, A. K., & Fausto, N. (2006). *Robbins and cotran pathologic basis of disease* (7th ed., pp. 48–85). Philadelphia, PA: Saunders Elsevier.

Mukherjee, P. K. (2007). *Quality control of herbal drugs–An approach to evaluation of botanicals* (1st ed.). New Delhi: Business Horizons.

Prakash, P., Prasad, K., Nitin, M., & Vijay Kumar, M. (2011). Evaluation of anti-inflammatory effect of *Calotropis procera* (AIT.) R.BR. root extract against different mediators of inflammation in albino rats. *International Research Journal of Pharmacy, 2*, 279–284.

Rang, H. P., Dale, M. M., Ritter, J. M., & Flower, R. J. (2008). Drugs used in the treatment of infections and cancer. *Rang and Dale's pharmacology* (6th ed., pp. 647–660). Philadelphia, PA: Elsevier.

Rayburn, E. R., Ezell, S. J., & Zhang, R. (2009). Anti-inflammatory agents for cancer therapy. *Molecular and Cellular Pharmacology, 1*, 29–43. http://dx.doi.org/10.4255/mcpharmacol

Romijn, J. C., Verkoelen, C. F., & Schroeder, F. H. (1988). Application of the MTT assay to human prostate cancer cell lines in vitro: Establishment of test conditions and assessment of hormone-stimulated growth and drug-induced cytostatic and cytotoxic effects. *The Prostate, 12*, 99–110. http://dx.doi.org/10.1002/(ISSN)1097-0045

Siriwatanametanon, N., Fiebich, B. L., Efferth, T., Prieto, J. M., & Heinrich, M. (2010). Traditionally used Thai medicinal plants: In vitro anti-inflammatory, anticancer and antioxidant activities. *Journal of Ethnopharmacology, 130*, 196–207. http://dx.doi.org/10.1016/j.jep.2010.04.036

Swingle, K. F., & Shideman, F. E. (1972). Phases of the inflammatory response to subcutaneous implantation of a cotton pellet and their modification by certain anti-inflammatory agents. *Journal of Pharmacology and Experimental Therapeutics, 183*, 226–234.

Thangam, C., & Dhananjayan, R. (2008). Anti-inflammatory potential of the seeds of *Carum capticum* Linn. *Indian Journal of Pharmacology, 35*, 388–391.

Thun, M. J., Henley, S. J., & Patrono, C. (2002). Nonsteroidal anti-inflammatory drugs as anticancer agents: Mechanistic, pharmacologic, and clinical issues. *JNCI Journal of the National Cancer Institute, 94*, 252–266. http://dx.doi.org/10.1093/jnci/94.4.252

Trease, G. E., & Evans, W. C. (1989). *Test book of pharmacognosy* (11th ed., pp. 176–180), London: Brailliare Tindall and Macmillian.

Vinegar, R., Schreibar, N., & Hugo, B. (1969). Biphasic development of carrageenan edema in rats. *Journal of Pharmacology and Experimental Therapeutics, 166*, 96–103.

Vogel, H. G. (2002). *Analgesic, anti-inflammatory and antipyretic activity in drug discovery and evaluation pharmacological assays* (2nd ed., pp. 759–767). New York, NY: Springer.

Winter, C. A., Risley, E. A., & Nuss, G. W. (1962). Carrageenin-induced edema in hind paw of the rat as an assay for antiinflammatory drugs. *Experimental Biology and Medicine, 111*, 544–547. http://dx.doi.org/10.3181/00379727-111-27849

World Health Organization. (2014). *World Cancer Report*.

9

Antibacterial activity of 3,6-di(pyridin-2-yl)-1,2,4,5-s-tetrazine capped Pd(0) nanoparticles against Gram-positive *Bacillus subtilis* bacteria

Sutapa Joardar[1,2]*, Shounak Ray[3], Suvendu Samanta[3] and Paramita Bhattacharjee[2]*

*Corresponding authors: Sutapa Joardar, Department of Biotechnology, Neotia Institute of Technology, Management and Science, Jhinga, Diamond Harbour Road, South 24 Parganas, Amira 743368, India; Department of Food Technology and Bio-chemical Engineering, Jadavpur University, 188, Raja S. C. Mallick Road, Kolkata 700 032, India; Paramita Bhattacharjee, Department of Food Technology and Bio-chemical Engineering, Jadavpur University, 188, Raja S. C. Mallick Road, Kolkata 700 032, India
E-mails: sutapajor@yahoo.co.in; pb@ftbe.jdvu.ac.in
Reviewing editor: Yasser Gaber, Beni-Suef University, Egypt
Additional information is available at the end of the article

Abstract: In this work, we report antibacterial activity of 3,6-di(pyridin-2-yl)-1,2,4,5-tetrazine (pytz) capped Pd(0) nanoparticles (TzPdNPs). The TzPdNPs were characterized by transmission electron microscopy (TEM) and powder X-ray diffraction techniques (PXRD). The antibacterial properties of TzPdNPs were studied using Gram-positive *B. subtilis* and Gram-negative *E. coli* by the plate count method. The antimicrobial activity of this TzPdNPs shows that the microbial growth has been fully inhibited only for Gram-positive *B. subtilis* bacteria. Morphological changes obtained by transmission electron microscope observation show that TzPdNPs can cause leakage and chaos of intracellular contents.

Subjects: Microbiology; Biology; Biotechnology

Keywords: palladium nanoparticles; antimicrobial activity; antibacterial growth kinetics transmission electron microscopy

1. Introduction
As a result of increasing microbial resistance to multiple antimicrobial agents and development of resistant strains, there is an increasing demand for novel antimicrobial materials with superior performance for disinfection applications. In this regard, nanoparticles are particularly effective, due to

ABOUT THE AUTHORS
Sutapa Joardar (SJ) is currently working as an assistant professor in the Department of Biotechnology, Neotia Institute of Technology, Management and Science, West Bengal, India. The focus of her research is application of nanostructured materials as antimicrobial agents. She is particularly interested about the antimicrobial properties of metallic nanoparticles immobilized on carrier such as zeolites, silicates, and polymer. Paramita Bhattacharjee (PB) is currently working as an assistant professor in the Department of Food Technology & Bio-Chemical Engineering, Jadavpur University, West Bengal, India. The focus of her research is supercritical fluid extraction, microencapsulation, gamma irradiation technology, technology of fats and oils, and technology of herbs and spices.

PUBLIC INTEREST STATEMENT
Infectious diseases remain among the top five causes of mortality in countries of all socioeconomic classes worldwide, resulting in 9.2 million deaths in 2013 (about 17% of all deaths). At any given time, an estimated 1.4 million people are affected by health care-associated infections (HAIs) in developed countries, around 5–10% of patients in hospitals get HAIs, but the risk is 200–2,000% higher in undeveloped countries. Therefore, antibacterial agents play a significant role in combating microbial infections. However, due to the wide use of antibiotics and the emergence of antibiotic-resistant bacterial strains, conventional antibiotics are becoming less effective, resulting in poor treatment efficacy and significantly increased cost of health care. Each year, in the United States alone, more than 2 million people with antibiotic-resistant infections are reported, leading to at least 23,000 deaths. Therefore, there is continuous demand to discover new antimicrobial agents to control microbial infections.

a high surface-to-volume ratio, which provides a large active surface in the contact with micro-organisms. It has been demonstrated that nanoparticles (NPs) have a wide variety of unique applications, including cell targeting (Manolova et al., 2008; Weissleder, Kelly, Sun, Shtatland, & Josephson, 2005), intravenous nucleic acid delivery (Dobson, 2006; Li & Szoka, 2007; Neuberger, Schöpf, Hofmann, Hofmann, & von Rechenberg, 2005), environmental remediation (Kamat & Meisel, 2003; Tratnyek & Johnson, 2006; Zhang et al., 2010), and catalysis (Witham et al., 2010). The antimicrobial function of metal ions and NPs has also long been recognized and exploited industrially for several purposes, including amendments to textiles and cosmetics, food processing, water treatment, and so on (Ilić et al., 2009; Mueller & Nowack, 2008). Commonly used antimicrobial nanocomposite materials include metal ions or NPs (silver (Tamboli et al., 2012; Vukoje et al., 2014), copper (Cady et al., 2011; Cioffi et al., 2005; Liang et al., 2012), gold (Boomi & Prabu, 2013; Mei et al., 2014), platinum (Ma et al., 2012), metal oxide (titanium dioxide (Zhang & Chen, 2009), zinc oxide (Liang et al., 2012), copper oxide (Ojas, Bhagat, Gopalakrishnan, & Arunachalam, 2008)), and organically modified nanoclay (Rhim, Hong, Park, & Ng, 2006), to name a few.

So far, the antimicrobial property of silver, copper, gold, and platinum NPs has been investigated thoroughly; however, potential of palladium nanoparticles (PdNPs) as antimicrobial agent remains a relatively undeveloped area. Pd is one of the most widely used transition metals for carbon–carbon and carbon–heteroatom cross-coupling reactions such as the Suzuki–Miyaura reaction (1995), Heck reaction (Heck, 1968a, 1968b, 1968c, 1968d; Heck & Nolley, 1972), Kumada reaction (Jana, Pathak, & Sigman, 2011), Sonogashira reaction (Sonogashira, 2002), Negishi reaction (Astruc, 2007), Stille reaction (Milstein & Stille, 1978), and Buchwald–Hartwig reaction (Widenhoefer & Buchwald, 1996). Pd has also been used as a catalyst to manufacture pharmaceuticals (Malleron, Fiaud, & Legros, 2000), degrade harmful environmental pollutants (Nutt, Hughes, & Wong, 2005), and as sensors for the detection of various analytes (Chang et al., 2008; Favier, Walter, Zach, Benter, & Penner, 2001; Yu et al., 2005). Additionally, Pd and Pd^{2+} ions also play fundamental roles in several biotechnological processes (Baccar, Adams, Abdelghani, & Obare, 2013). Baccar et al. developed a non-enzymatic biosensor using various sizes of PdNPs to detect hydrogen peroxide in milk (2013). The anti-tumor, -viral, -malarial, -fungal and -bacterial activities of various Pd(II) complexes of N, O, S donor ligands, and drugs are long known (Baccar et al., 2013; Garoufis, Hadjikakou, & Hadjikakou, 2009). Size-dependent antimicrobial effects of PdNPs have also been reported recently (Adams, Walker, Obare, Docherty, & van Raaij, 2014; Garoufis, Hadjikakou, & Hadjiliadis, 2005).

In the present study, we report the activity of 3,6-di-(pyridin-2-yl)-1,2,4,5-tetrazine (pytz) capped PdNPs (TzPdNPs) against bacterial growth. To investigate the antibacterial activity of TzPdNPs, the viability of Gram-positive *Bacillus subtilis* was investigated in the presence of TzPdNPs.

2. Experimental

2.1. Materials and methods
The synthesis of 3,6-di(pyridin-2-yl)-1,2,4,5-tetrazine (pytz) capped Pd(0) nanoparticles (TzPdNPs) was carried out according to the method reported earlier (Das, Samanta, Ray, & Biswas, 2015). The antimicrobial activities were investigated against bacterial strains Gram-positive *Bacillus subtilis* (MTCC 441) and Gram-negative *Escherichia coli* (MTCC 2939).

2.2. Instrumentation
Powder X-ray diffraction (PXRD) patterns were obtained on a Philips PW 1140 parallel beam X-ray diffractometer with Bragg–Bretano focusing geometry and monochromatic CuKα radiation (λ = 1.540598 Å). Transmission electron micrographs (TEM) images were collected using JEOL JEM-2100 microscope working at 200 kV. TEM samples were prepared by sonicating an aliquot of the sample in ethanol for 15 min and a single drop of this suspension was drop-casted onto carbon-coated 300 mesh copper grids. Grids were allowed to air-dry prior to imaging.

2.3. Antimicrobial test for TzPdNPs

The antibacterial activities of TzPdNPs were measured by serial dilution method. The 50 mg of TzPdNPs was suspended in 6 mL of Muller-Hinton broth and inoculated with 1 mL of 18-h stock culture of *Bacillus subtilis*. Aliquots of each sample were diluted and 200 µL were plated on sterile Petri dishes, covered with Muller-Hinton agar. The plates were incubated for 20 h at 37°C. After the incubation period, the colony-forming units (CFU) of each plate were determined. Antimicrobial activity of pytz was also tested by the same method to compare its activity with TzPdNPs.

2.4. Minimum inhibitory concentration (MIC)

Bacillus subtilis was cultured in Muller-Hinton broth at 37°C on a shaker incubator at 200 rpm for 6 h. Then the concentration of micro-organism of 1×10^8 CFU mL^{-1} dilution was fixed at 0.1, optical density at 600 nm with addition of medium and then again diluted to 1×10^6 CFU mL^{-1} with medium. We mixed bacterial suspension (1×10^6 CFU mL^{-1}, 100 µL) with the solutions of TzPdNPs with different concentrations in medium and the final volume of the solutions was adjusted to 10 mL with medium. The solutions were shaken at 37°C on a shaker incubator at 200 rpm for 24 h. The bacterial viability was determined by measuring optical density at 600 nm ($OD_{600\,nm}$). Simultaneously, aliquots of each solution were diluted and plated on Muller-Hinton agar. The plates were incubated at 37°C for 20 h and colonies were counted and compared to those on control plates to calculate changes in the cell growth inhibition. The percentage of cell growth reduction (*R*, %) was calculated using the following equation:

$$R = (C_0 - C)/C_0 \times 100 \tag{1}$$

where C_0 is the number of CFU from the control sample and C is the number of CFU from treated samples. Each concentration was prepared and measured in triplicate, and all experiments were repeated at least thrice in parallel.

2.5. Antibacterial growth kinetics

Bacterial suspension (~1×10^7 CFU mL^{-1}) was inoculated with different concentrations (63–250 µg mL^{-1}) of TzPdNPs in Muller-Hinton broth. The mixtures were shaken at 37°C on a shaker incubator at 200 rpm for 0–36 h. The initial time of addition of the bacteria was taken as zero. The growth curves resulted from bacteria only, and TzPdNPs incubated with bacteria were plotted against time. Aliquots were drawn from each of the mixtures at definite time intervals and diluted further. Afterward, 200 µL of the final solutions were plated on Muller-Hinton agar plates immediately. The plates were incubated at 37°C for 24 h, and bacterial colonies were counted. A plot of colony-forming unit (CFU) in logarithmic scale versus time ($\log_{10}$ CFU mL^{-1} vs time) was then plotted to determine the bactericidal kinetics.

2.6. Characterization of treated bacteria by TEM

The size and morphology of TzPdNPs treated and untreated bacteria were examined by TEM. Prior to microscopy analysis, bacterial samples were prepared using procedures described previously in literature (Hao, Jayawardana, Chen, & Yan, 2015). *Bacillus subtilis* was inoculated for 10 h in Muller-Hinton broth at 37°C and at 200 rpm until an OD_{600} of 0.5 was attained. The bacteria cell suspension (30 mL) was then harvested, centrifuged, and redispersed in pH 7.4 PBS buffer. TzPdNPs (300 µg mL^{-1}) were added to bacteria and the mixture was incubated at 37°C for 10 h while shaking at 200 rpm. The mixture was then centrifuged at 2,000 rcf for 8 min, and the supernatant was discarded. A drop of bacteria cell suspension was placed onto a Cu grid followed by overnight drying.

3. Results and discussion

3.1. Synthesis and characterization of the TzPdNPs

Synthesis of 3,6-di(pyridin-2-yl)-1,2,4,5-s-tetrazine (pytz) capped Pd(0) nanoparticles (TzPdNPs) was achieved by reacting aqueous solution of Na$_2$PdCl$_4$ with 3,6-di(pyridin-2-yl)-1,4-dihydro-1,2,4,5-tetrazine (H$_2$pytz) in ethanol under sonication. The crystallinity and purity of TzPdNPs were examined (Figure 1(a))

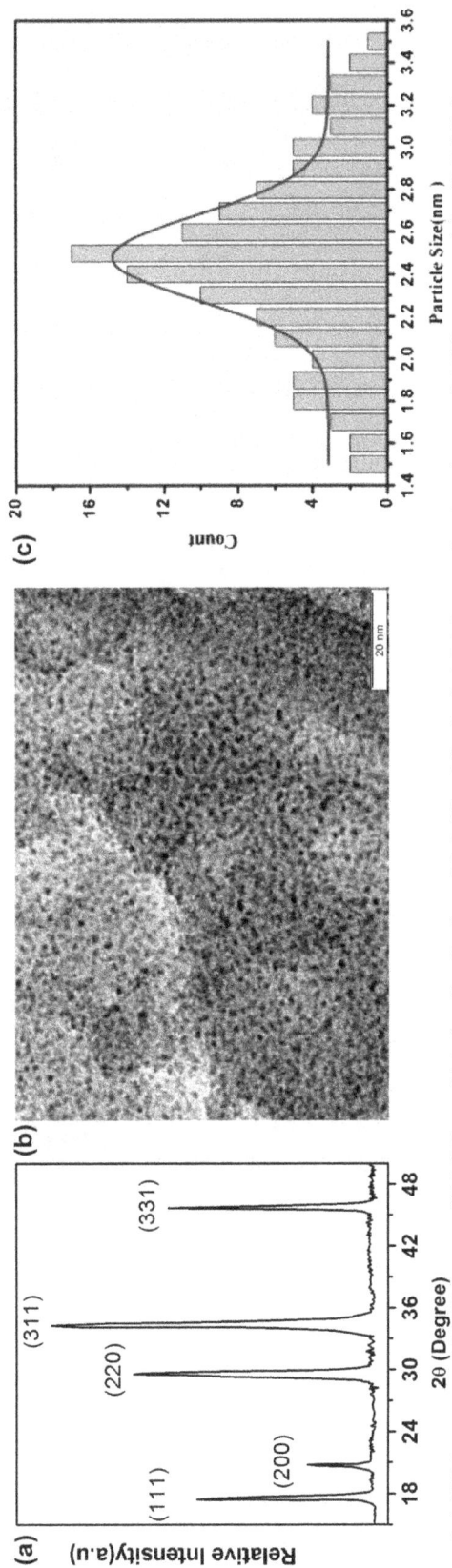

Figure 1. (a) XRD pattern, (b) TEM image of TzPdNPs and (c) size distribution of 3,6-di(pyridin-2-yl)-1,2,4,5-tetrazine (pytz) capped Pd(0) nanoparticles (TzPdNPs).

by the PXRD technique. The XRD pattern of TzPdNPs shows the inter planar d-spacing of XRD peaks correspond to (111), (200), (220), and (311) planes of Pd(0) with fcc structure (JCPDS No. 01–1310), without any impurity phase. The morphology and size distribution of the prepared nanoparticles were elucidated from the TEM. Figure 1(b) exhibits morphological images of the prepared palladium nanoparticles. The shape of the particles is quasi-spherical and the size of the particles varies from 1.5 to 3.5 nm and the average size observed was 2.5 ± 0.2 nm (Figure 1(c)). It can also be noted that, despite the small size, the particles are separated from one another and no significant aggregation is observed.

3.2. Antibacterial activity

The antibacterial activities of these TzPdNPs nanoparticles have been evaluated toward Gram-positive B. Subtilis and Gram-negative E. coli. TzPdNPs (50 mg) were dispersed in 6 mL of Mueller-Hinton Broth and inoculated with 1 mL of freshly cultured bacteria. After an incubation time of 18 h, sample was diluted and plated on Mueller-Hinton agar plates. The plates were incubated for 20 h. Colonies were counted and recorded for dilutions containing between 30 and 300 colonies. The Mueller-Hinton agar plates with the B. subtilis exhibited no growth of bacterial colonies for 10^9–10^7 CFU mL^{-1} concentrations of bacteria, indicating excellent antibacterial activity against Gram-positive bacteria (Figure 2). Antibacterial activities of TzPdNPs were then compared with pytz and a control solution was prepared with the bacteria but without any nanoparticles. Solutions of bacteria with TzPdNPs, pytz, and without any nanoparticles were diluted to 10^5 CFU mL^{-1} after 18 h incubation and plated on Mueller-Hinton agar plates. As shown in Figure 2, after 20 h of incubation, dense bacterial colonies were observed on the control and pytz Mueller-Hinton agar B. subtilis (Figures 3(a) and (b)). However, the Mueller-Hinton agar plates with the TzPdNPs exhibited no growth of bacterial colonies, indicating again excellent antibacterial activity of TzPdNPs against B. subtilis (Figure 3(c)).

MIC values of TzPdNPs were evaluated by growth inhibition of B. subtilis by varying the concentration of TzPdNPs and measuring their OD$_{600 nm}$ after 24 h. The original bacterial concentration was 1×10^6 CFU mL^{-1}. Aliquots of each sample were diluted and plated on Mueller-Hinton agar, incubated for 24 h, and colonies were counted. The MIC of TzPdNPs to inhibit proliferation of B. subtilis was

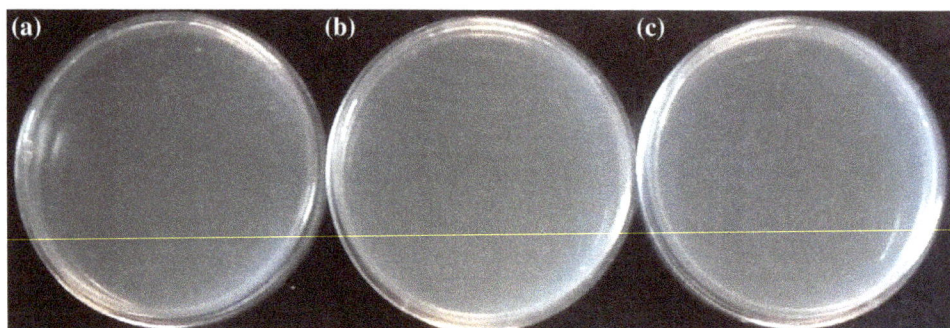

Figure 2. Antibacterial activities of TzPdNPs with (a) 10^9 (b) 10^{87} CFU mL^{-1} *Bacillus subtilis*.

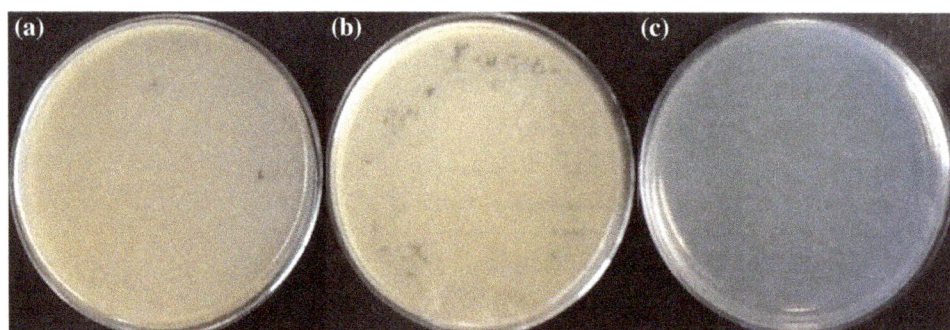

Figure 3. Photographs of colonies of B. subtilis (a) control, (b) after treatment with pytz and (c) treated with TzPdNPs.

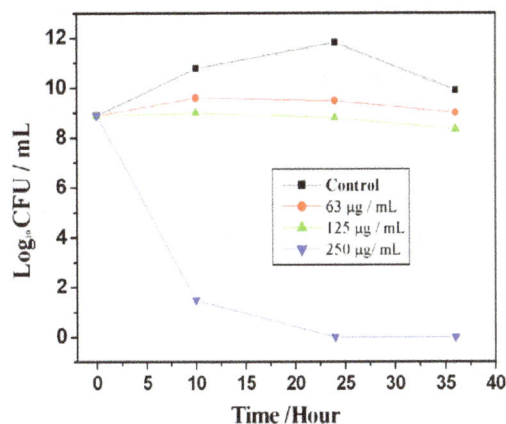

Figure 4. Concentration effects of TzPdNPs on the growth of *B. subtilis*.

determined to be 62.5 μg mL^{-1}. Percentage of the cell growth reduction was calculated to be 95% for *B. subtilis* after 24 h of incubation. The effect of TzPdNPs nanoparticles on the growth kinetics of *B. subtilis* in liquid media was studied (Figure 4) further. Freshly cultured bacteria were added with various concentrations of TzPdNPs solution; the optical density at 600 nm (OD$_{600\,nm}$) based on the turbidity of the cell suspension was measured to monitor bacterial growth. Each solution was also diluted, plated on Mueller-Hinton agar, incubated for 24 h, and bacterial colonies were counted. The growth of both the bacteria was inhibited as the concentration of TzPdNPs increased, but the rates of bacterial growth inhibition were different. The growth of *B. subtilis* started to decrease when the concentration of TzPdNPs was 62.5 μg mL^{-1} and at a concentration of 250 μg mL^{-1}, the growth can be completely inhibited (Figure 4). It can also be observed that no colony was formed for bacteria in the plate until 36 h.

Next, we explored antibacterial mechanisms of TzPdNPs. The exact mechanisms of NPs toxicity against bacteria are not understood completely. Generally, NPs are able to bind to the membrane of bacteria by electrostatic interaction and disrupt the integrity of the bacterial membrane (Thill et al., 2006). Nanotoxicity is generally triggered by the induction of oxidative stress by formation of reactive oxygen species (ROS), following the administration of NPs (Nel et al., 2009; Soenen, Rivera-Gil, Montenegro, Parak, & De Smedt, 2011). On the other hand, rather different mechanism has been proposed particularly for silver nanoparticles. Previous studies suggest that silver nanoparticles (AgNPs) release Ag$^+$ ions which interact strongly with thiol groups of intracellular enzymes and proteins, leading to degeneration (Agnihotri, Mukherji, & Mukherji, 2013; Kim, Lee, Cha, Kim, & Kang, 2007; Xiu, Zhang, Puppala, Colvin, & Alvarez, 2012). The activities of enzymes and proteins are closely related to physiological functions of bacteria (Eckhardt et al., 2013; Xiu et al., 2012). Therefore, it may be suggested that Ag$^+$ could affect the physiological functions of intracellular enzymes and proteins, leading to inhibition of activity and cellular damage.

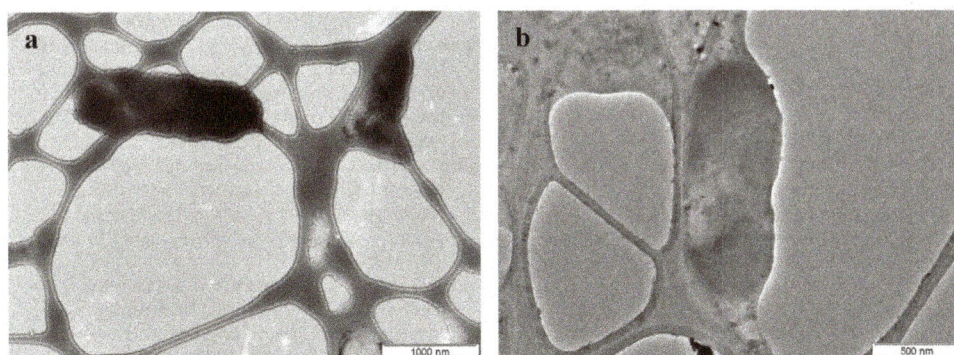

Figure 5. TEM images of *B. subtilis* before (a) and after (b) inoculation with TzPdNPs for 6 h.

To clarify the antibacterial mechanism of the action of TzPdNPs, we investigated ultrastructural damage of *B. subtilis* before and after incubation with TzPdNPs, examining TEM images. Untreated bacteria exhibited a smooth surface and distinct nucleoid structures (Figure 5(a)). In contrast, the bacterial cell surface became rough after incubation with TzPdNPs for 12 h. Moreover, numerous blebs were formed with cytoplasmic release (Figure 5(b)). The penetration of positively charged nanoparticles into negatively charged bacterial cytomembranes via electrostatic interactions accelerates cell division, leading to the formation of blebs. Literature reports have shown that electrostatic attraction between negatively charged bacterial cells and positively charged nanoparticles is crucial for the activity of nanoparticles as bactericidal materials (Hamouda & Baker, 2000; Stoimenov, Klinger, Marchin, & Klabunde, 2002). In addition, pytz molecules on nanoparticle probably provide a lipophilic segment compatible with the lipid bilayer of the bacterial cytoplasmic membrane (Beney & Gervais, 2001) and can cause decreased membrane fluidity and disrupt the cytoplasmic membrane (Silvestro, Weiser, & Axelsen, 2000). These two factors act together to damage the bacterial cytomembrane, resulting in leakage of intracellular content. It is also possible that the nanoparticles interact with the intracellular enzymes and proteins to inhibit their activity. Therefore, positively charged nanoparticles increase the permeability of the cytoplasmic bacterial membrane due to the presence of external pytz. Subsequently, these nanoparticles penetrate the cells and strongly associate with intracellular enzymes and proteins, resulting in cell death.

4. Conclusions

Synthesis of 3,6-di(pyridin-2-yl)-1,2,4,5-tetrazine (pytz) capped PdNPs was achieved by a simple method. Synthesized TzPdNPs have been characterized by powder X-ray diffraction and transmission electron microscope techniques. A systematic evaluation of the antibacterial activity of the 3,6-di(pyridin-2-yl)-1,2,4,5-tetrazine (pytz) capped Pd(0) nanoparticles (TzPdNPs) was carried out and showed excellent bactericidal properties against Gram positive *B. subtilis* bacteria. TzPdNPs showed effective antimicrobial behavior, which caused 100% mortality against *B. subtilis* bacterial strains within 24 h. This material reported here can be used for different applications in biomaterials.

Acknowledgment
We are thankful to Dr P. Biswas, Department of Chemistry, Indian Institute of Engineering Science and Technology, for helpful suggestions during work and for preparation of the manuscript.

Funding
The authors received no direct funding for this research.

Competing interests
The authors declare no competing interests.

Author details
Sutapa Joardar[1,2]
E-mail: sutapajor@yahoo.co.in
Shounak Ray[3]
E-mail: shounak.ray.89@gmail.com
Suvendu Samanta[3]
E-mail: samanta.suvendu88@gmail.com
Paramita Bhattacharjee[2]
E-mail: pb@ftbe.jdvu.ac.in
[1] Department of Biotechnology, Neotia Institute of Technology, Management and Science, Jhinga, Diamond Harbour Road, South 24 Parganas, Amira 743368, India.
[2] Department of Food Technology and Bio-Chemical Engineering, Jadavpur University, 188, Raja S. C. Mallick Road, Kolkata 700 032, India.
[3] Department of Chemistry, Indian Institute of Engineering Science and Technology, Shibpur, Howrah 711 103, India.

Cover image
Source: Authors.

References
Adams, C. P., Walker, K. A., Obare, S. O., Docherty, K. M., & van Raaij, M. J. (2014). Size-dependent antimicrobial effects of novel palladium nanoparticles. *PLoS ONE, 9*, e85981. http://dx.doi.org/10.1371/journal.pone.0085981
Agnihotri, S., Mukherji, S., & Mukherji, S. (2013). Immobilized silver nanoparticles enhance contact killing and show highest efficacy: Elucidation of the mechanism of bactericidal action of silver. *Nanoscale, 5*, 7328–7340. http://dx.doi.org/10.1039/c3nr00024a
Astruc, D. (2007). Palladium nanoparticles as efficient green homogeneous and heterogeneous carbon–carbon coupling precatalysts: A unifying view. *Inorganic Chemistry, 46*, 1884–1894. http://dx.doi.org/10.1021/ic062183h
Baccar, H., Adams, C. P., Abdelghani, A., & Obare, S. O. (2013). Chronoamperometric-based detection of hydrogen peroxide using palladium nanoparticles. *International Journal of Nanotechnology, 10*, 563–576. http://dx.doi.org/10.1504/IJNT.2013.053525

Beney, L., & Gervais, P. (2001). Influence of the fluidity of the membrane on the response of microorganisms to environmental stresses. *Applied Microbiology and Biotechnology, 57*, 34–42.

Boomi, P., & Prabu, H. G. (2013). Synthesis, characterization and antibacterial analysis of polyaniline/Au-Pd nanocomposite. *Colloids and Surfaces A: Physicochemical and Engineering Aspects, 429*, 51–59. http://dx.doi.org/10.1016/j.colsurfa.2013.03.053

Cady, N. C., Behnke, J. L., & Strickland, A. D. (2011). Copper-based nanostructured coatings on natural cellulose: Nanocomposites exhibiting rapid and efficient inhibition of a multi-drug resistant wound pathogen, A. baumannii, and mammalian cell biocompatibility *in vitro*. *Advanced Functional Materials, 21*, 2506–2514. http://dx.doi.org/10.1002/adfm.201100123

Chang, Z., Fan, H., Zhao, K., Chen, M., He, P., & Fang, Y. (2008). Electrochemical DNA biosensors based on palladium nanoparticles combined with carbon nanotubes. *Electroanalysis, 20*, 131–136. http://dx.doi.org/10.1002/(ISSN)1521-4109

Cioffi, N., Torsi, L., Ditaranto, N., Tantillo, G., Ghibelli, L., Sabbatini, L., ... Traversa, E. (2005). copper nanoparticle/ polymer composites with antifungal and bacteriostatic properties. *Chemistry of Materials, 17*, 5255–5262. http://dx.doi.org/10.1021/cm0505244

Das, S., Samanta, S., Ray, S., & Biswas, P. (2015). 3,6-Di(pyridin-2-yl)-1,2,4,5-tetrazine capped Pd(0) nanoparticles: A catalyst for copper-free Sonogashira coupling of aryl halides in aqueous medium. *RSC Advances, 5*, 75263–75267. http://dx.doi.org/10.1039/C5RA13252E

Dobson, J. (2006). Magnetic nanoparticles for drug delivery. *Drug Development Research, 67*, 55–60. http://dx.doi.org/10.1002/(ISSN)1098-2299

Eckhardt, S., Brunetto, P. S., Gagnon, J., Priebe, M., Giese, B., & Fromm, K. M. (2013). Nanobio silver: Its interactions with peptides and bacteria, and its uses in medicine. *Chemical Reviews, 113*, 4708–4754. http://dx.doi.org/10.1021/cr300288v

Favier, F., Walter, E. C., Zach, M. P., Benter, T., & Penner, R. M. (2001). Hydrogen sensors and switches from electrodeposited palladium mesowire arrays. *Science, 293*, 2227–2231. http://dx.doi.org/10.1126/science.1063189

Garoufis, A., Hadjikakou, S. K., & Hadjiliadis, N. (2005). In M. Gielen & E. R. T. Tiekink (Eds.), *Metals in medicine, Palladium (Pd), in metallotherapeutic drugs and metal-based diagnostic agents: The use of metals in medicine* (pp. 399–415). John Wiley & Sons.

Garoufis, A., Hadjikakou, S. K., & Hadjikakou, N. (2009). Palladium coordination compounds as anti-viral, anti-fungal, anti-microbial and anti-tumor agents. *Coordination Chemistry Reviews, 253*, 1384–1397. http://dx.doi.org/10.1016/j.ccr.2008.09.011

Hamouda, T., & Baker, Jr., J. R. (2000). Antimicrobial mechanism of action of surfactant lipid preparations in enteric Gram-negative bacilli. *Journal of Applied Microbiology, 89*, 397–403. http://dx.doi.org/10.1046/j.1365-2672.2000.01127.x

Hao, N., Jayawardana, K. W., Chen, X., & Yan, M. (2015). One-step synthesis of amine-functionalized hollow mesoporous silica nanoparticles as efficient antibacterial and anticancer materials. *ACS Applied Materials & Interfaces, 7*, 1040–1045. http://dx.doi.org/10.1021/am508219g

Heck, R. F. (1968a). The arylation of allylic alcohols with organopalladium compounds. A new synthesis of 3-aryl aldehydes and ketones. *Journal of the American Chemical Society, 90*, 5526–5531. http://dx.doi.org/10.1021/ja01022a035

Heck, R. F. (1968b). Allylation of aromatic compounds with organopalladium salts. *Journal of the American Chemical Society, 90*, 5531–5534. http://dx.doi.org/10.1021/ja01022a036

Heck, R. F. (1968c). The palladium-catalyzed arylation of enol esters, ethers, and halides. A new synthesis of 2-aryl aldehydes and ketones. *Journal of the American Chemical Society, 90*, 5535–5538. http://dx.doi.org/10.1021/ja01022a037

Heck, R. F. (1968d). The addition of alkyl- and arylpalladium chlorides to conjugated dienes. *Journal of the American Chemical Society, 90*, 5542–5546.

Heck, R. F., & Nolley, J. P. (1972). Palladium-catalyzed vinylic hydrogen substitution reactions with aryl, benzyl, and styryl halides. *The Journal of Organic Chemistry, 37*, 2320–2322. http://dx.doi.org/10.1021/jo00979a024

Ilić, V., Šaponjić, Z., Vodnik, V., Potkonjak, B., Jovančić, P., Nedeljković, J., & Radetić, M. (2009). The influence of silver content on antimicrobial activity and color of cotton fabrics functionalized with Ag nanoparticles. *Carbohydrate Polymers, 78*, 564–569.

Jana, R., Pathak, T. P., & Sigman, M. S. (2011). Advances in transition metal (Pd,Ni,Fe)-catalyzed cross-coupling reactions using alkyl-organometallics as reaction partners. *Chemical Reviews, 111*, 1417–1492. http://dx.doi.org/10.1021/cr100327p

Kamat, P. V., & Meisel, D. (2003). Nanoscience opportunities in environmental remediation. *Comptes Rendus Chimie, 6*, 999–1007. http://dx.doi.org/10.1016/j.crci.2003.06.005

Kim, Y. H., Lee, D. K., Cha, H. G., Kim, C. W., & Kang, Y. S. (2007). Synthesis and characterization of antibacterial Ag–SiO$_2$ nanocomposite. *The Journal of Physical Chemistry C, 111*, 3629–3635. http://dx.doi.org/10.1021/jp068302w

Li, W., & Szoka, Jr., F. C. (2007). Lipid-based nanoparticles for nucleic acid delivery. *Pharmaceutical Research, 24*, 438–449. http://dx.doi.org/10.1007/s11095-006-9180-5

Liang, X., Sun, M., Li, L., Qiao, R., Chen, K., Xiao, Q., & Xu, F. (2012). Preparation and antibacterial activities of polyaniline/Cu$_{0.05}$Zn$_{0.95}$O nanocomposites. *Dalton Transactions, 41*, 2804–2811. http://dx.doi.org/10.1039/c2dt11823h

Ma, S., Izutani, N., Imazato, S., Chen, J. H., Kiba, W., Yoshikawa, R., ... Ebisu, S. (2012). Assessment of bactericidal effects of quaternary ammonium-based antibacterial monomers in combination with colloidal platinum nanoparticles. *Dental Materials Journal, 31*, 150–156. http://dx.doi.org/10.4012/dmj.2011-180

Malleron, J. L., Fiaud, J. C., & Legros, J. Y. (2000). *Handbook of palladium-catalyzed organic reactions*. London: Academic Press.

Manolova, V., Flace, A., Bauer, M., Schwarz, K., Saudan, P., & Bachmann, M. F. (2008). Nanoparticles target distinct dendritic cell populations according to their size. *European Journal of Immunology, 38*, 1404–1413. http://dx.doi.org/10.1002/(ISSN)1521-4141

Mei, L., Zhang, X., Wang, Y., Zhang, W., Lu, Z., Luo, Y., ... Li, C. (2014). Multivalent polymer–Au nanocomposites with cationic surfaces displaying enhanced antimicrobial activity. *Polymer Chemistry, 5*, 3038–3044. http://dx.doi.org/10.1039/c3py01578e

Milstein, D., & Stille, J. K. (1978). A general, selective, and facile method for ketone synthesis from acid chlorides and organotin compounds catalyzed by palladium. *Journal of the American Chemical Society, 100*, 3636–3638. http://dx.doi.org/10.1021/ja00479a077

Miyaura, N., & Suzuki, A. (1995). Palladium-catalyzed cross-coupling reactions of organoboron compounds. *Chemical Reviews, 95*, 2457–2483. http://dx.doi.org/10.1021/cr00039a007

Mueller, N. C., & Nowack, B. (2008). Exposure modeling of engineered nanoparticles in the environment. *Environmental Science & Technology, 42*, 4447–4453. http://dx.doi.org/10.1021/es7029637

Nel, A. E., Mädler, L., Velegol, D., Xia, T., Hoek, E. M. V., Somasundaran, P., ... Thompson, M. (2009). Understanding biophysicochemical interactions at the nano-bio interface. *Nature Materials, 8*, 543–557. http://dx.doi.org/10.1038/nmat2442

Neuberger, T., Schöpf, B., Hofmann, H., Hofmann, M., & von Rechenberg, B. (2005). Superparamagnetic nanoparticles for biomedical applications: Possibilities and limitations of a new drug delivery system. *Journal of Magnetism and Magnetic Materials, 293*, 483–496. http://dx.doi.org/10.1016/j.jmmm.2005.01.064

Nutt, M. O., Hughes, J. B., & Wong, M. S. (2005). Designing Pd-on-Au bimetallic nanoparticle catalysts for trichloroethene hydrodechlorination. *Environmental Science & Technology, 39*, 1346–1353. http://dx.doi.org/10.1021/es048560b

Ojas, M., Bhagat, M., Gopalakrishnan, C., & Arunachalam, K. D. (2008). Ultrafine dispersed CuO nanoparticles and their antibacterial activity. *Journal of Experimental Nanoscience, 3*, 185–193.

Rhim, J.-W., Hong, S.-I., Park, H.-M., & Ng, P. K. W. (2006). Preparation and characterization of chitosan-based nanocomposite films with antimicrobial activity. *Journal of Agricultural and Food Chemistry, 54*, 5814–5822. http://dx.doi.org/10.1021/jf060658h

Silvestro, L., Weiser, J. N., & Axelsen, P. H. (2000). Antibacterial and antimembrane activities of Cecropin A in *Escherichia coli*. *Antimicrobial Agents and Chemotherapy, 44*, 602–607.

Soenen, S. J., Rivera-Gil, P., Montenegro, J.-M., Parak, W. J., & De Smedt, S. C. (2011). Cellular toxicity of inorganic nanoparticles: Common aspects and guidelines for improved nanotoxicity evaluation. *Nano Today, 6*, 446–465. http://dx.doi.org/10.1016/j.nantod.2011.08.001

Sonogashira, K. J. (2002). Development of Pd–Cu catalyzed cross-coupling of terminal acetylenes with sp^2-carbon halides. *Journal of Organometallic Chemistry, 653*, 46–49. http://dx.doi.org/10.1016/S0022-328X(02)01158-0

Stoimenov, P. K., Klinger, R. L., Marchin, G. L., & Klabunde, K. J. (2002). Metal oxide nanoparticles as bactericidal agents. *Langmuir, 18*, 6679–6686. http://dx.doi.org/10.1021/la0202374

Tamboli, M. S., Kulkarni, M. V., Patil, R. H., Gade, W. N., Navale, S., & Kale, B. B. (2012). Nanowires of silver–polyaniline nanocomposite synthesized via *in situ* polymerization and its novel functionality as an antibacterial agent. *Colloids*

and Surfaces B: Biointerfaces, 92*, 35–41. http://dx.doi.org/10.1016/j.colsurfb.2011.11.006

Thill, A., Zeyons, O., Spalla, O., Chauvat, F., Rose, J., Auffan, M., & Flank, A. M. (2006). Cytotoxicity of CeO_2 nanoparticles for *Escherichia coli*. Physico-chemical insight of the cytotoxicity mechanism. *Environmental Science & Technology, 40*, 6151–6156. http://dx.doi.org/10.1021/es060999b

Tratnyek, P. G., & Johnson, R. L. (2006). Nanotechnologies for environmental cleanup. *Nano Today, 1*, 44–48. http://dx.doi.org/10.1016/S1748-0132(06)70048-2

Vukoje, I. D., Džunuzović, E. S., Vodnik, V. V., Dimitrijević, S., Ahrenkiel, S. P., & Nedeljković, J. M. (2014). Synthesis, characterization, and antimicrobial activity of poly(GMA-co-EGDMA) polymer decorated with silver nanoparticles. *Journal of Materials Science, 49*, 6838–6844. http://dx.doi.org/10.1007/s10853-014-8386-x

Weissleder, R., Kelly, K., Sun, E. Y., Shtatland, T., & Josephson, L. (2005). Cell-specific targeting of nanoparticles by multivalent attachment of small molecules. *Nature Biotechnology, 23*, 1418–1423. http://dx.doi.org/10.1038/nbt1159

Widenhoefer, R. A., & Buchwald, S. L. (1996). Formation of palladium bis(amine) complexes from reaction of amine with palladium tris(o-tolyl)phosphine mono(amine) complexes. *Organometallics, 15*, 3534–3542. http://dx.doi.org/10.1021/om9603169

Witham, C. A., Huang, W., Tsung, C. K., Kuhn, J. N., Somorjai, G. A., & Toste, F. D. (2010). Converting homogeneous to heterogeneous in electrophilic catalysis using monodisperse metal nanoparticles. *Nature Chemistry, 2*, 36–41. http://dx.doi.org/10.1038/nchem.468

Xiu, Z.-M., Zhang, Q.-B., Puppala, H. L., Colvin, V. L., & Alvarez, P. J. J. (2012). Negligible particle-specific antibacterial activity of silver nanoparticles. *Nano Letters, 12*, 4271–4275. http://dx.doi.org/10.1021/nl301934w

Yu, S., Welp, U., Hua, L. Z., Rydh, A., Kwok, W. K., & Wang, H. H. (2005). Fabrication of palladium nanotubes and their application in hydrogen sensing. *Chemistry of Materials, 17*, 3445–3450. http://dx.doi.org/10.1021/cm048191i

Zhang, H., & Chen, G. (2009). Potent antibacterial activities of Ag/TiO_2 nanocomposite powders synthesized by a one-pot sol–gel method. *Environmental Science & Technology, 43*, 2905–2910. http://dx.doi.org/10.1021/es803450f

Zhang, D., Wei, S., Kaila, C., Su, X., Wu, J., Karki, A. B., ... Guo, Z. (2010). Carbon-stabilized iron nanoparticles for environmental remediation. *Nanoscale, 2*, 917–919.

10

Cottonseed yield and its quality as affected by mineral nutrients and plant growth retardants

Zakaria M. Sawan[1]*

*Corresponding author: Zakaria M. Sawan, Agricultural Research Center, Cotton Research Institute, Ministry of Agriculture and Land Reclamation, 9 Gamaa street, 12619 Giza, Egypt
E-mail: zmsawan@hotmail.com
Reviewing editor: Raffaele Dello Ioio, Universita degli Studi di Roma La Sapienza, Italy
Additional information is available at the end of the article

Abstract: Seed quality is one of the most important factors for stand establishment in cotton (*Gossypium* sp.), and the use of good quality seeds is therefore essential to obtain an optimum plant population. Conditions prevailing during seed formation can affect the quality of seed produced, and hence crop establishment in the next growing season. These conditions can affect the germination of the seeds and the ability of the seedlings to emerge from soil, these being the most critical stages during the life cycle of cotton plant. Field experiments were conducted to investigate the effect of nitrogen (N), phosphorus (P), potassium (K), foliar application of zinc (Zn) and calcium (Ca), the use of plant growth retardants (PGR's) [e.g. 1,1-dimethyl piperidinium chloride (MC); 2-chloroethyl trimethyl ammonium chloride (CC); or succinic acid 2,2-dimethyl hydrazide (SADH)], during square initiation and boll setting stage, on growth, seed yield, seed viability, and seedling vigor of cotton.

Subjects: Bioscience; Environment & Agriculture; Environmental Studies & Management

Keywords: calcium; phosphorus; plant growth retardants; potassium; zinc

ABOUT THE AUTHOR

Effect of different rates and application systems of nitrogen fertilization and indole-3-butyric acid on Egyptian cotton growth and yield. Effect of organic fertilizer and micro-nutrient treatments on cotton agronomy and fiber characteristics of Giza 69. Effect of concentration and time of application of the defoliant Harvade on the lint, seed, protein and oil yields, and oil properties of cottonseed. Effect of 1-naphthalene acetic acid and Kinetin on yield and fiber properties, seed, protein, oil, and fatty acids of Egyptian cotton. Plant growth retardants, plant nutrients, and cotton production. Plant density; plant growth retardants: its direct and residual effects on cotton yield and fiber properties. Plant nutrition and plant growth retardants: their effects on cottonseed, protein, oil yields, and oil properties. Direct and Residual affects of Plant Nutrition's and Plant Growth Retardants, on Cottonseed. Studying the nature relationships between climatic factors and cotton production by different applied statistical methods.

PUBLIC INTEREST STATEMENT

Stand establishment of cotton seedlings is one of the most critical stages in cotton production. Cotton-seed quality is affected, to a large extent, by the indeterminate growth habit of the cotton plant, which allows seed to set and develop across an extended period of time. Seed vigor and viability are important components influencing seedling establishment, crop growth, and productivity. Any factor that negatively affects seed vigor and viability during seed development will have adverse consequences on crop production. Plant nutrition using a balanced fertilization programmer with both macro- and micro-nutrients has become very important in the production of high quality seed. Plant growth retardants (PGR's) represent diverse chemistries and mode of action, and provide numerous possibilities for altering crop growth and development, provide farmers with a new management tool for controlling undesirable vegetative growth, and to balance vegetative and reproductive growth as well as to improve yield and its quality.

1. Introduction

Sowing is a critical time in the life cycle of any crop and the seeds are frequently exposed to adverse conditions that may compromise the establishment of seedlings in the field (de Figueiredo e Albuquerque, 2003). Stand establishment of cotton seedlings is one of the most critical stages in cotton production. Cotton-seed quality is affected, to a large extent, by the indeterminate growth habit of the cotton plant, which allows seed to set and develop across an extended period of time. Seed vigor and viability are important components influencing seedling establishment, crop growth, and productivity. Any factor (biotic and/or environmental) that negatively affects seed vigor and viability during seed development will have adverse consequences on crop production, especially when seeds are sown under environmentally stressful conditions (Welch, 1995). Both size and number of seeds, produced by maternal plants, are most likely determined by their nutritional status at the time of flowering and bud initiation. Furthermore, the most important single determinant of mineral nutrient reserves in seeds is the mineral nutrient availability to the maternal plant during reproductive development, with increasing supplies of a particular mineral nutrient enhancing the nutrient concentration in the mature seed (Fenner, 1992).

Plant nutrition using a balanced fertilization programmer with both macro- and micro-nutrients has become very important in the production of high quality seed. Many management practices and breeding efforts have allowed plants to partition more carbohydrates into bolls and less into vegetative growth. Mineral nutritional status of plants has a considerable impact on partitioning of carbohydrates and dry matter between shoots and roots. Often, the number of sink organs is the yield component that is affected mostly by mineral nutrients. The positive effect of mineral nutrient supply on the number of sink organs may result not only from an increase in mineral nutrient supply, but also from an increase in photosynthate supply to the sink sites or from hormonal effects (Borowski, 2001).

1.1. Nitrogen

In cotton culture, N have the most necessity role in production inputs, which controls growth and prevents abscission of squares and bolls, essential for photosynthetic activity (Reddy, Reddy, Padjung, & Hodges, 1996), and stimulates the mobilization and accumulation of metabolites in newly developed bolls, thus increasing their number and weight. Additionally, with a dynamic crop like cotton, excess N serves to delay maturity, promote vegetative tendencies, and usually results in lower yields (Rinehardt, Edmisten, Wells, & Faircloth, 2004). Therefore, errors made in N management that can impact the crop can be through either deficiencies or excesses. With a dynamic crop like cotton, excess N serves to delay maturity, promote vegetative tendencies, and usually results in lower yields (McConnell, Baker, & Frizzell, 1996; Rinehardt et al., 2004). Therefore, errors made in N management that can impact the crop can be through either deficiencies or excesses. If an N deficiency is developing in a cotton crop, it is not particularly difficult to diagnose and correct. Excess N fertility levels, which, can be damaging to final crop productivity, are subtler to detect, and are difficult to correct (Silvertooth & Norton, 1997).

1.2. Phosphorus

Phosphorus (P) is the second most limiting nutrient in cotton production after nitrogen Response to P fertilizer, however, is often difficult to predict, even with soil test-based applications (Bronson, Onken, Keeling, Booker, & Torbert, 2001). The high soil pH (>7.6) and the high quantities of $CaCO_3$ result in precipitation of P, which reduces the soluble P supply. Its deficiency tends to limit the growth of cotton plants, especially when plants are deprived of phosphorus at early stages than later stages of growth (Hearn, 1981). P is also involved in cell division and development of meristematic tissues (Russell, 1973). Moreover, on a whole-plant scale, P plays a decisive role in carbon assimilate transport and metabolic regulation (Bisson, Cretenet, & Jallas, 1994). Phosphorus deficiencies lead to a reduction in the rate of leaf expansion and photosynthesis per unit leaf area (Rodríguez et al., 1998). The high soil pH (>7.6) and the high quantities of $CaCO_3$ result in precipitation of P, which reduces the soluble P supply. Sasthri, Thiagarajan, Srimathi, Malarkodi, and Venkatasalam (2001) found that application of 2% diammonium phosphate to cotton plants increased seed yield, seed germination, root length, vigor index, and dry matter production.

1.3. Potassium

The physiological role of K during fruit formation and maturation periods is mainly expressed in carbohydrate metabolism and translocation of metabolites from leaves and other vegetative organs to developing bolls. K increases the photosynthetic rates of crop leaves, CO_2 assimilation and facilitating carbon movement (Sangakkara, Frehner, & Nosberger, 2000). At least 60 enzymes are known to be activated by this ion. The enzyme pyruvate kinas (more correctly referred to as ATP: pyruvate phosphotransferase), which participates in glycolysis (Glass ADM: Plant Nutrition, 1989). The high concentration of K^+ is thought to be essential for normal protein synthesis. Potassium role in this process is considered to be the maintenance of a proper association between t RNA molecules and ribosomes during the translation of mRNA (Glass ADM: Plant Nutrition, 1989). Potassium also acts as an activator for several enzymes involved in carbohydrates metabolism. The requirement of cotton for K increases with the beginning of bud formation stage. A greater accumulation of sugars and starch in leaves under K-deficient conditions adversely affects development of bolls due to deficiency of metabolites. K deficiency during the reproductive period can limit the accumulation of crop biomass (Colomb, Bouniols, & Delpech, 1995), markedly changes the structure of fruit-bearing organs, and decreases yield and quality. Pettigrew (1999) stated that the elevated carbohydrate concentrations remaining in source tissue, such as leaves, appear to be part of the overall effect of K deficiency in reducing the amount of photosynthate available for reproductive sinks and thereby producing the changes in yield and quality seen in cotton.

1.4. Calcium

Ca is essential in cell nucleus matrix. It activates enzymes, particularly those that are membrane-bound (Rensing & Cornelius, 1980). Calcium is important in membrane permeability, maintenance of cell integrity, and in ion uptake. Calcium deficiency may also decrease the basipetal transport of auxin (dela Fuente & Leopold, 1973). Addicott and Lyon (1973) listed Ca deficiency as one of the causes of abscission and suggested this plus the role of Ca in the middle lamella (Ca pectates) as the possible reason. It is thought that Ca is important in the formation of cell membranes and lipid structures. Ma and Sun (1997), suggested that Ca might be involved in light signal transduction chain for phototropism. Ca also plays an important role in plant growth as a major component of the middle lamella (calcium pectate). A likely reason was that Ca deficiency affected translocation of carbohydrates, causing accumulation in the leaves and a decline in stems and roots.

1.5. Zinc

Although only small amounts of Zn are removed from the field by a cotton crop (0.5 oz per bale), Zn is critical for several key enzymes in the plant (Sharma, Sharma, Bisht, & Nautiyal, 1982). Zinc influences electron transfer reactions, including those of the Kreb cycle, and thereby affecting the plant's energy production. Zinc binds tightly to Zn-containing essential metabolites in vegetative tissues, e.g. Zn-activated enzymes such as carbonic anhydrase (Welch, 1995). Zn deficiency has been shown to affect growing sink organs; it adversely affects the development and viability of pollen grains (Sharma, Chatterjee, Agarwala, & Sharma, 1990). Zinc deficiency occurs on high-pH soils, particularly where topsoil has been removed in preparing fields for irrigation and thereby exposing the Zn-deficient subsoil. Also, Zn deficiencies have occurred where high rates of P are applied. The high P rates in the plant interfere with the utilization of Zn (Oosterhuis, Hake, & Burmester, July 1991).

1.6. Plant growth retardants

Plant growth retardants (PGR's) represent diverse chemistries and mode of action, and provide numerous possibilities for altering crop growth and development (Cothren, 1994). PGR's [e.g. 1,1-dimethyl piperidinium chloride (MC); 2-chloroethyl trimethyl ammonium chloride (CC); or succinic acid 2,2-dimethyl hydrazide (SADH)]. provide farmers with a new management tool for controlling undesirable vegetative growth. An objective for using PGR in cotton is to balance vegetative and reproductive growth as well as to improve yield and its quality (Zhao & Oosterhuis, 2000). Visual growth-regulating activity of MC, CC, or SADH is similar (Stuart, Isbell, Wendt, & Abernathy, 1984;

von Heyendorff-Scheel, Schott, & Rittig, 1983), being expressed as reduced plant height and width (shortened stem and branch internodes and leaf petioles), influence leaf chlorophyll concentration, structure and CO_2 assimilation, and thicker leaves.

In Egypt, soil fertilization is the primary limiting factor affecting growth and production under intensive land use for two or more crops per year. Furthermore, recently released varieties have high yielding ability, which largely depends on ensuring the plant's essential nutritional requirements (e.g. N, P, K, Ca; Zn). Considerable interest also exists in using PGR for cotton production because of their potential for altering crop growth and seed development (Cothren, 1994). All environmental factors and their interactions that influence plant growth can potentially influence the complicated and dynamic processes that control their seed initiation, development, and seed nutrient reserves. These factors can modify the ultimate vigor and viability of seeds (Sawan, Fahmy, & Yousef, 2011). The objectives of this study were to evaluate the effects of N and P, and K fertilization and foliar application of chelated Ca and Zn nutrients, and the PGR's (e.g. MC, CC, or SADH) during square initiation and boll setting stage and to identify the best combination of these production treatments in order to improve seed yield, seed weight, and seed quality (as measured by seed viability, seedling vigor, and cool germination test) of Egyptian cotton (*Gossypium barbadense*).

2. Methods and measurements
Field experiments were conducted at the Agricultural Research Center (ARC), in Giza (30°N, 31°: 28′E and 19 m altitude). The soil type was a clay loam with an alluvial substratum. Average textural properties (Kilmer & Alexander, 1940) and chemical properties (Chapman & Pratt, 1961) of soil in both seasons are reported in Table 1. Range and mean values of the climatic factors recorded during the growing seasons are presented in Table 2. These data were obtained from the Agricultural Meteorological Station of the, ARC, Giza, Egypt. No rainfall occurred during the two growing seasons (Sawan, Fahmy, & Yousef, 2009). The experiments were arranged as a randomized complete block

Table 1. Physical and chemical properties of the soil used in I and II seasons

Season	I	II
Soil texture		
Clay (%)	43.0	46.5
Silt (%)	28.4	26.4
Fine sand (%)	19.3	20.7
Coarse sand (%)	4.3	1.7
Soil texture	Clay	Loam
Chemical analysis		
Organic matter (%)	1.8	1.9
Calcium carbonate (%)	3.0	2.7
Total soluble salts (%)	0.13	0.13
pH (1:2.5)	8.1	8.1
Total nitrogen (%)[a]	0.12	0.12
Available nitrogen (mg kg^{-1} soil)[b] (1% K_2SO_4, extract)	50.0	57.5
Available phosphorus (mg kg^{-1} soil) ($NaHCO_3$ 0.5 N, extract)	15.7	14.2
Available potassium (mg kg^{-1} soil) (NH_4OAC 1 N, extract)	370.0	385.0
Total sulfur (mg kg^{-1} soil)	21.3	21.2
Calcium (meq/100 g) (with Virsen, extract)	0.2	0.2

Note: The Physical analysis (soil fraction) added to the organic matter, calcium carbonate and total soluble salts to a sum of about 100% (Sawan et al., 2009).

[a]Total nitrogen, i.e. organic N + inorganic N.

[b]Available nitrogen, i.e. NH_4^+ and NO_3^-.

Weather variables	Season I		Season II		Overall date (two seasons)	
	Range	Mean	Range	Mean	Range	Mean
Max temp (°C)	20.8–44.0	32.6	24.6–43.4	32.7	20.8–44.0	32.6
Min temp (°C)	10.4–24.5	19.4	12.0–24.3	19.3	10.4–24.5	19.3
Max-min temp (°C)	4.7–23.6	13.2	8.5–26.8	13.4	4.7–26.8	13.3
Sunshine (h d^{-1})	0.3–12.9	11.1	1.9–13.1	11.2	0.3–13.1	11.1
Max hum (%)	48–96	79.5	46–94	74.7	46–96	77.2
Min hum (%)	6–48	30.1	8–50	33.0	6–50	31.5
Wind speed (m s^{-1})	0.9–11.1	5.2	1.3–11.1	5.0	0.9–11.1	5.1

Table 2. Range and mean values of the weather variables recorded during the growing seasons (April–October)

Source: Sawan et al. (2009).

design in a factorial arrangement. The plot size was 1.95 m. (3 ridges, its ridge was 65 cm width) × 4 m (length), including three ridges (beds). Hills were spaced 25 cm apart on one side of the ridge (16 hills per ridge), and seedlings were thinned to two plants hill^{-1} 6 weeks after sowing (AS), providing plant density of 123,000 plants ha^{-1}. Total irrigation amount during the growing season (surface irrigation) was about 6,000 m^3 ha^{-1}. The first irrigation was applied 3 weeks AS, and the second one was 3 weeks later. Thereafter, the plots were irrigated every 2 weeks until the end of the season, thus providing a total of nine irrigations. In every experiment, fertilization along with pest and weed management was applied as needed during the growing season, according to local practices performed at the experimental station.

2.1. Experiments

2.1.1. Influence of N and Foliar-applied PGR's and Zn on cotton seed yield, viability, and seedling vigor

2.1.1.1. Materials. A field experiment was conducted at the Agricultural Research Center, Ministry of Agriculture in Giza, Egypt using the cotton cultivar Giza 75 (*G. barbadense* L.) in the two seasons I and II. Each experiment included 16 treatments which were the combinations of:

(1) Two N rates (farmer rate of 107 kg ha^{-1} N as control, and a higher rate of 161 kg ha^{-1} N) applied as ammonium nitrate (MH$_4$NO$_3$, 33.5% N) in two equal doses, six and eight weeks AS.

(2) Three PGR's: (1) 1,1-dimethyl piperidinium chloride (MC); (2) 2-chloroethyl trimethyl ammonium chloride (CC); and (3) succinic acid 2,2-dimethyl hydrazide (SADH). Each was foliar-sprayed once after 75 days after sowing (DAS) (during square initiation and boll formation stages) at a rate of 300 ppm. The volume of solution used for each one was 960 l ha^{-1}. Water alone (0 PGR) was foliar-sprayed at 960 l ha^{-1} as the control or check treatment.

(3) Two Zn rates (0 as control or 50 ppm Zn) in the chelated form (ethylene diamine tetra-acetic acid, EDTA) were applied. Each was foliar-sprayed two times, i.e. 80 and 95 DAS. The volume of solution used for each rate was 960 l ha^{-1} (Sawan, Gregg, & Yousef, 1998).

2.2.1. Effect of P, Zn, and Ca on cotton seed yield, viability, and seedling vigor

2.2.1.1. Materials. A field experiment was conducted at the ARC in Giza, Egypt, using the cotton cultivar Giza 75 in the two seasons I and II. The experiment included 16 treatments, combinations of:

(1) Two P rates (44 or 74 kg ha^{-1} P$_2$O$_5$), were applied (banded into soil) as calcium super phosphate (15% P$_2$O$_5$) three weeks AS, just before the first irrigation. General Farmer practice of applying 44 kg ha^{-1} P$_2$O$_5$ was used as the control treatment.

(2) Two Zn rates (0 as control or 40 ppm Zn) were applied as the chelated form. Each rate was foliar-sprayed twice, at 75 and 90 DAS. Volume of solution applied was 960 l ha⁻¹ (actual Zn applied in each spray was 0 or 38.4 g ha⁻¹).

(3) Four Ca rates (0 as control, 20, 40 or 60 ppm Ca) were applied as the chelated form. Each rate was foliar-sprayed twice, at 80 and 95 DAS. Volume of solution applied was 960 l ha⁻¹ (actual Ca applied in each spray was 0, 19.2, 38.4, or 57.6 g ha⁻¹) (Sawan, Gregg, & Yousef, 1999).

2.3.1. Effect of N, K, and PGR's on cotton growth, seed yield, seed viability, and seedling vigor

2.3.1.1. Materials. A field experiment was conducted at the ARC in Giza, Egypt using the cotton cultivar Giza 86 in the two seasons I and II. The factors studied were N fertilization, foliar application of potassium, and the PGR MC. Two N rates, 95.2 and 142.8 kg N ha⁻¹, were applied as ammonium nitrate at two equal doses, six and eight weeks AS. Each application in the form of side dressing beside each hill was followed immediately by irrigation. Four K rates (0, 0.38, 0.77, 1.15 kg ha⁻¹ K_2O) were applied in a volume 960 l ha⁻¹ as K sulfate (K_2SO_4, "48% K_2O") twice during the reproductive phase. The first application occurred 70 DAS during square initiation and the second at 95 DAS during boll development. The foliar PGR MC was applied twice during reproductive phase, both times delivered in of 960 l H_2O ha⁻¹. The first application (0.048 kg a.i. ha⁻¹) occurred 75 DAS and the second (0.024 kg a.i. ha⁻¹) 90 DAS. Control plots received no MC. The K_2O and MC were both applied to the leaves with uniform coverage using a knapsack sprayer (Sawan et al., 2009).

2.4.1. Effect of K, Zn, and P on seed yield, seed viability, and seedling vigor of cotton

2.4.1.1. Materials. A field experiment was conducted at the ARC in Giza, Egypt using the cotton cultivar Giza 86 in the two seasons I and II. Each experiment included 16 treatments, which were the combinations of two K rates (0.0 and 47 kg ha⁻¹ K), two Zn rates (0.0 or 58 g ha⁻¹ Zn) (0 or 57.6 g of Zn ha⁻¹) and four P rates (0.0, 576, 1,152, and 1,728 g ha⁻¹ P) (0, 576, 1,152, and 1,728 g ha⁻¹ P). The K was applied as K sulfate (K_2SO_4, "48% K_2O") eight weeks AS (as a band close to the seed ridge) and the application was followed immediately by irrigation. Zn was applied to the foliage in chelated form two times (70 and 85 DAS). Foliar application of P (calcium super phosphate, 15% P_2O_5) was made twice (80 and 95 DAS). The Zn and P were both applied to the leaves with uniform coverage at a solution volume of 960 l ha⁻¹, using a knapsack sprayer (Sawan et al., 2011).

2.4.1.2. Measurements. At harvest, bolls of 10 randomly chosen plants from each plot were harvested (handpicking) and laboratory-ginned to determine seed yield in g per plant. Total seed cotton yield of each plot (including the 10-plant sub-sample) was lab-ginned to determine seed yield in kg ha⁻¹. A random sample of 100 g of seeds from each plot was taken to determine seed weight (weight of 100 seed in g) and to evaluate seed quality in terms for seed viability, cool germination test performance, and seedling vigor (Sawan et al., 2009).

2.4.1.3. Seed viability. Germination was evaluated using the International Rules of Seed Testing (International Seed Testing Association, 1976) in the Seed Research Unit, Central Administration of Seed, ARC, Giza, Egypt. Aluminum dishes 17 cm in diameter and 3 cm deep were used, and the sand substratum was sieved/washed/sterilized and kept moistened to 50% of water-holding capacity. Fifty seeds were planted in each dish in sand depressions made with a standard puncher and then covered with a top layer of 2 cm of loose moist sand. Each of the four replicates of each treatment included two dishes for each replicate. Dishes were then incubated at 30 ± 1°C for 12 days. The following parameters were measured:

(1) First germination count (germination velocity): percentage of seeds that sprouted after four days of incubation.

(2) Second germination count: percentage of seeds that sprouted after eight days of incubation. This count was used to calculate germination rate index (GRI).

(3) Total germination capacity (final count): total percentage of normal seedlings after 12 days of incubation.

(4) GRI: was calculated according to Bartlett (1937) as follows:

$$GRI = \frac{a + (a + b) + (a + b + c)}{n(a + b + c)}$$

where $n = 3$ is the number of times counts were taken (Sawan et al., 2009).

2.4.1.4. Cool germination test performance. In this test, the germination chamber was maintained at a constant temperature of $18 \pm 1°C$ with sufficient humidity to prevent drying of the paper towel substratum (Bird & Reyes, 1967). Two hundred seeds (four replicates of 50 seeds each) were tested per field treatment. Four paper towels represented each of the four replicates of each treatment. In each of the four replicates, the 50 seeds were randomly placed on moist towels, as usually practiced in the standard germination test. Two towels were placed over the seeds before rolling. The towels were moistened, but not so wet that by pressing, a film of water formed around the finger. Rolled towel tests were then set upright in wire mesh baskets in the germinator. Additional moisture was not needed during the test period. Two counts of germination were made for germination on the fourth and seventh day under test conditions (Sawan et al., 2009).

2.4.1.5. Seedling vigor. Aluminum dishes, similar to those described regarding the seed viability test, were used to evaluate seedling vigor. Two dishes represented each of the four replicates of each treatment; in each dish, 50 seeds were planted. Ten seedlings were randomly taken from each dish after eight days of incubation at $30 \pm 1°C$ to measure the following seedling vigor characters: (1) length (in cm) of hypocotyl, radicle, and entire seedling; and (2) fresh and dry weights (g) of 10 seedlings. The 10 seedlings were weighed immediately to record fresh weight, and then oven-dried for 72 h at 85°C to determine dry weight (Sawan et al., 2009).

2.4.1.6. Statistical analysis. Data for the studied characters observed were analyzed as a factorial experiment arranged in a randomized complete block design, and combined statistical analysis for the two years had been done, according to Snedecor and Cochran (1980). The Least Significant Difference (LSD) test (*t*-test) at the 0.05 significance level was used to examine differences among treatment means (Sawan et al., 2009).

3. Analyzed data for measurements

3.1. Experiments

3.1.1. Influence of N and foliar-applied PGR and Zn on cotton seed yield, viability, and seedling vigor

3.1.1.1. Seed yield. Seed yield plant^{-1} and yield ha^{-1} significantly increased by raising the N-rate in both years (Table 3) (Sawan et al., 1998). Nitrogen is an important nutrient which control growth and prevents abscission of squares and bolls, essential for photosynthetic activity (Reddy et al., 1996) and stimulate the mobilization and accumulation of metabolites in newly developed bolls and thus their number and weight are increased. Abdel-Malak, Radwan, and Baslious (1997) stated that cotton yield was higher when *N* was applied at a rate of 190 kg ha^{-1} than at the rate of 143 kg ha^{-1}.

Treatment	Cotton seed yield (g plant⁻¹)		Cotton seed yield (kg ha⁻¹)		Seed weight (g 100 seed⁻¹)	
	I	II	I	II	I	II
N-rate (kg ha⁻¹)						
107 (Control)	17.55	18.90	1,854.5	1,960.8	10.11	10.46
161	19.17**	20.55**	2,026.3**	2,131.0**	1,0.25**	10.63**
L.S.D. 0.05	0.944	0.790	107.80	86.07	0.096	0.083
0.01	1.261	1.055	143.99	114.97	0.129	0.111
Plant growth retardants (ppm)						
0 (Control)	17.12	18.29	1,810.6	1,893.8	10.05	10.43
MC, 300	19.12*	20.55**	2,021.7*	2,130.2**	10.27*	10.62**
CC, 300	18.79*	20.36**	1,985.6*	2,110.5**	10.24*	10.59**
SADH, 300	18.41	19.74*	1,943.8	2,049.1*	10.17	10.55*
L.S.D. 0.05	1.335	1.118	152.45	121.73	0.137	0.117
0.01	n.s.	1.493	n.s.	162.59	n.s.	0.156
Zn-rate (ppm)						
0 (Control)	17.70	18.85	1,870.9	1,954.0	10.12	10.48
50	19.02**	20.59**	2,010.0*	2,137.8**	10.24*	10.61**
L.S.D. 0.05	0.944	0.790	107.80	86.07	0.096	0.083
0.01	1.261	1.055	n.s.	114.97	n.s.	0.111

Table 3. Effect of N-rate and foliar application of plant growth retardants and Zn on seed yield plant⁻¹ and seed yield ha⁻¹, and seed weight

Note: n.s.: Not significant.

*Significant at 5% level.

**Significant at 1% level (Sawan et al., 1998).

Palomo Gil and Chávez González (1997) applied N at a rate ranging from 40 to 200 kg ha⁻¹ to cotton plants and found highest yield was associated with high rates of applied N. Similar results were obtained by Sarwar Cheema, Akhtar, and Nasarullah (2009) and Saleem, Bilal, Awais, Shahid, and Anjum (2010) when N was applied at 120 kg ha⁻¹. In both years, all three PGR increased seed yield plant⁻¹ and ha⁻¹, compared to untreated control. In I season, only MC and CC produced statistically significant increases, while in II season, increases were significant with all tested PGR. The highest numerically increase in seed yield was with MC, followed in order by CC, and SADH. Such increases may be due to increased photosynthetic activity of leaves when these substances are applied. Increased photosynthesis greatly increased flowering, boll retention, and boll weight (Kler, Raj, & Dhillon, 1989; Mauney, Fry, & Guinn, 1978). Abdel-Al (1998) indicated that cotton yield significantly increased with MC treatment at a rate 11.90 ml (formulation) ha⁻¹ at the beginning of flowering. Pípolo, Athayde, Pípolo, and Parducci (1993) found that spraying cotton plants at an age of 70 d after emergence with CC at rates ranging from 25 to 100 g ha⁻¹ resulted in yield increases. Sawan and Gregg (1993) stated that application of CC and SADH, at rates ranging from 250 to 700 ppm 105 DAS increased cotton seed yield ha⁻¹. Similar results were obtained by Sarwar Cheema et al. (2009). Zinc application significantly increased seed yield plant⁻¹ and ha⁻¹ in both years, as compared with untreated plants. Zinc is required in the synthesis of tryptophan, which is a precursor of IAA synthesis which is the hormone that inhibits abscission of squares and bolls. Also, this nutrient has favorable effect on the photosynthetic activity of leaves and plant metabolism (Li, Ma, Wang, & Tai, 2004), which might account for higher accumulation of metabolites in reproductive organs (bolls). Similar results were obtained by Basilious, Abdel Malak, and Abdel Kader (1991), by foliar spraying of 0.314 kg ha⁻¹ Zn at flowering stage, Gomaa (Gomaa, 1991) when cotton was sprayed with 0.4 kg ha⁻¹ $ZnSO_4$. Zeng (1996) stated that application of Zn to cotton plants on calcareous soil increased yield by 7.8–25.7% and Ibrahim, Bekheta, El-Moursi, and Gaafar (2009).

3.1.1.2. Seed weight. Seed weight significantly increased by adding the high N-rate in both years (Table 3). This may be partially due to enhanced photosynthetic activity (Abdel-Malak et al., 1997). Similar findings were obtained by Palomo Gil and Chávez González (1997). Application of PGR increased seed weight over the untreated control in both years. The increase was significant for MC and CC in season I. In season II, the increase was significant for all the three PGR's. Spraying plants with MC produced the highest numerically seed weight. Increased seed weight as a result of MC, CC, or SADH applications may be due to increase in photosynthetic activity, which stimulates photosynthetic activity and dry matter accumulation, and in turn increases formation of fully mature seed and increases seed weight. These agree with previous works of Kler et al. (1989) when CC was applied at 58 ppm after 75 DAS, Sawan and Sakr (1990) when MC was applied at 10–100 ppm once after 90 D or twice after 90 and 110 DAS, Sawan and Gregg (1993) when CC or SADH was sprayed at 250, 500 or 750 ppm after 105 DAS. Results are confirmed by those of Carvalho et al. (1994) by applying MC and CC and Abdel-Al (1998), by applying MC. Seed weight significantly increased with Zn application, compared with the control in both years. This may be due to its favorable effects on photosynthetic activity, improving mobilization of photosynthesis and directly affect seed weight. In this connection, Ibrahim et al. (2009) noted that seed weight increased due to the application of Zn.

3.1.1.3. Seed viability, seedling vigor, and cool germination test. Seed viability, seedling vigor, and cool germination test performance significantly increased by addition the high N-rate, as compared to the low rate. Also, the same pattern was true with the application of the three PGR or Zn in both years (Tables 4 and 5) (Sawan et al., 1998). However, the exceptions of this trend were in case of germination velocity, second germination count, cool germination test performance (four and seven day counts), radicle seedling length and seedling fresh and dry weights, which were not significantly increased when SADH was applied in season I. Results from MC application were numerically higher than from CC or SADH. GRI was not significantly affected in either year by N-rate, PGR, or Zn. The stimulatory effect of raising N-rate on these characters may be attributed to the increase in seed weight. These favorable effects of increased N-rate on seed viability and seedling vigor agreed with the results obtained by Sawan, Maddah El-Din, and Gregg (1989). No information on the residual effects on N-rate on cool germination was noted in the available literature. Beneficial residual effects of PGR on seed viability, seedling vigor, and cool germination test performance may be due to their favorable effects on seed weight. Bartee and Kreig (1974) reported that organic and inorganic materials available to germination seedlings were greater in high-density seed than in low-density seed. Sawan, Basyony, McCuistion, and El-Farra (1993) found that application of CC or SADH (at 250, 500, or 750 ppm as foliar spray after 105 DAS) increased seed viability, seedling vigor, and cool germination test performance. Zinc's stimulation of seed viability, seedling vigor, and cool germination performance, which are important in stand establishment, may also be associated with increased seed weight and changed composition, as Zn has favorable effects on metabolism of nucleic acids, proteins, vitamins, and growth substances (Vlasynk et al., 1978). These are manifested in metabolites formed in plant tissues, and directly influence growth and development processes. Favorable effects of Zn on seed viability and seedling vigor were mentioned by Sawan et al. (1989) when Zn was applied at 12.5 ppm three times as foliar spray after 70, 85, and 100 DAS. No information on residual effects of Zn on cool germination test performance was found. The interaction between N-rate and application of PGR or zinc showed no significant effect on the investigated characters.

3.2.1. Effect of P, Zn, and Ca on cotton seed yield, viability, and seedling vigor

3.2.1.1. Seed yield. Seed yield plant^{-1} and plot^{-1}, significantly increased when P was applied at the highest rate with the application of high P-rate in both years (Table 6) (Sawan et al., 1999). Phosphorus as a constituent of cell nuclei is essential for cell division and development of meristematic tissue, and hence it should have a stimulating effect on the plants, increasing the number of flowers and bolls per plant. Further, P has a well-known impact in photosynthesis as well as synthesis of nucleic acids, proteins, lipids, and other essential compounds (Guinn, 1984), all of which are major factors affecting boll weight and consequently cottonseed. These results are confirmed by those of Abdel-Malak et al. (1997), Ibrahim et al. (2009), and Saleem, Shakeel, Bilal, Shahid, and Anjum (2010).

Table 4. Effect of N-rate and foliar application of plant growth retardants and Zn on seed viability and cool germination test performance

Treatment	Germination velocity (%)		Second germination count (%)		Germination capacity (%)		Germination rate index (GRI unit)		Cool germination test performance (%)			
									4-day count		7-day count	
	I	II	I	II	I	II	I	II	I	II	I	II
N-rate (kg ha⁻¹)												
0 (Control)	72.81	75.38	79.31	81.12	80.56	83.19	0.655	0.656	33.19	33.94	67.44	69.12
161	75.69**	79.06**	84.44**	85.62**	85.81**	87.81**	0.653	0.655	35.75**	37.31**	71.81**	74.25**
L.S.D. (0.05)	1.802	1.668	2.229	1.838	2.038	1.662	n.s.	n.s.	1.707	1.702	1.730	1.626
(0.01)	2.407	2.228	2.978	2.455	2.722	2.220	n.s.	n.s.	2.281	2.273	2.311	2.172
Plant growth retardants (ppm)												
0 (Control)	71.88	74.62	79.12	80.62	80.25	82.88	0.654	0.655	32.38	33.25	67.50	69.25
MC, 300	75.62*	78.75**	83.75*	85.12**	84.88*	87.12**	0.654	0.655	35.62*	37.12*	70.75*	73.12**
CC, 300	75.12*	78.38**	82.88*	84.50**	84.25*	86.38**	0.654	0.656	35.12*	36.38*	70.38*	72.62**
SADH, 300	74.38	77.12*	81.75	83.25*	83.38*	85.62*	0.654	0.655	34.75	35.75*	69.88	71.75*
L.S.D. (0.05)	2.548	2.359	3.153	2.600	2.883	2.350	n.s.	n.s.	2.415	2.407	2.447	2.300
(0.01)	n.s.	3.152	n.s.	3.472	n.s.	3.139	n.s.	n.s.	n.s.	n.s.	n.s.	3.072
Zn-rate (ppm)												
0 (Control)	73.31	75.81	80.56	82.06	81.88	84.12	0.654	0.655	33.56	34.44	68.56	69.94
50	75.19*	78.62**	83.19*	84.69**	84.50*	86.88**	0.654	0.656	35.38*	36.81**	70.69*	73.44**
L.S.D. (0.05)	1.802	1.668	2.229	1.838	2.038	1.662	n.s.	n.s.	1.707	1.702	1.730	1.626
(0.01)	n.s.	2.228	n.s.	2.455	n.s.	2.220	n.s.	n.s.	n.s.	2.273	n.s.	2.172

Note: n.s.: Not significant.
*Significant at 5% level.
**Significant at 1% level (Sawan et al., 1998).

Table 5. Effect of N-rate and foliar application of plant growth retardants and Zn on seedling vigor

Treatment	Hypocotyl length (cm)		Radicle Length (cm)		Seedling length (cm)		Seedling fresh weight (g 10 seedling⁻¹)		Seedling dry weight (g 10 seedling⁻¹)	
	I	II	I	II	I	II	I	II	I	II
N-rate (kg ha⁻¹)										
107 (Control)	7.59	7.58	15.76	15.91	23.35	23.49	6.89	7.20	0.594	0.597
161	7.80**	7.87**	16.26**	16.52**	24.06**	24.38**	7.27**	7.61**	0.622**	0.636**
L.S.D. 0.05	0.149	0.162	0.348	0.310	0.484	0.429	0.206	0.180	0.0167	0.0156
0.01	0.199	0.217	0.465	0.414	0.647	0.573	0.275	0.240	0.0224	0.0209
Plant growth retardants (ppm)										
0 (Control)	7.50	7.46	15.58	15.72	23.08	23.18	6.83	7.14	0.587	0.593
MC, 300	7.79*	7.86**	16.25*	16.46**	24.04*	24.32**	7.23*	7.60**	0.621*	0.630**
CC, 300	7.76*	7.82**	16.16*	16.38**	23.92*	24.21**	7.17*	7.48**	0.615*	0.624**
SADH, 300	7.73*	7.76*	16.06	16.29*	23.78*	24.05**	7.10	7.40*	0.609	0.619*
L.S.D. 0.05	0.210	0.230	0.492	0.439	0.685	0.607	0.291	0.254	0.0237	0.0221
0.01	n.s.	0.307	n.s.	0.586	n.s.	0.811	n.s.	0.340	n.s.	0.0296
Zn-rate (ppm)										
0 (Control)	7.62	7.61	15.82	15.99	23.43	23.60	6.97	7.28	0.599	0.606
50	7.77*	7.84**	16.20*	16.43**	23.98*	24.27**	7.19*	7.53**	0.616*	0.627**
L.S.D. 0.05	0.149	0.162	0.348	0.310	0.484	0.429	0.206	0.180	0.0167	0.0156
0.01	n.s.	0.217	n.s.	0.414	n.s.	0.573	n.s.	0.240	n.s.	0.0209

Note: n.s.: Not significant.

*Significant at 5% level.

**Significant at 1% level (Sawan et al., 1998).

Application of Zn significantly increased Seed yield plant⁻¹ and plot⁻¹, as compared with the untreated control. This may be due to its favorable effect on photosynthetic activity, which improves mobilization of photosynthates and directly influences of boll weight (Glass ADM: Plant Nutrition, 1989). Also, Zn enhances the activity of tryptophan synthesis, which is involved in the synthesis of the growth control compound IAA, the major hormone that inhibits abscission of squares and bolls. The application of Zn increased the number of retained bolls plant⁻¹. Similar results were obtained by Alikhanova and Tursunov (1988) by application of Zn at 2.5–7.5 kg ha⁻¹, by Sawan et al. (Sawan et al., 1989) when cotton was sprayed with Zn at 12.5 ppm and by Gomaa (1991) when cotton was sprayed with 0.952 kg Zn SO₄ ha⁻¹. Results are confirmed by those of Zeng (1996) and Ibrahim et al. (2009). The three concentrations of Ca applied significantly exceeded the control (Table 6) (Sawan et al., 1999). In general, the highest Ca concentration (60 ppm) was better than the other two. The role of Ca in increasing seed yield can possibly be ascribed to its involvement in the process of photosynthesis and the translocation of carbohydrates to young bolls. Calcium deficiency depressed the rate of photosynthesis (rate of CO_2 fixation). Guinn (Guinn, 1984) stated that Ca deficiency would cause carbohydrates to accumulate in leaves and not in young bolls. The results obtained agree with those reported by Shui and Meng (1990) and Wright et al. (1995).

3.2.1.2. Seed weight. Application of P at the high rate of 74 kg ha⁻¹ P_2O_5 and Zn at the concentration of 40 ppm, both significantly increased seed weight relative to the control in the two seasons (Table 6) (Sawan et al., 1999). A possible explanation for increased seed weight due to the application of P at the higher rate is that this nutrient activated biological reactions in the cotton plants, particularly CO_2 fixation and the synthesis of sugar, amino acids, protein, lipids, and other organic compounds. It also increased the translocation of assimilates from photosynthetic organs to the sink (Kosheleva,

Table 6. Effect of P-rate and foliar application of Zn and Ca on seed yield plant⁻¹ and seed yield plot⁻¹, and seed weight

Treatment	Cotton seed yield (g plant⁻¹)		Cotton seed yield (kg plot⁻¹)		Seed weight (g 100 seed⁻¹)	
	I	II	I	II	I	II
P_2O_5 rate (kg ha⁻¹)						
44 (Control)	17.04	17.93	1.302	1.345	10.06	10.32
74	18.80**	20.03**	1.439**	1.505**	10.26**	10.54**
L.S.D. (0.05)	0.451	0.697	0.0385	0.0511	0.091	0.075
(0.01)	0.603	0.931	0.0515	0.0683	0.121	0.101
Zn rate (ppm)						
0 (Control)	17.30	18.10	1.322	1.358	10.11	10.36
40	18.54**	19.87**	1.419**	1.492**	10.21*	10.50**
L.S.D. (0.05)	0.451	0.697	0.0385	0.0511	0.091	0.075
(0.01)	0.603	0.931	0.0515	0.0683	n.s.	0.101
Ca rate (ppm)						
0 (Control)	16.61	17.79	1.270	1.333	10.03	10.30
20	17.96**	18.92*	1.368**	1.419*	10.18*	10.43*
40	18.46**	19.43**	1.411**	1.460**	10.21*	10.47**
60	18.65**	19.80**	1.432**	1.488**	10.22*	10.51**
L.S.D. (0.05)	0.638	0.986	0.0545	0.0723	0.129	0.107
(0.01)	0.853	1.317	0.0728	0.0966	n.s.	0.143

Note: n.s.: Not significant.
*Significant at 5% level.
**Significant at 1% level (Sawan et al., 1999).

Bakhnova, Semenova, & Mil'Kevich, 1984). Similar results were obtained by El-Debaby, Hammam, and Nagib (1995). Application of Zn significantly increased seed index, compared to the control. This may be due to its favorable effect on photosynthetic activity. Zinc improves mobilization of photosynthates and directly influences boll weight that coincide directly with increased seed index. These results are confirmed by those obtained by Ibrahim et al. (2009). Calcium applied at all rates significantly increased seed index over the control. The highest rate of Ca (60 ppm) showed the highest numerical value of seed index. Similar results were obtained by Ibrahim et al. (2009).

3.2.1.3. Seed viability, seedling vigor, and cool germination test. Seed viability, seedling vigor, and cool germination test performance were generally significantly increased by addition of P at high rate and by application of Zn at 40 ppm and Ca at different concentrations in both years (Tables 7 and 8) (Sawan et al., 1999). This may be attributed to the fact that P is required for production of high quality seed, since it occurs in coenzymes involved in energy transfer reactions. Energy is tapped in photosynthesis in the form of adenosine triphosphate (ATP) and nicotinamide adenine dinucleotide phosphate (NADP). This energy is then used in photosynthetic fixation of CO_2 and in the synthesis of lipids and other essential organic compounds (Taiz & Zeiger, 1991). Also, this could be attributed to the increase in total photo assimilates (e.g. lipids) and the translocated assimilates to the sink as a result of applying zinc. The highest Ca-concentration (60 ppm) significantly increased total germination count. Ochiai (1977) notes that Ca^{2+} can bridge phosphate and carboxylate groups of phospholipids and proteins; that it increases hydrophobicity of membranes; that it generally increases membrane stability and reduces water permeability. Certain hydrolase's acting on macromolecular substrates (e.g. some @-amylases, phospholipase, and nucleases) requires Ca^{2+} for activity. Although Ca^{2+} probably has a structural role rather than catalytic (binds at sites other than catalytic, changing enzyme conformation), it is also possible that macromolecular substrates such as starch might require Ca^{2+} for bridge complexes (Scott & Peterson, 1979). Phosphorus-rates or application of Zn or Ca at different concentrations had no significant effect in either year on GRI.

3.3.1. Effect of N, K, and PGR on cotton growth, seed yield, seed viability, and seedling vigor

3.3.1.1. Plant growth and mineral contents. From Table 9, it follows that there were significant effects under the high N-rate regime (142.8 kg ha^{-1} N) on growth and nutrient content of cotton plants (105 DAS) compared with the lower rate (95.2 kg ha^{-1} N) (Sawan et al., 2009). These findings coincide with the fact that N is an essential nutrient in building a plant dry matter as well as many energy-rich compounds (ATP), which regulate photosynthesis. Under N deficiency, a considerably larger proportion of dry matter (photosynthates) is partitioned to roots than shoots, leading to reduced shoot/root dry weight ratios (Engels & Marschner, 1995). Shrivastava, Tomar, and Singh (1993) found that an increase in N level (from 0 up to 120 kg N ha^{-1}) application caused an increased uptake of N, P, and K. Perumai (1999) stated that when cotton was given 0–120 kg N ha^{-1}, an increase in an N level (from 0 up to 120 kg ha^{-1} N) significantly increased aboveground biomass production. Bronson et al. (2001) found that petiole NO$_3$-N readings were positively related to N rate when applied up to 112 kg N ha^{-1}. According to the N-status in our experimental soil (Table 1), it was classified as medium fertile for N. The K applied at all the three K concentrations (0.38, 0.77, 1.15 kg of K$_2$O ha^{-1}) significantly enhanced growth, N and K uptake of cotton plants as compared to control (0 kg ha^{-1} K$_2$O). In this connection, (Fan, Yuzhang, and Chaojun (1999) found that K content in petioles and total dry matter production in cotton increased by application of K. Gormus (2002) indicated that the 0 kg ha^{-1} K$_2$O plots (untreated control) had lower leaf K concentrations compared with the plots with 80, 160, and 240 kg ha^{-1} K$_2$O. According to the K-status in our experimental soil (Table 1), it classified as medium fertile for K. MC significantly increased dry matter yield and N and K uptake of cotton plants compared with plots not treated with MC. Hodges, Reddy, and Reddy (1991) stated that application of MC increased canopy grass photosynthesis of cotton within 48 h, suggesting a direct effect of MC on photosynthesis. Zhao and Oosterhuis (1999) stated that MC application improved leaf photosynthetic rate compared with the untreated control.

3.3.1.2. Seed yield. An increase in N dose from 95.2 to 142.8 kg ha^{-1} N increased seed yield per plant and per ha (by 13.08 and 13.03%) (Table 10) (Sawan et al., 2009). There is an optimal relationship between the nitrogen content in the plant and CO$_2$ assimilation, where decreases in CO$_2$ fixation are well documented for N-deficient plants. Nitrogen deficiency is associated with elevated levels of ethylene (which increase boll shedding), suggesting ethylene production in response to N-deficiency stress (Legé, Cothren, & Morgan, 1997). N is also an essential nutrient in creating plant dry matter, as well as many energy-rich compounds which regulate photosynthesis and plant production (Wu, Wu, & Xu, 1998), thus influencing boll development, increasing the number of bolls plant^{-1} and boll weight. Similar findings were obtained by McConnell & Mozaffari (2004) when N fertilizer was applied at 120 kg ha^{-1}, and Saleem, Bilal et al., (2010) when N fertilizer was applied at 120 kg ha^{-1}. Also, similar results were obtained by Sarwar Cheema et al. (2009). On the other hand, Boquet (2005) reported that increasing N from 90 to 157 kg ha^{-1} did not result in increased cotton yield in irrigated or rainfed cotton. All the three K concentrations (0.38, 0.77, 1.15 kg ha^{-1} K$_2$O) significantly increased seed yield per plant (by 10.28–16.45%) and per ha (by 10.02–16.26%) compared to the untreated control. This could be attributed to the fact that K significantly enhanced growth and N and K uptake of the plants (Table 9) (Sawan et al., 2009). These increases could be due to the favorable effects of this nutrient on yield components such as number of opened bolls plant^{-1}, boll weight, or both, leading to higher cotton yield. Zeng (1996) indicated that K fertilizer affects abscission and reduced boll shedding and it certainly affects yield. Pettigrew (1999) stated that, the elevated carbohydrate concentrations remaining in source tissue, such as leaves, appear to be part of the overall effect of K deficiency in reducing the amount of photosynthate available for reproductive sinks and thereby producing changes in boll weight. Cakmak, Hengeler, and Marschner (1994) found that, the K nutrition had pronounced effects on carbohydrate partitioning by affecting either the phloem export of photosynthates (sucrose) or growth rate of sink and/or source organs. Results obtained here confirmed those obtained by Aneela, Muhammad, and Akhtar (2003) when applying 200 kg K$_2$O ha^{-1}, Pervez, Ashraf, and Makhdum (2004) under 62.5, 125, 250 kg K ha ha^{-1}, and Pettigrew, Meredith, and Young (2005) under K fertilizer (112 kg ha^{-1}). MC significantly increased seed yield plant^{-1} (by 9.68%), and ha^{-1} (by 9.72%) compared with untreated plants. This could be attributed to the fact that MC

Table 7. Effect of P-rate and foliar application of Zn and Ca on seed viability and cool germination test performance

| Treatment | First germination count (%) | | Second germination count (%) | | Total germination count (%) | | Germination rate index (GRI unit) | | Cool germination test performance (%) | | | |
| | | | | | | | | | 4-day count | | 7-day count | |
	I	II	I	II	I	II	I	II	I	II	I	II
P_2O_5 rate (kg ha^{-1})												
44 (Control)	73.19	75.44	79.25	82.31	80.62	83.69	0.656	0.655	33.06	34.00	68.19	70.06
74	76.38**	78.94**	84.00**	85.62**	85.38**	86.88**	0.654	0.656	35.81**	37.88**	71.56**	73.12**
L.S.D. (0.05)	1.703	1.588	1.719	1.579	1.675	1.504	n.s.	n.s.	1.555	1.548	1.837	1.626
(0.01)	2.275	2.121	2.297	2.109	2.238	2.009	n.s.	n.s.	2.077	2.068	2.454	2.171
Zn rate (ppm)												
0 (Control)	73.75	76.12	80.50	82.75	81.94	84.19	0.655	0.655	33.56	34.88	68.81	70.44
40	75.81*	78.25**	82.75*	85.19**	84.06*	86.38**	0.655	0.656	35.31*	37.00**	70.94*	72.75**
L.S.D. (0.05)	1.703	1.588	1.719	1.579	1.675	1.504	n.s.	n.s.	1.555	1.548	1.837	1.626
(0.01)	n.s.	2.121	n.s.	2.109	n.s.	2.009	n.s.	n.s.	n.s.	2.068	n.s.	2.171
Ca rate (ppm)												
0 (Control)	72.12	74.25	78.88	81.62	80.25	82.75	0.655	0.655	32.62	.75	67.50	69.12
20	74.75*	77.38**	81.50*	83.88*	82.62*	85.38*	0.656	0.656	34.25	36.38*	70.20*	71.50*
40	75.88**	78.25**	82.75**	84.62**	83.75**	85.88**	0.656	0.656	35.12	36.62*	70.75*	72.62**
60	76.38**	78.88**	83.38**	85.75**	85.38**	87.12**	0.654	0.656	35.75*	37.00*	71.12*	73.12**
L.S.D. (0.05)	2.409	2.246	2.432	2.233	2.369	2.127	n.s.	n.s.	2.199	2.190	2.598	2.299
(0.01)	3.218	2.999	3.248	2.983	3.165	2.841	n.s.	n.s.	n.s.	n.s.	n.s.	3.071

Note: n.s.: Not significant.

*Significant at 5% level.

**Significant at 1% level (Sawan et al., 1999).

Table 8. Effect of P-rate and foliar application of Zn and Ca on seedling vigor

Treatment	Hypocotyl length (cm)		Radicle length (cm)		Seedling length (cm)		Seedling fresh weight (g 10 seedling⁻¹)		Seedling dry weight (g 10 seedling⁻¹)	
	I	II	I	II	I	II	I	II	I	II
P_2O_5 rate (kg ha⁻¹)										
44 (Control)	7.53	7.59	15.72	15.86	23.25	23.45	6.99	7.30	0.596	0.601
74	7.80**	7.92**	16.30**	16.55**	24.10**	24.47**	7.23**	7.52**	0.620**	0.625**
L.S.D. (0.05)	0.182	0.149	0.325	0.344	0.487	0.468	0.135	0.163	0.0133	0.0112
(0.01)	0.243	0.199	0.434	0.459	0.651	0.625	0.181	0.217	0.0178	0.0149
Zn rate (g ha⁻¹)										
0.0 (Control)	7.57	7.64	15.79	15.95	23.36	23.59	7.02	7.33	0.600	0.605
40	7.76*	7.86**	16.22*	16.46**	23.98*	24.32**	7.20*	7.50*	0.615*	0.622**
L.S.D. (0.05)	0.182	0.149	0.325	0.344	0.487	0.468	0.135	0.163	0.0133	0.0112
(0.01)	n.s.	0.199	n.s.	0.459	n.s.	0.625	n.s.	n.s.	n.s.	0.0149
Ca rate (ppm)										
0 (Control)	7.44	7.53	15.53	15.70	22.98	23.23	6.95	7.16	0.592	0.597
20	7.69	7.77*	16.08*	16.26*	23.77*	24.03*	7.12	7.40*	0.609	0.614*
40	7.74*	7.81*	16.16*	16.38**	23.90*	24.19**	7.16*	7.52**	0.612*	0.618*
60	7.78*	7.89**	16.26*	16.49**	24.04*	24.38**	7.22*	7.58**	0.618*	0.624**
L.S.D. (0.05)	0.257	0.210	0.459	0.486	0.689	0.662	0.191	0.230	0.0189	0.0158
(0.01)	n.s.	0.281	n.s.	0.649	n.s.	0.884	n.s.	0.308	n.s.	0.0211

Note: n.s.: Not significant.
*Significant at 5% level.
**Significant at 1% level (Sawan et al., 1999).

Table 9. Mean effects of N-rate and foliar application of potassium and the plant growth MC on dry matter yield and uptake of N and K by cotton plants (season II, sampled 105 days after planting)

Treatments	Dry matter yield (g plant⁻¹)	N		K	
		Conc./d.m. (%)	Uptake (mg plant⁻¹)	Conc./d.m. (%)	Uptake (mg plant⁻¹)
N rate (kg ha⁻¹)					
95.2	39.8ᵇ	3.12ᵇ	1,242.7ᵇ	2.10ᵇ	836.4ᵇ
142.8	47.9ᵃ	3.26ᵃ	1,560.6ᵃ	2.46ᵃ	1,177.6ᵃ
L.S.D. 0.05	3.4	0.11	77.2	0.14	46.7
K_2O rate (kg ha⁻¹)					
0	38.2ᵇ	3.08ᵇ	1,177.2ᶜ	2.04ᵇ	779.7ᵈ
0.38	43.0ᵇ	3.16ᵇ	1,358.2ᵇ	2.26ᵃ	971.4ᶜ
0.77	45.3ᵃ	3.22ᵇ	1,457.7ᵃᵇ	2.38ᵃ	1,077.4ᵇ
1.152	48.9ᵃ	3.30ᵃ	1,615.0ᵃ	2.44ᵃ	1,194.1ᵃ
L.S.D. 0.05	4.8	0.16	109.5	0.20	68.9
MC rate (kg ha⁻¹)					
0	40.2ᵇ	3.09ᵇ	1,241.9ᵇ	2.18ᵇ	876.1ᵇ
0.048 + 0.024	47.5ᵃ	3.29ᵃ	1,563.4ᵃ	2.38ᵃ	1,131.0ᵃ
L.S.D. 0.05	3.4	0.11	77.2	0.14	46.7

Note: Values followed by the same letter are not significantly different from each other at 0.05 levels (Sawan et al., 2009).

Table 10. Mean effects of N-rate and foliar application of potassium and the plant growth retardant MC on seed yield plant^{-1} and seed yield ha^{-1}, and seed weight

Treatments	Cotton seed yield (g plant^{-1})	Cotton seed yield (kg ha^{-1})	Seed weight (g 100 seed^{-1})
N rate (kg ha^{-1})			
95.2	19.11[b]	1,862.4[b]	10.09[b]
142.8	21.61[a]	2,105.0[a]	10.32[a]
L.S.D. 0.05	0.864	78.78	0.075
K$_2$O rate (kg ha^{-1})			
0	18.48[b]	1,804.3[c]	10.03[c]
0.38	20.38[a]	1,985.1[b]	10.19[b]
0.77	21.05[a]	2,047.7[ab]	10.27[ab]
1.15	21.52[a]	2,097.6[a]	10.32[a]
L.S.D. 0.05	1.222	111.42	0.107
MC rate (kg ha^{-1})			
0	19.42[b]	1,891.8[b]	10.13[b]
0.048 + 0.024	21.30[a]	2,075.6[a]	10.27[a]
L.S.D. 0.05	0.864	78.78	0.075

Note: Values followed by the same letter are not significantly different from each other at 0.05 levels (Sawan et al., 2009).

significantly controlled new growth and N and K uptake of cotton plants compared to plots not treated with MC (Table 9) (Sawan et al., 2009). Such increases could be due to the fact that, the application of MC restrict vegetative growth and thus enhance reproductive organs by allowing plants to direct more energy toward the reproductive structure (Pípolo et al., 1993). This means that bolls on treated cotton would have a larger photo synthetically supplied sink of carbohydrates and other metabolites than did those on untreated cotton (Wang, Yin, & Sun, 1995). Results agreed with those obtained by Ram, Prasad, and Pachauri (2001) when MC was applied at 50 ppm, Mekki (1999) when MC was applied at 100 ppm, and Kumar, Patil, and Chetti (2004). Also, similar results were obtained by Sarwar Cheema et al. (2009).

3.3.1.3. Seed weight. Seed weight significantly increased with an increase in N from 95.2 to 142.8 kg ha^{-1} (Table 10). This may be due to increased photosynthetic activity, which increases accumulation of metabolites, with direct impact on seed weight (Table 9) (Sawan et al., 2009). This may be due to increased photosynthetic activity that increases accumulation of metabolites, with direct impact on seed weight. Reddy et al. (1996), in a pot experiment under natural environmental conditions, where 20-day old cotton plants received 0, 0.5, 1.5, or 6 mM NO$_3$, found that, net photosynthetic rates, stomatal conductance and transpiration were positively correlated with leaf N concentration. Similar findings were reported by Palomo Gil, Godoy Avila, and Chávez González (1999), when N was applied at 40–200 kg ha^{-1}, and Ali and El-Sayed (2001), when N was applied at 95–190 kg ha^{-1}. The K application, at all the three concentrations, increased seed weight compared with a control. Results from K application were more effective and significant when applied at the high concentration (1.15 kg ha^{-1} K$_2$O) than those produced from the low concentration (0.38 kg ha^{-1} K$_2$O). This may be due to its favorable effects on nutrient uptake, photosynthetic activity, improving its mobilization (Table 9) (Sawan et al., 2009), which directly influence boll weight and increased seed weight (Pettigrew, 1999). Ghourab, Wassel, and Raya (2000) and Ibrahim et al. (2009) reported that, the application of K fertilizer MC significantly increased seed weight as compared to untreated control. Increased seed weight as a result of MC applications may be due to an increase in photosynthetic activity, which stimulates photosynthetic activity, and dry matter accumulation (Bednarz & Oosterhuis, 1999; Kumar et al., 2004), and in turn increases the formation of fully-mature seed and thus increases seed weight. Similar results to the present study were obtained by Ghourab et al. (2000) and Lamas (2001).

Table 11. Mean effects of N-rate and foliar application of potassium and the plant growth retardant MC on seed viability and cool germination test performance						
Treatments	**First germination count (%)**	**Second germination count (%)**	**Total germination Capacity (%)**	**Germination rate index (GRI unit)**	**Cool germination test performance (%)**	
					4-day count	**7-day count**
N rate (kg ha⁻¹)						
95.2	74.56ᵇ	81.25ᵇ	82.91ᵇ	0.655	32.56ᵇ	70.03ᵇ
142.8	78.13ᵃ	84.84ᵃ	86.00ᵃ	0.656	35.00ᵃ	72.50ᵃ
L.S.D. 0.05	1.373	1.329	1.206	n.s.	1.077	1.160
K_2O rate (kg ha⁻¹)						
0	72.88ᶜ	0.81ᶜ	81.63ᶜ	0.654	31.81ᵇ	68.88ᵇ
0.38	75.88ᵇ	82.69ᵇ	84.31ᵇ	0.655	33.81ᵃ	71.25ᵃ
0.77	77.75ᵃᵇ	84.25ᵃᵇ	85.31ᵃᵇ	0.656	34.38ᵃ	72.19ᵃ
1.15	78.88ᵃ	85.44ᵃ	86.56ᵃ	0.656	35.13ᵃ	72.75ᵃ
L.S.D. 0.05	1.942	1.879	1.705	n.s.	1.523	1.641
MC rate (kg ha⁻¹)						
0	74.84ᵇ	81.56ᵇ	83.19ᵇ	0.655	32.91ᵇ	70.28ᵇ
0.048 + 0.024	77.84ᵃ	84.53ᵃ	85.72ᵃ	0.656	34.66ᵃ	72.25ᵃ
L.S.D. 0.05	1.373	1.329	1.206	n.s.	1.077	1.160

Values followed by the same letter are not significantly different from each other at 0.05 levels (Sawan et al., 2009).

3.3.1.4. Seed viability, seedling vigor, and cool germination test performance. Seed viability, seedling vigor, and cool germination test performance were significantly increased by the high N rate as compared to the low rate and with the application of the three K concentrations and the PGR MC (Tables 11 and 12) (Sawan et al., 2009). Results from K application were more effective and significant when applied at the high concentration (1.15 kg ha⁻¹ K_2O) than those produced from the low concentration (0.38 kg ha⁻¹ K_2O) concerning its effect on seed viability and hypocotyl and entire seedling length. The GRI was not significantly affected by N rate, three K concentrations and the PGR MC. Effects of adding the high N-rate, application of potassium at different concentrations and MC on seed viability, seedling vigor and the cool germination test performance, which are important in stand establishment, may be attributed to the higher synthesis of assimilates (Table 9) in the recovered leaves which were deviated toward bolls (Sawan et al., 2009). This caused an increase in seed weight associated with its changed composition (Taiz & Zeiger, 1991; Welch, 1995; Wiatrak, Wright, & Marois, 2006). These effects are manifested in metabolites formed in plant tissues, which have a direct impact through utilization in growth and development processes. This may be reflected in distinct changes in seed quality and weight. Speed, Krieg, and Jividen (1996) stated that seed density was positively associated with germination capability at 15°C. Gadallah (2000) indicated that viability, germination, and seedling emergence were directly related to seed density for the *G. barbadense* cultivars. He pointed out that high seed density of all cultivars usually exhibited faster and more uniform rates of radicle emergence than low seed density. Seedlings from heavier seeds had greater accumulation of fresh and dry weight, and quality and vigor indices were also higher than those from light seeds. N is essential for plant protein synthesis including chlorophyll, which is indispensable for photosynthesis and enzymes which act as catalysts in biochemical reactions, and seed reserve proteins (Bisson et al., 1994). Maiya, Basave Gowda, Gouda, and Khadi (2001) indicated that the large- and medium-sized seeds (>4.75 mm) recorded higher field emergence (>65%) and produced vigorous seedlings compared to smaller seeds (<4.75 mm). The shoot and root length of seedlings increased with an increase in seed size. Thus, heavier cottonseed has a higher growth potential than lighter seed. Potassium application has favorable effects as an activator of several enzymes involved in carbohydrate metabolism and on the metabolism of nucleic acids, proteins, vitamins,

Table 12. Mean effects of N-rate and foliar application of potassium and the plant growth retardant MC on seedling vigor

Treatments	Hypocotyl length (cm)	Radicle length (cm)	Seedling length (cm)	Seedling fresh weight (g 10 seedling^{-1})	Seedling dry weight (g 10 seedling^{-1})
N rate (kg ha^{-1})					
95.2	7.31[b]	16.24[b]	23.56[b]	6.81[b]	0.637[b]
142.8	7.62[a]	16.69[a]	24.31[a]	7.04[a]	0.655[a]
L.S.D. 0.05	0.149	0.239	0.377	0.101	0.0085
K$_2$O rate (kg ha^{-1})					
0	7.18[c]	15.91[b]	23.09[c]	6.71[b]	0.626[b]
0.38	7.45[b]	16.48[a]	23.93[b]	6.94[a]	0.647[a]
0.77	7.55[ab]	16.70[a]	24.24[ab]	6.99[a]	0.653[a]
1.15	7.69[a]	16.79[a]	24.48[a]	7.07[a]	0.657[a]
L.S.D. 0.05	0.211	0.338	0.532	0.143	0.0120
MC rate (kg ha^{-1})					
0	7.33[b]	16.28[b]	23.61[b]	6.84[b]	0.638[b]
0.048 + 0.024	7.60[a]	16.66[a]	24.26[a]	7.02[a]	0.654[a]
L.S.D. 0.05	0.149	0.239	0.377	0.101	0.0085

Note: Values followed by the same letter are not significantly different from each other at 0.05 levels (Sawan et al., 2009).

and growth substances (Bednarz & Oosterhuis, 1999; Bisson et al., 1994). This may be reflected in distinct changes in seed weight and quality. Vasudevan, Virupakshappa, Venugopal, and Bhaskar (1997) studied the response of sunflower to K, P, and Zn along with recommended doses of N, P and K, and found that those nutrients increased the seed-quality parameters such as higher germination after accelerated ageing, higher speed of germination, higher shoot and root lengths, higher vigor index and seedling-growth rate compared with the control. No information on residual effects of K on cool germination test performance was found in the available literature on cotton plants. Beneficial residual effects of MC on seed viability, seedling vigor, and cool germination test performance may be due to their favorable effects on seed weight (Kumar et al., 2004). Wang et al. (1995) stated that the application of MC (at 50 or 100 mg kg^{-1}) to the cotton plants at squaring decreased the partitioning of assimilates to the main stem, the branches and their growing points, and increased partitioning to the reproductive organs and roots. Also, they indicated that, from flower to boll setting, MC application was very effective in promoting the partitioning of assimilates into reproductive organs (seeds). Lamas and Athayde (1999) indicated that seedling emergence and seedling dry matter increased with increasing MC rate, when applied at 50, 75, 100, or 125 g ha^{-1}. Significant effects for the main treatments and years were detected, on all studied characters, with one exception for the GRI. Replications within years showed significant effects only on cotton seed yield plant^{-1}

Table 13. Effect of interaction between K rate and foliar application of Zn on dry matter yield and uptake of K, Zn, and P by cotton plants (season II, sampled 105 days after planting)

K rate (kg ha^{-1})	Dry matter yield (g plant^{-1})		K uptake (mg plant^{-1})		Zn uptake (µg plant^{-1})		P uptake (mg plant^{-1})	
	Zn rate (g ha^{-1})							
	0.0	58	0.0	58	0.0	58	0.0	58
0.0	26.51[d]	35.90[b]	1442.4[bc]	1705.2[b]	1381[b]	2191[a]	94.44[b]	98.81[b]
47	33.18[c]	46.33[a]	1686.6[b]	2307.7[a]	2052[a]	2109[a]	102.88[b]	138.56[a]
LSD (0.05)	1.649		245.8		503		23.72	

Note: Values followed by the same letter are not significantly different from each other at 0.05 levels (Sawan et al., 2011).

Table 14. Effect of interaction between K rate and foliar application of P on chlorophyll and uptake of K and Zn and by cotton plants (season II, sampled 105 days after planting)

P rate (g ha⁻¹)	Chlorophyll (mg l⁻¹)		K uptake (mg plant⁻¹)		Zn uptake (µg plant⁻¹)	
	K rate (kg ha⁻¹)					
	0.0	47	0.0	47	0.0	47
0.0	3.859[f]	4.389[e]	1,533.6[c]	1,687.6[bc]	1,828[bc]	1,402[c]
576	5.002[d]	5.120[d]	1,485.9[c]	1,551.0[c]	1,668[bc]	1,970[b]
1,152	6.386[c]	7.002[b]	2,003.8[ab]	1,820.5[b]	2,120[ab]	2,505[a]
1,728	7.355[b]	7.839[a]	2,050.5[ab]	2,151.5[a]	1,989[b]	1,984[b]
L.S.D. (0.05)	0.450		245.8		503	

Note: Values followed by the same letter are not significantly different from each other at 0.05 levels (Sawan et al., 2011).

Table 15. Effect of interaction between Zn rate and foliar application of P on uptake of K and Zn by cotton plants (season II, sampled 105 days after planting)

P rate (g ha⁻¹)	K uptake (mg plant⁻¹)		Zn uptake (µg plant⁻¹)	
	Zn rate (g ha⁻¹)			
	0.0	58	0.0	58
0.0	1,658.4[bc]	1,879.0[b]	2,188[ab]	1,551[c]
576	1,613.1[c]	1,895.0[b]	1,760[c]	2,139[ab]
1,152	1,495.1[c]	2,041.3[ab]	1,836[bc]	1,569[c]
1,728	1,528.6[c]	2,173.4[a]	2,020[abc]	2,402[a]
L.S.D. (0.05)	245.8		503	

Note: Values followed by the same letter are not significantly different from each other at 0.05 levels (Sawan et al., 2011).

and ha⁻¹, seed weight and first germination count. Boman and Westerman (1994) found that no significant N × MC rate interactions on growth and yield of cotton when the plants received 0 to 227 kg ha⁻¹ N and were sprayed at early flowering with 0, 25, or 50 g ha⁻¹ MC.

3.4.1. Effect of K, Zn, and P on seed yield, seed viability, and seedling vigor of cotton

3.4.1.1. Effects of interactions among treatments. There were no significant interactions among K, Zn and P with respect to quantitative and qualitative characters under investigation, except for the following significant interaction effects (Sawan et al., 2011):

Two factor interactions:

- between K and Zn for dry matter yield of cotton plants (shoots) at 105 DAS, as well as for K, Zn, and P uptake (Table 13);
- between K and P for total chlorophyll concentration, as well as for K, and Zn uptake (Table 14);
- between Zn and P for K and Zn uptake (Table 15).

Three factor interactions:

- between K, Zn, and P for total chlorophyll concentration, as well as for Zn uptake (Table 16). Application of the high K rate combined with Zn and or P application increased dry matter yield of cotton plants (shoots) at 105 DAS, as well as total chlorophyll concentration and K, Zn, and P uptake, over that obtained with the high K rate or Zn and or P alone.

Table 16. Effect of interactions between K rate, foliar application of Zn; P on chlorophyll and uptake of Zn by cotton plants (season II, sampled 105 days after planting)				
Treatment			**Chlorophyll (mg l⁻¹)**	**Zn uptake (µg plant⁻¹)**
K rate (kg ha⁻¹)	**Zn rate (g ha⁻¹)**	**P rate (g ha⁻¹)**		
0.0	0.0	0.0	3.163[i]	1,488[ghi]
		576	3.965[h]	1,253[ij]
		1,152	4.543[gh]	1,124[j]
		1,728	4.652[g]	1,661[g]
	58	0.0	4.555[g]	2,168[ef]
		576	4.813[g]	1,550[gh]
		1,152	5.461[f]	2,212[de]
		1,728	5.588[f]	2,279[cde]
47	0.0	0.0	5.950[ef]	2,888[a]
		576	6.550[de]	2,419[bcd]
		1,152	6.812[d]	1,978[f]
		1,728	6.650[d]	1,478[ghi]
	58	0.0	6.823[cd]	1,353[hij]
		576	7.454[bc]	2,591[b]
		1,152	7.899[b]	2,000[f]
		1,728	9.027[a]	2,490[bc]
L.S.D. (0.05)			0.636	251

Note: Values followed by the same letter are not significantly different from each other at 0.05 levels (Sawan et al., 2011).

3.5.1. Effects of main treatments

3.5.1.1. Plant growth and mineral content. Dry matter yield (shoots) at 105 DAS; total chlorophyll concentration, as well as K, Zn, and P content was determined to study the effect of applied K and foliar application of Zn and P on cotton growth and mineral uptake (Table 17). Soil-applied K was associated with higher dry matter yield, higher chlorophyll concentration and higher contents of K, Zn and P (Sawan et al., 2011). Hiremath and Hunsigi (1995) found that K application [presumably to soil] was associated with increased total dray matter and increased K concentration in petioles. Gormus (2002) found that K content in petioles and total dry matter production increased by K applied to cotton plants. Gormus (2002) indicated that control plots had lower leaf K concentrations, compared with the other plots, when applying K at the rates of 66.4, 132.8, and 199.2 kg ha⁻¹. Aneela et al. (2003) indicated that the K content significantly increased with increasing fertilizer K levels and was highest at 166 kg ha⁻¹. The P content increased significantly with K application and was highest at 83 kg ha⁻¹. According to the K-status in the experimental soil (Table 1) (Sawan et al., 2011), it is classified as medium fertile for K. Foliar application of Zn was associated with higher dry matter yield, higher chlorophyll concentration, and higher contents of K, Zn, and P. This stimulation is due to a low level of available Zn in the soil (Table 1). Because the pH value of the soil site was more than 2 units higher than 6, applied Zn almost certainly would give a profitable response (Benton, Wolf, & Mills, 1991). Cakmak (2000) has speculated that Zn deficiency stress may inhibit the activities of a number of antioxidant enzymes, resulting in extensive oxidative damage to membrane lipids, proteins, chlorophyll, and nucleic acids. Applied at different concentrations significantly enhanced dry matter yield, chlorophyll concentration as well as Zn and K content in cotton plants. The highest increase in dry matter yield was obtained from the highest P application rate (1,728 g ha⁻¹). In chlorophyll biosynthesis, P is required as a pyridoxal phosphate (Ambrose & Easty, 1977). The importance of P and Zn nutrition for Egyptian cotton (Sawan et al., 2011) was also confirmed by Mahmoud, Abdel

Treatments	Dry matter yield (g plant^{-1})	Chlorophyll (mg l^{-1})	K uptake (mg plant^{-1})	Zn uptake (μg plant^{-1})	P uptake (mg plant^{-1})
K rate (kg ha^{-1})					
0.0	29.85[b]	4.593[b]	1,564.5[b]	1,717[b]	98.66[b]
47	41.12[a]	7.146[a]	2,006.4[a]	2,150[a]	119.19[a]
L.S.D. (0.05)	1.166	0.225	122.91	251	11.86
Zn rate (g ha^{-1})					
0.0	31.21[b]	5.286[b]	1,573.8[b]	1,786[b]	97.13[b]
58	39.76[a]	6.453[a]	1,997.2[a]	2,081[a]	120.72[a]
L.S.D. (0.05)	1.166	0.225	122.91	251	11.86
P rate (g ha^{-1})					
0.0	33.50[c]	5.123[c]	1,768.7[a]	1,974[a]	91.50[b]
576	34.56[c]	5.696[b]	1,754.1[a]	1,954[a]	108.56[a]
1152	36.27[b]	6.178[a]	1,768.1[a]	1,829[a]	116.19[a]
1728	37.59[a]	6.479[a]	1,851.0[a]	1,977[a]	119.44[a]
L.S.D. (0.05)	1.649	0.318	n.s.	n.s.	16.77

Table 17. Mean effects of K and foliar application of Zn and P on dry matter yield, chlorophyll and uptake of K, Zn, and P by cotton plants (season II, sampled 105 days after planting)

Note: Values followed by the same letter are not significantly different from each other at 0.05 levels, and n.s.: not significant (Sawan et al., 2011).

Aziz, and Ashoub (1985) who found a significant relationship between Zn uptake and P uptake by plants. This reflects the positive relationship that exists between the two elements in the nutrition of cotton plants. These results can be seen in the sense that both K and Zn are necessary for the biosynthesis of chlorophyll (Amberger, 1974). Therefore, the factors responsible for the green color of tissues (NPK and minor-elements) are themselves stimulators for chlorophyll biosynthesis. Data (Table 17) also reveal that the uptake of P by cotton plants was increased significantly by the application of K, Zn, and P treatments, individually (Sawan et al., 2011). More and Agale (1993) indicated that when applying P to cotton plants in the range of 10.9–32.7 kg P ha^{-1}, plant uptake increased with increasing P fertilization, while total dry matter yield increased with increasing P levels up to 21.8 kg ha^{-1}. Deshpande and Lakhdive (1994) found that P application (10.9–21.8 kg ha^{-1}) increased P uptake and content in stem, leaf, reproductive parts, and seed. Ahmad et al. (2000) pointed out that P deficiency reduced biomass.

3.5.1.2. Seed yield. Seed yield plant^{-1}, as well as plot^{-1}, significantly increased when K was applied (by as much as 14.72, 13.13; 14.69, 13.26%, respectively) in both seasons (Table 18) (Sawan et al., 2011). K would have a favorable impact on yield components, including number of opened bolls plant^{-1} and boll weight, leading to a higher cotton yield. Guinn (1985) suggested that growth, flowering, and boll retention decrease when the demand for photosynthate increases and exceeds the supply. The obtained results of total chlorophyll (a; b) confirmed these findings (Table 17) (Sawan et al., 2011). This means that an increase in photosynthesis should permit more bolls to be set before cutout. Sangakkara et al. (2000) indicated that, K increases the photosynthetic rates of crop leaves, CO_2 assimilation and facilitates carbon movement. Also, the role of K suggests that it affects abscission and reduced boll shedding and it certainly affects yield (Zeng, 1996). Mullins, Schwab, and Burmester (1999) evaluated cotton yield in relation to long-term surface application of K at 49.8–149.4 kg ha^{-1} and found that K application increased yield. Results obtained here were similar to those of Aneela et al. (2003), Gormus, (2002), Ibrahim et al. (2009), Pervez et al. (2004), Pettigrew et al., (2005), Sawan, Hafez, Basyony, and Alkassas (2007), and Sawan, Mahmoud, and El-Guibali (2006). Application of Zn significantly increased seed yield plant^{-1}, and seed yield plot^{-1}, as compared

Table 18. Effect of K-rate and foliar application of Zn and P on seed yield plant⁻¹ and seed yield plot⁻¹, and seed weight

Treatments	Cotton seed yield (g plant⁻¹)		Cotton seed yield (kg plot⁻¹)		Seed weight (g 100 seed⁻¹)	
	I	II	I	II	I	II
K rate (kg ha⁻¹)						
0.0	19.16[b]	18.35[b]	1.457[b]	1.395[b]	10.12[b]	9.89[b]
47	21.98[a]	20.76[a]	1.671[a]	1.580[a]	10.29[a]	10.04[a]
L.S.D. (0.05)	1.14	1.28	0.095	0.098	0.09	0.08
Zn rate (g ha⁻¹)						
0.0	19.57[b]	18.73[b]	1.489[b]	1.425[b]	10.15[b]	9.92[b]
58	21.57[a]	20.38[a]	1.639[a]	1.549[a]	10.26[a]	10.00[a]
L.S.D. (0.05)	1.14	1.28	0.095	0.098	0.09	0.08
P rate (g ha⁻¹)						
0.0	18.55[b]	17.82[b]	1.413[b]	1.358[b]	10.08[b]	9.86[b]
576	20.63[a]	19.30[ab]	1.566[a]	1.467[ab]	10.21[a]	9.95[ab]
1,152	21.21[a]	20.35[a]	1.611[a]	1.546[a]	10.26[a]	10.00[a]
1,728	21.88[a]	20.74[a]	1.667[a]	1.578[a]	10.29[a]	10.04[a]
L.S.D. (0.05)	1.62	1.81	0.134	0.138	0.12	0.11

Note: Values followed by the same letter are not significantly different from each other at 0.05 levels (Sawan et al., 2011).

with the untreated control (by 10.22, 8.81; 10.07, 8.70%, respectively) in the two seasons (Table 18) (Sawan et al., 2011). Zn could have a favorable effect on photosynthetic activity of leaves, which improves mobilization of photosynthesis coincidence the increases total chlorophyll (a; b) as shown in Table 17 and directly influences boll weight. Further, Zn is required in the synthesis of tryptophan, a precursor of indole-3-acetic acid, which is the major hormone inhibits abscission of squares and bolls. Thus the number of retained bolls plant⁻¹ and consequently seed yield ha⁻¹ would be increased (Rathinavel & Dharmalingam, 2000). Similar results were obtained by Rathinavel and Dharmalingam (2000), by soil application of 50 kg ha⁻¹ $ZnSO_4$, and Ibrahim et al. (2009). Applying P significantly increased seed yield plant⁻¹ and plot⁻¹ in both seasons (Table 18), as compared to the untreated plants, when treatment rate was increased from 576 up to 1728 g ha⁻¹ (by 11.21–17.95, 8.31–16.39; 10.83–17.98, 8.03–16.20%, respectively). Generally seed yield plant⁻¹ and plot⁻¹ were the greatest when the highest P-concentration (1728 g ha⁻¹) was applied. Such results reflect the pronounced improvement of yield components due to application of P which is possibly ascribed to its involvement in photosynthesis and translocation of carbohydrates to young bolls (Rodríguez et al., 1998). Phosphorus as a constituent of cell nucleus is essential for cell division and the development of meristematic tissue and hence it would have a stimulating effect on increasing the number of flowers and bolls plant⁻¹ (Russell, 1973). These results agree with that reported by Katkar, Turkhede, Solanke, Wankhade, and Sakhare (2002) Ibrahim et al. (2009) and Saleem, Shakeel, et al. (2010).

3.5.1.3. Seed weight. Seed weight significantly increased with applying K in both years (Table 18). A possible explanation for the increased seed weight due to the application of K may be due in part to its favorable effects on photosynthetic activity rate of crop leaves and CO_2 assimilation, which improves mobilization of photosynthates and directly influences boll weight which in turn directly affect seed weight (Ghourab et al., 2000). Result was similar to those obtained by Sabino, da Silva, and Kondo, (1999); Sawan et al. (2006, 2007), i.e. an increase in seed weight due to K application. Application of Zn significantly increased seed weight (Table 18) coinciding with the increased total chlorophyll (a; b) (Table 17) compared to the control in both seasons. The increased seed weight might be due to an increased photosynthesis activity resulting from the application of Zn (Welch, 1995) which improves mobilization of photosynthates and the amount of photosynthate available for reproductive sinks and

Table 19. Effect of K-rate and foliar application of Zn and P on seed viability and cool germination test performance

Treatments	Germination velocity (%)		Second germination count (%)		Total germination capacity (%)		Germination rate index (GRI unit)		Cool germination test performance (%)			
									4-day count		7-day count	
	I	II	I	II	I	II	I	II	I	II	I	II
K rate (kg ha⁻¹)												
0.0	75.19b	74.19b	82.19b	81.13b	83.63b	82.44b	0.655a	0.655a	32.88b	32.38b	70.56b	68.69b
47	79.13a	77.00a	87.81a	84.63a	89.13a	86.00a	0.654a	0.655a	36.06a	35.44a	72.88a	71.25a
L.S.D. (0.05)	1.85	1.97	2.11	1.71	2.07	1.81	n.s.	n.s.	1.49	1.74	1.64	1.94
Zn rate (g ha⁻¹)												
0.0	75.81b	74.44b	83.38b	81.81b	84.63b	83.19b	0.655a	0.654a	33.13b	32.75b	70.81b	68.88b
58	78.50a	76.75a	86.83a	83.94a	88.13a	85.25a	0.654a	0.655a	35.81a	35.06a	72.63a	71.06a
L.S.D. (0.05)	1.85	1.97	2.11	1.71	2.07	1.81	n.s.	n.s.	1.49	1.74	1.64	1.94
P rate (g ha⁻¹)												
0.0	74.25b	73.38b	81.63b	80.13c	82.88b	81.38c	0.654a	0.655a	32.25b	32.13b	69.50b	68.00b
576	76.88a	75.13ab	84.75a	82.63b	86.25a	84.00b	0.654a	0.654a	34.50a	33.38ab	71.88a	69.25ab
1,152	78.13a	76.63a	86.25a	83.25ab	87.75a	84.88ab	0.654a	0.655a	35.00a	34.63a	72.38a	70.75a
1,728	79.38a	77.25a	87.38a	85.50a	88.63a	86.63a	0.655a	0.655a	36.13a	35.50a	73.13a	71.88a
L.S.D. (0.05)	2.61	2.78	2.98	2.42	2.92	2.56	n.s.	n.s.	2.11	2.46	2.32	2.74

Note: Values followed by the same letter are not significantly different from each other at 0.05 levels, and n.s.: not significant (Sawan et al., 2011).

Table 20. Effect of K-rate and foliar application of Zn and P on seedling vigor									
Treatments	**Hypocotyl length (cm)**		**Radicle length (cm)**		**Seedling length (cm)**		**Seedling fresh weight (g 10 seedling⁻¹)**		**Seedling dry weight (g 10 seedling⁻¹)**	
	I	**II**	**I**	**II**	**I**	**II**	**I**	**II**	**I**	**II**
K rate (kg ha⁻¹)										
0.0	7.28ᵇ	7.12ᵇ	16.26ᵇ	15.82ᵇ	23.54ᵇ	22.94ᵇ	6.86ᵇ	6.66ᵇ	0.640ᵇ	0.621ᵇ
47	7.71ᵃ	7.41ᵃ	16.73ᵃ	16.34ᵃ	24.44ᵃ	23.75ᵃ	7.10ᵃ	6.88ᵃ	0.658ᵃ	0.641ᵃ
L.S.D. (0.05)	0.22	0.20	0.35	0.36	0.56	0.54	0.14	0.15	0.011	0.012
Zn rate (g ha⁻¹)										
0.0	7.34ᵇ	7.16ᵇ	16.30ᵇ	15.90ᵇ	23.63ᵇ	23.05ᵇ	6.88ᵇ	6.69ᵇ	0.642ᵇ	0.623ᵇ
58	7.66ᵃ	7.37a	16.70ᵃ	16.27ᵃ	24.36ᵃ	23.63ᵃ	7.08ᵃ	6.85ᵃ	0.657ᵃ	0.638ᵃ
L.S.D. (0.05)	0.22	0.20	0.35	0.36	0.56	0.54	0.14	0.15	0.011	0.012
P rate (g ha⁻¹)										
0.0	7.12ᵇ	6.98ᵇ	15.91ᵇ	15.58ᵇ	23.02ᵇ	22.56ᵇ	6.77ᵇ	6.55ᵇ	0.632ᵇ	0.614ᵇ
576	7.51ᵃ	7.25ᵃᵇ	16.50ᵃ	16.10ᵃ	24.01ᵃ	23.35ᵃ	7.00ᵃ	6.78ᵃ	0.649ᵃ	0.630ᵃᵇ
1152	7.65ᵃ	7.33ᵃ	16.74ᵃ	16.26ᵃ	24.39ᵃ	23.59ᵃ	7.04ᵃ	6.84ᵃ	0.653ᵃ	0.637ᵃ
1728	7.72ᵃ	7.48ᵃ	16.84ᵃ	16.40ᵃ	24.56ᵃ	23.87ᵃ	7.11ᵃ	6.91ᵃ	0.662ᵃ	0.643ᵃ
L.S.D. (0.05)	0.32	0.28	0.50	0.51	0.80	0.76	0.20	0.21	0.016	0.017

Note: Values followed by the same letter are not significantly different from each other at 0.05 levels (Sawan et al., 2011).

thereby influences boll weight, factors that coincide with increased in seed weight. A similar result (in seed weight due to Zn application) was obtained by Rathinavel and Dharmalingam (2000). The same pattern was found (Table 18) with regard to application of the three P concentrations, which also brought about a significant increment in seed weight over the control in both seasons. This increase was significant for all P concentrations in the first season and for P at 1,152 and 1,728 g ha⁻¹ in the second season. Spraying plants with P at 1,728 g ha⁻¹ produced the highest seed weight. This increased seed weight may be due to the fact that P activated the biological reaction in cotton plant, particularly photosynthesis fixation of CO_2 and synthesis of sugar, and other organic compounds (Welch, 1995; Wiatrak et al., 2006). This indicates that treated cotton bolls had larger photosynthetically supplied sinks for carbohydrates and other metabolites than untreated bolls.

3.5.1.4. Seed viability, seedling vigor, and cool germination test. Seed viability, cool germination test performance, and seedling vigor were significantly increased by application of K (47 kg ha⁻¹), Zn (58 g ha⁻¹) and P (576, 1,152; 1,728 g ha⁻¹), as compared with the untreated plants (Tables 19 and 20) (Sawan et al., 2011). The differences between the effects of the three P rates were not significant, with the exception of the 1,728 g ha⁻¹ dose, which significantly increased seed viability, seedling vigor and the cool germination test performance as compared with P applied at 576 g ha⁻¹. Application of K, Zn, or P had no effect on GRI. Beneficial residual effects of K addition and application of Zn, and P at different concentrations on stimulating the seed viability, seedling vigor and cool germination test performance may be attributed to their favorable effects on increased seed weight associated with changes in the metabolism of nucleic acids, proteins, vitamins, and growth substances (Sharma et al., 1990; Taiz & Zeiger, 1991; Welch, 1995; Wiatrak, Wright, Marois, Koziara, & Pudelko, 2005). At least 60 enzymes are known to be activated by this ion. The enzyme pyruvate kinas (more correctly referred to as ATP: pyruvate phosphotransferase), which participates in glycolysis (Glass ADM: Plant Nutrition, 1989). The high concentration of K^+ is thought to be essential for normal protein synthesis. Potassium role in this process is considered to be the maintenance of a proper association between t RNA molecules and ribosomes during the translation of mRNA (Glass ADM: Plant Nutrition, 1989). Potassium also acts as an activator for several enzymes involved in carbohydrates metabolism. The requirement of cotton for K increases with the beginning of bud formation stage. A greater accumulation of sugars and starch in leaves under K-deficient conditions adversely affects development of bolls due to deficiency of metabolites. The significance of the correlation coefficients between seed weight and the various performance measures are as follows: 0.9487**–0.9893** with seed viability, 0.8912**–0.9437** with seedling vigor; 0.6660**–0.9550** with cool germination test performance. Wang, Yang, Zhou, and Xu (2003) studied the influence of physical characteristics of cotton seed on germination and emergence rates and found that germination rate and seed weight had an extreme positive correlation. Also found that seed weight had a positive correlation with emergence rate.

4. Conclusion

From the findings of this study, it seems rational to recommended application of N at a rate of 161 of kg ha^{-1}, spraying of cotton plants with plant PGR, and application of Zn in comparison with the ordinary cultural practices adopted by Egyptian cotton producers, it is quite apparent that applications of such PGR, Zn, and increased N fertilization rates could bring about better impact on seed yield and seedling characters studied (Sawan et al., 1998).

Under the conditions of this study, it can be concluded that addition of P at 74 kg ha^{-1} P_2O_5 and spraying cotton plants with Zn at 40 ppm and also with Ca at 60 ppm can be recommended to improve cotton seed yield, viability, and seedling vigor (Sawan et al., 1999).

Application of N at the rate of 142.8 kg ha^{-1} and application of K (foliar, at the rate of 1.15 kg ha^{-1} K_2O) and MC (at the rate of 0.048 + 0.024 kg ha^{-1} MC) should help achieve higher cotton seed productivity and quality (seed viability and seedling vigor) in comparison with the usual cultural practices adopted by Egyptian cotton procedures (Sawan et al., 2009).

From the findings of this study, the addition of K at 47 kg ha^{-1}, spraying cotton plants with Zn twice (at 57 g ha^{-1}), and also with P twice (especially the P concentration of 1728 g ha^{-1}) along with the soil fertilization used P at sowing time have been proven beneficial to the quality and yield of cotton plants. These combinations appeared to be the most effective treatments, affecting cottonseed productivity and quality (as indicated by better seed viability, seedling vigor, and cool germination test performance) (Sawan et al., 2011).

In comparison with the ordinary cultural practices adopted by Egyptian cotton producers, it is apparent that the applications of such treatments could produce an improvement in cottonseed yield and quality.

Funding
The author received no direct funding for this research.

Competing Interests
The author declare no competing interest

Author details
Zakaria M. Sawan[1]
E-mail: zmsawan@hotmail.com
[1] Agricultural Research Center, Cotton Research Institute, Ministry of Agriculture and Land Reclamation, 9 Gamaa street, 12619 Giza, Egypt.

References
Abdel-Al, M. H. (1998). Response of Giza 85 cotton cultivar to the growth regulators Pix and Atonic. *Egyptian Journal of Agricultural Research, 76*, 1173–1181.

Abdel-Malak, K. I., Radwan, F. E., & Baslious, S. I. (1997). Effect of row width, hill spacing and nitrogen levels on seed cotton yield of Giza 83 cotton cultivar. *Egyptian Journal of Agricultural Research, 75*, 743–752.

Addicott, F. T., & Lyon, J. L. (1973). Physiological ecology of abscission. In T. T. Kozlowski (Ed.), *Shedding of plants parts* (pp. 85–124). New York, NY: Academic Press. http://dx.doi.org/10.1016/B978-0-12-424250-0.50008-7

Ahmad, Z., Gill, M. A., Qureshi, R. H., Aslam, M., Iqbal, J., & Nawas, S. (2000). Inter-cultivar variations of phosphorus-deficiency stress tolerance in cotton. *Tropical Agricultural Research, 12*, 119–127.

Ali, S. A., & El-Sayed, A. E. (2001). Effect of sowing dates and nitrogen levels on growth, earliness and yield of Egyptian cotton cultivar Giza 88. *Egyptian Journal of Agricultural Research, 79*, 221–232.

Alikhanova, O. I., & Tursunov, D. (1988). Efficiency of trace element fertilizers applied to cotton on reclaimed grey-brown stony soils in the Khodzha-Bakirgan massif. *Agrokhimiya, 2*, 72–77.

Amberger, A. (1974). Micronutrients, dynamics in the soil and function in plant metabolism. *Iron Proceedings Egypt Botanic Social Workshop 1 Cairo*, 81–90.

Ambrose, E. J., & Easty, D. M. (1977). *Cell Biology*. London: The English Language Book Society and Longman.

Aneela, S., Muhammad, A., & Akhtar, M. E. (2003). Effect of potash on boll characteristics and seed cotton yield in newly developed highly resistant cotton varieties. *Pakistan Journal of Biological Sciences, 6*, 813–815.

Bartee, S. N., & Krieg, D. R. (1974). Cotton seed density: Associated physical and chemical properties of ten cultivars. *Agronomy Journal, 66*, 433–435. http://dx.doi.org/10.2134/agronj1974.00021962006600030028x

Bartlett, M. S. (1937). Some examples of statistical methods of research in agriculture and applied biology. *Supplement to the Journal of the Royal Statistical Society, 4*, 137–183. http://dx.doi.org/10.2307/2983644

Basilious, S. I., Abdel Malak, K. K. I., & Abdel Kader, A. E. M. (1991). Response of cotton "Giza 83" to some micronutrients as affected by time of application of nitrogen levels. *Assiut Journal of Agricultural Sciences, 22*, 351–366.

Bednarz, C. W., & Oosterhuis, D. M. (1999). Physiological changes associated with potassium deficiency in cotton. *Journal of Plant Nutrition, 22*, 303–313. http://dx.doi.org/10.1080/01904169909365628

Benton, J. J., Wolf, B., & Mills, H. A. (1991). *Plant analysis handbook*. Georgia, ATL: Micro-Macro Pub.

Bird, L. S., & Reyes, A. A. (1967). Effects of cottonseed quality on seed and seedling characteristics. *In Proceedings Beltwide Cotton Conferences*, 199–206.

Bisson, P., Cretenet, M., & Jallas, E. (1994). Nitrogen, phosphorus and potassium availability in the soil-physiology of the assimilation and use of these nutrients by the plant. Challenging the Future. In G. A. Constable & N. W. Forrester (Eds.), *Proceedings of the World Cotton Research Conference-1, Brisbane Australia, February 14–17* (pp. 115–124). Melbourne: CSIRO.

Boman, R. K., & Westerman, R. L. (1994). Nitrogen and mepiquat chloride effects on the production of Nonrank, irrigated, short-season cotton. *Journal of Production Agriculture, 7*, 70–75. http://dx.doi.org/10.2134/jpa1994.0070

Boquet, D. J. (2005). Cotton in ultra-narrow row spacing plant density and nitrogen fertilizer rates. *Agronomy Journal, 97*, 279–287. http://dx.doi.org/10.2134/agronj2005.0279

Borowski, E. (2001). The effect of nitrogenous compounds on the growth, photosynthesis and phosphorus uptake of sunflowers. *Annales Universitatis Mariae Curie-Sklodowska. Sectio EEE Horticultura, 9*, 23–31.

Bronson, K. F., Onken, A. B., Keeling, J. W., Booker, J. D., & Torbert, H. A. (2001). Nitrogen response in cotton as affected by tillage system and irrigation level. *Soil Science Society of America Journal, 65*, 1153–1163. http://dx.doi.org/10.2136/sssaj2001.6541153x

Cakmak, I. (2000). Possible roles of zinc in protecting plant cells from damage by reactive oxygen species. *New Phytologist, 146*, 185–205. http://dx.doi.org/10.1046/j.1469-8137.2000.00630.x

Cakmak, I., Hengeler, C., & Marschner, H. (1994). Changes in phloem export of sucrose in leaves in response to phosphorus, potassium and magnesium deficiency in bean plants. *Journal of Experimental Botany, 45*, 1251–1257. http://dx.doi.org/10.1093/jxb/45.9.1251

Carvalho, L. H., Chiavegato, E. J., Cia, E., Kondo, J. I., Sabino, J. C., Pettinelli Júnior, A., ... & Gallo, P. B. (1994). Plant growth regulators and pruning in the cotton crop. *Bragantia, 53*, 247–254. http://dx.doi.org/10.1590/S0006-87051994000200014

Chapman, H. D., & Pratt, P. F. (1961). *Methods of analysis for soils, plants and waters* (pp. 60–61 and 150–179). Los Angeles, CA: University of California, Division of Agricultural Science.

Colomb, B., Bouniols, A., & Delpech, C. (1995). Effect of various phosphorus availabilities on radiation-use efficiency in sunflower biomass until anthesis. *Journal of Plant Nutrition, 18*, 1649–1658. http://dx.doi.org/10.1080/01904169509365010

Cothren, J. T. (1994). Use of growth regulators in cotton production. Challenging the Future. In G. A. Constable & N. W. Forrester (Eds.), *Proceedings of the World Cotton Research Conference-1, Brisbane Australia, February 14–17* (pp. 6–24). Melbourne: CSIRO.

Deshpande, R. M., & Lakhdive, B. A. (1994). Effect of plant growth substances and phosphorus levels on yield and phosphorus uptake by cotton. *PKV Research Journal, 18*, 118–121.

de Figueiredo e Albuquerque, M. C. (2003). Effect of the type of environmental stress on the emergence of sunflower (*Helianthus annuus* L.), soybean (*Glycine max* (L.) Merril) and maize (*Zea mays* L.) seeds with different levels of vigor. *Seed Science and Technology, 31*, 465–479. http://dx.doi.org/10.15258/sst

dela Fuente, R. K., & Leopold, A. C. (1973). A role for calcium in Auxin transport. *Plant Physiology, 51*, 845–847. http://dx.doi.org/10.1104/pp.51.5.845

El-Debaby, A. S., Hammam, G. Y., & Nagib, M. A. (1995). Effect of planting date, N and P application levels on seed index, lint percentage and technological characters of Giza 80 cotton cultivar. *Annals of Agricultural Science, Moshtohor, 33*, 455–464.

Engels, C., & Marschner, H. (1995). Plant uptake and utilization of nitrogen. In P. E. Bacan (Ed.), *Nitrogen fertilization in the environment* (pp. 41–81). New York, NY: Marcel Dekker.

Fan, S., Yuzhang, X., & Chaojun, Z. (1999). Effects of nitrogen, phosphorus and potassium on the development of cotton bolls in summer. *Acta Gossypii Sinica, 11*, 24–30.

Fenner, M. (1992). Environmental influences on seed size and composition. *Horticultural Reviews, 13*, 183–213.

Gadallah, F. M. (2000). Seed density in relation to germination and seedling quality in cotton (*Gossypium barbadense* L.). *Alexandria Journal of Agricultural Research, 45*, 119–137.

Ghourab, M. H. H., Wassel, O. M. M., & Raya, N. A. A. (2000). Response of cotton plants to foliar application of (Pottasin-P)™ under two levels of nitrogen fertilizer. *Egyptian Journal of Agricultural Research, 78*, 781–793.

Glass ADM: Plant Nutrition. (1989). *An introduction to current concepts*. Boston, MA: Jones and Bartlett.

Gomaa, M. A. (1991). Response of cotton plant to phosphatic and zinc fertilization. *Annals of Agricultural Science, Moshtohor, Egypt, 29*, 1051–1061.

Gormus, O. (2002). Effects of rate and time of potassium application on cotton yield and quality in Turkey. *Journal of Agronomy and Crop Science, 188*, 382–388. http://dx.doi.org/10.1046/j.1439-037X.2002.00583.x

Guinn, G. (1984). Boll abscission in cotton. In U. S. Gupta (Ed.), *Crop physiology: Advancing frontiers* (pp. 177–225). New Delhi: Oxford & IBH.

Guinn, G. (1985). Fruiting of cotton. III. Nutritional stress and cutout. *Crop Science, 25*, 981–985. http://dx.doi.org/10.2135/cropsci1985.0011183X002500060020x

Hearn, A. B. (1981). Cotton nutrition. *Field Crop Abstracts, 34*, 11–34.

Hiremath, G. M., & Hunsigi, G. (1995). Effect of nitrogen and potash levels on concentration of nitrogen and potassium in petiole of two cotton hybrids (*Gossypium* sp.). Karnataka. *Karnataka Journal of Agricultural Science, 8*, 99–101.

Hodges, H. F., Reddy, V. R., & Reddy, K. R. (1991). Mepiquat chloride and temperature effects on photosynthesis and respiration of fruiting cotton. *Crop Science, 31,* 1302–1308. http://dx.doi.org/10.2135/cropsci1991.0011183X003100050044x

Ibrahim, M. E., Bekheta, M. A., El-Moursi, A., & Gaafar, N. A. (2009). Effect of arginine, prohexadione-Ca, some macro and micro-nutrients on growth, yield and fiber quality of cotton plants. *World Journal of Agricultural Sciences, 5,* 863–870.

International Seed Testing Association. (1976). International rules for seed testing. *Seed Science and Technology, 4,* 51–177.

Katkar, R. N., Turkhede, A. B., Solanke, V. M., Wankhade, S. T., & Sakhare, B. A. (2002). Effect of foliar sprays of nutrients and chemicals on yield and quality of cotton under rainfed condition. *Research Crops, 3,* 27–29.

Kilmer, V. J., & Alexander, L. T. (1940). Methods of making mechanical analysis of soils. *Soil Science, 68,* 15.

Kler, D. S., Raj, D., & Dhillon, G. S. (1989). Modification of micro-environment with cotton canopy for reduced abscission and increased seed yield. *Environment and Ecology, 7,* 800–802.

Kosheleva, L. L., Bakhnova, K. V., Semenova, T. A., & Mil'Kevich, Z. A. (1984). Effect of phosphorus nutrition on metabolism of young fiber flax plants in relation to assimilate distribution in them. *Referativnyi Zhurnal, 6,* 533.

Kumar, K. A. K., Patil, B. C., & Chetti, M. B. (2004). Effect of plant growth regulators on biophysical, biochemical parameters and yield of hybrid cotton. *Karnataka Journal of Agricultural Science, 16,* 591–594.

Lamas, F. M. (2001). Comparative study of mepiquat chloride and chlormequat chloride application in cotton. *Pesquisa Agropecuária Brasileira, 36,* 265–272. http://dx.doi.org/10.1590/S0100-204X2001000200008

Lamas, F. M., & Athayde, M. L. F. (1999). Effect of mepiquat chloride and thidiazuron on some characteristics of cotton seed. *Pesquisa Agropecuária Brasileira, 34,* 2015–2019. http://dx.doi.org/10.1590/S0100-204X1999001100006

Legé, K. E., Cothren, J. T., & Morgan, P. W. (1997). Nitrogen fertility and leaf age effects on ethylene production of cotton in a controlled environment. *Plant Growth Regulation, 22,* 23–28. http://dx.doi.org/10.1023/A:1005854625641

Li, L.-L., Ma, Z.-B., Wang, W.-L., & Tai, G.-Q. (2004). Effect of spraying nitrogen and zinc at seedling stage on some physiological characteristics and yield of summer cotton. *Journal of Henan Agricultural University, 38,* 33–35.

Ma, L.-G., & Sun, D.-Y. (1997). The involvement of calcium in the light signal transduction chain for phototropism in sunflower seedling. *Biologia Plantarum, 39,* 569–574.

Mahmoud, M. H., Abdel Aziz, I., & Ashoub, M. A. (1985). The relationship between phosphorous and zinc and cotton plant. *Annals Agricultural Science, Faculty of Agriculture, Ain Shams Univeristy, Cairo, Egypt, 30,* 1011–1030.

Maiya, M. R., Basave Gowda, Gouda, M. S., & Khadi, B. M. (2001). Effect of size grading on recovery and quality in naturally coloured cotton. *Seed Research, 29,* 248–250.

Mauney, J. R., Fry, K. E., & Guinn, G. (1978). Relationship of photosynthetic rate to growth and fruiting of cotton, soybean, sorghum and sunflower. *Crop Science, 18,* 259–263. http://dx.doi.org/10.2135/cropsci1978.0011183X001800020016x

McConnell, J. S., Baker, W. H., & Frizzell, B. S. (1996). Distribution of residual nitrate-N in longterm fertilization studies of an alfisol cropped for cotton. *Journal of Environment Quality, 25,* 1389–1394. http://dx.doi.org/10.2134/jeq1996.00472425002500060032x

McConnell, J. S., & Mozaffari, M. (2004). Yield, petiole nitrate, and node development responses of cotton to early season nitrogen fertilization. *Journal of Plant Nutrition, 27,* 1183–1197.

Mekki, B. B. (1999). Effect of mepiquat chloride on growth, yield and fiber properties of some Egyptian cotton cultivars. *Arab Universities Journal of Agricultural Science, 7,* 455–466.

More, S. D., & Agale, B. N. (1993). Phosphate balance studies in irrigated cotton. *Journal of the Indian Society of Soil Science, 41,* 498–500.

Mullins, G. L., Schwab, G. J., & Burmester, C. H. (1999). Cotton response to surface applications of potassium fertilizer: A 10-year summary. *Journal of Production Agriculture, 12,* 434–440. http://dx.doi.org/10.2134/jpa1999.0434

Ochiai, E.-L. (1977). *Bioinorganic chemistry* (p. 515). Boston, MA: Allyn & Bacon.

Oosterhuis, D., Hake, K., & Burmester, C. (1991, July). Leaf feeding insects and mites. *Cotton Physiology Today, 2,* 1–7.

Palomo Gil, A., & Chávez González, J. F. (1997). Response of the early cotton cultivar CIAN 95 to nitrogen fertilizer application. *ITEA. Producción Vegetal, 93,* 126–132.

Palomo Gil, A., Godoy Avila, S., & Chávez González, J. F. (1999). Reductions in nitrogen fertilizers use with new cotton cultivars: Yield, yield components and fiber quality. *Agrociencia, 33,* 451–455.

Perumai, N. K. (1999). Effect of different nitrogen levels on morpho-physiological characters and yield in rainfed cotton. *Indian Journal of Plant Physiology, 4,* 65–67.

Pervez, H., Ashraf, M., & Makhdum, M. I. (2004). Influence of potassium rates and sources on seed cotton yield and yield components of some elite cotton cultivars. *Journal of Plant Nutrition, 27,* 1295–1317.

Pettigrew, W. T. (1999). Potassium deficiency increases specific leaf weights of leaf glucose levels in field-grown cotton. *Agronomy Journal, 91,* 962–968. http://dx.doi.org/10.2134/agronj1999.916962x

Pettigrew, W. T., Meredith, W. R., Jr., & Young, L. D. (2005). Potassium fertilization effects on cotton lint yield, yield components, and reniform nematode populations. *Agronomy Journal, 97,* 1245–1251. http://dx.doi.org/10.2134/agronj2004.0321

Pípolo, A. E., Athayde, M. L. F., Pípolo, V. C., & Parducci, S. (1993). Comparison of different rates of chlorocholine chloride applied to herbaceous cotton. *Pesquisa Agropecuária Brasileira, 28,* 915–923.

Ram, P., & Prasad, M., & Pachauri, D. K. (2001). Effect of nitrogen, chlormequat chloride and FYM on growth yield and quality of cotton (*Gossypium hirsutum* L.). *Annals of Agricultural Research, 22,* 107–110.

Rathinavel, K., & Dharmalingam, C. (2000). Paneersel vam S: Effect of micronutrient on the productivity and quality of cotton seed cv. TCB 209 (*Gossypium barbadense* L.). Madrase. *Madras Agricultural Journal, 86,* 313–316.

Reddy, A. R., Reddy, K. R., Padjung, R., & Hodges, H. F. (1996). Nitrogen nutrition and photosynthesis in leaves of Pima cotton. *Journal of Plant Nutrition, 19,* 755–770. http://dx.doi.org/10.1080/01904169609365158

Rensing, L., & Cornelius, G. (1980). Biological membranes as components of oscillating systems. *Biologische Rundschau, 18,* 197–209.

Rinehardt, J. M., Edmisten, K. L., Wells, R., & Faircloth, J. C. (2004). Response of ultra-narrow and conventional spaced cotton to variable nitrogen rates. *Journal of Plant Nutrition, 27,* 743–755. http://dx.doi.org/10.1081/PLN-120030379

Rodríguez, D., Zubillaga, M. M., Ploschuk, E., Keltjens, W., Goudriaan, J., & Lavado, R. (1998). Leaf area expansion and assimilate prediction in sunflower growing under low phosphorus conditions. *Plant Soil, 202,* 133–147. http://dx.doi.org/10.1023/A:1004348702697

Russell, E. W. (1973). *Soil condition and plant growth* (p. 448). London: The English Language Book Society and Longman.

Sabino, N. P., da Silva, N. M., & Kondo, J. I. (1999). *Components of production and fiber quality of cotton as a function of potassium and gypsum* (pp. 703–706). Campina Grande: Empresa Brasileira de Pesquisa Agropecuáia, Embrapa Algodão.

Saleem, M. F., Bilal, M. F., Awais, M., Shahid, M. Q., & Anjum, S. A. (2010). Effect of nitrogen on seed cotton yield and fiber quality of cotton (*Gossypium hirsutum* L.). *The Journal Animal Plant Science, 20*, 23–27.

Saleem, M. F., Shakeel, A., Bilal, M. F., Shahid, M. Q., & Anjum, S. A. (2010). Effect of different phosphorus levels on earliness and yield of cotton cultivars. *Soil Environ, 29*, 128–135.

Sangakkara, U. R., Frehner, M., & Nosberger, J. (2000). Effect of soil moisture and potassium fertilizer on shoot water potential, photosynthesis and partitioning of carbon in mungbean and cowpea. *Journal of Agronomy and Crop Science, 185*, 201–207. http://dx.doi.org/10.1046/j.1439-037x.2000.00422.x

Sarwar Cheema, M., Akhtar, M., & Nasarullah, M. (2009). Effect of foliar application of mepiquat chloride under varying nitrogen levels on seed cotton yield and yield components. *Journal of Agricultural Research, 47*, 381–388.

Sasthri, G., Thiagarajan, C. P., Srimathi, P., Malarkodi, K., & Venkatasalam, E. P. (2001). Foliar application of nutrient on the seed yield and quality characters of nonaged and aged seeds of cotton cv. MCU5. *Madras Agricultural Journal, 87*, 202–206.

Sawan, Z. M., Basyony, A. E., McCuistion, W. L., & El-Farra, A. A. (1993). Effect of plant population densities and application of growth retardants on cottonseed yield and quality. *Journal of the American Oil Chemists' Society, 70*, 313–317. http://dx.doi.org/10.1007/BF02545314

Sawan, Z. M., Fahmy, A. H., & Yousef, S. E. (2009). Direct and residual effects of nitrogen fertilization, foliar application of potassium and plant growth retardant on Egyptian cotton growth, seed yield, seed viability and seedling vigor. *Acta Ecologica Sinica, 29*, 116–123. http://dx.doi.org/10.1016/j.chnaes.2009.05.008

Sawan, Z. M., Fahmy, A. H., & Yousef, S. E. (2011). Effect of potassium, zinc and phosphorus on seed yield, seed viability and seedling vigor of cotton (*Gossypium barbadense* L.). *Archives of Agronomy and Soil Science, 57*, 75–90. http://dx.doi.org/10.1080/03650340903222328

Sawan, Z. M., & Gregg, B. R. (1993). Influence of growth retardants and plant density on cotton yield and fiber properties. *Ann Bot (Rome), 51*, 33–42.

Sawan, Z. M., Gregg, B. R., & Yousef, S. E. (1998). Influence of nitrogen fertilisation and foliar-applied plant growth retardants and zinc on cotton seed yield, viability and seedling vigour. *Seed Science and Technology, 26*, 393–404.

Sawan, Z. M., Gregg, B. R., & Yousef, S. E. (1999). Effect of phosphorus, chelated zinc and calcium on cotton seed yield, viability and seedling vigour. *Seed Science and Technology, 27*, 329–337.

Sawan, Z. M., Hafez, S. A., Basyony, A. E., & Alkassas, A. R. (2007). Nitrogen, potassium and plant growth retardant effects on oil content and quality of cotton seed. *Grasas Y Aceites, 58*, 243–251.

Sawan, Z. M., Maddah El-Din, M. S., & Gregg, B. (1989). Effect of nitrogen fertilisation and foliar application of calcium and micro-elements on cotton seed yield, viability and seedling vigor. *Seed Science and Technology, 17*, 421–431.

Sawan, Z. M., Mahmoud, M. H., & El-Guibali, A. H. (2006). Response of yield, yield components, and fiber properties of Egyptian cotton (*Gossypium barbadense* L.) to nitrogen fertilization and foliar-applied potassium and mepiquat chloride. Journal of Cotton. *Science, 10*, 224–234.

Sawan, Z. M., & Sakr, R. A. (1990). Response of Egyptian cotton (*Gossypium barbadense*) yield to 1,1-dimethyl piperidinium chloride (Pix). *The Journal of Agricultural Science, 114*, 335–338. http://dx.doi.org/10.1017/S0021859600072725

Scott, M. G., & Peterson, R. L. (1979). The root endodermis in *Ranunculus acris*. I. Structure and ontogeny. *Canadian Journal of Botany, 57*, 1040–1062. http://dx.doi.org/10.1139/b79-129

Sharma, C. P., Sharma, P. N., Bisht, S. S., & Nautiyal, B. D. (1982). Zinc deficiency induced changes in cabbage. P. 601-606. In A. Scaife (Ed.), *Proceeding of the 9th Plant Nutrition Colloquy, Warwick, England* (pp. 22–27). Slough: Aug. Commonwealth Agric. Bur. Farnham House.

Sharma, P. N., Chatterjee, C., Agarwala, S. C., & Sharma, C. P. (1990). Zinc deficiency and pollen fertility in maize (*Zea mays*). *Plant and Soil, 124*, 221–225. http://dx.doi.org/10.1007/BF00009263

Shrivastava, U. K., Tomar, S. P. S., & Singh, D. (1993). Effect of N and Zn fertilization on uptake of nutrients and dry matter production. *Indian Society Cotton Improvement, 18*, 71–74.

Shui, J.-G., & Meng, S.-F. (1990). Effects of lime application on cotton yield in red soil fields. *China Cottons, 1*, 26–29.

Silvertooth, J. C., & Norton, E. R. (1997). *Evaluation of a feedback approach to nitrogen and Pix application, Cotton, A College of Agriculture Report* (Series P-112, 1998, pp. 469–475). Tucson, AZ: University of Arizona.

Snedecor, G. W., & Cochran, W. G. (1980). *Statistical Methods* (7th ed., p. 305). Ames, Iowa, USA: Iowa State University Press.

Speed, T. R., Krieg, D. R., & Jividen, G. (1996). Relationship between cotton seedling cold tolerance and physical and chemical properties. In *Proceedings Beltwide Cotton Conferences, Nashville, TN, USA, January 9–12* (pp. 1170–1171). Memphis, TN: National Cotton Council.

Stuart, B. L., Isbell, V. R., Wendt, C. W., & Abernathy, J. R. (1984). Modification of cotton water relations and growth with mepiquat chloride. *Agronomy Journal, 76*, 651–655. http://dx.doi.org/10.2134/agronj1984.000219620076000 40034x

Taiz, L., & Zeiger, E. (1991). *Plant physiology: Mineral nutrition* (pp. 100–119). Redwood City, CA: The Benjamin Cummings.

Vasudevan, S. N., Virupakshappa, K., Venugopal, N., & Bhaskar, S. (1997). Response of sunflower (*Helianthus annuus*) to phosphorus, sulphur, micronutrients and humic acid under irrigated conditions on red sandy-loam soil. *Indian Journal of Agricultural Science, 67*, 110–112.

Vlasynk, P. A., Zhidkov, V. A., Ivchenkov, V. I., Kibalenko, A. P., Klimovitskaya, Z. M., Okrim, M. F., & Rudakova, E. V. (1978). Trace elements in nutrient metabolism and productivity of plants. *Fiziologia Biokimya kul'turnykh Rastenii, 10*, 350–359.

von Heyendorff-Scheel, R. C., Schott, P. E., & Rittig, F. R. (1983). Mepiquat chloride, a plant growth regulator for cotton. *Zeitschrift fur Pflanzenkrankheiten und Pflanzenschutz, 90*, 585–590.

Wang, Z.-L., Yin, Y.-P., & Sun, X.-Z. (1995). The effect of DPC (N, N-dimethyl piperidinium chloride) on the $^{14}CO_2$-assimilation and partitioning of ^{14}C assimilates within the cotton plants interplanted in a wheat stand. *Photosynthetica, 31*, 197–202.

Wang, Y.-Q., Yang, W.-H., Zhou, D.-Y., & Xu, H.-X. (2003). Study on the relationship between the nutrients of cottonseed and germination and emergence percentage. *Cotton Science, 15*, 109–112.

Welch, R. M. (1995). Micronutrient nutrition of plants. *Critical Reviews in Plant Sciences, 14*, 49–82. http://dx.doi.org/10.1080/07352689509701922

Wiatrak, P. J., Wright, D. L., Marois, J. J., Koziara, W., & Pudelko, J. A. (2005). Tillage and nitrogen application on cotton following wheat. *Agronomy Journal, 97*, 288–293. http://dx.doi.org/10.2134/agronj2005.0288

Wiatrak, P. J., Wright, D. L., & Marois, J. J. (2006). Development and yields of cotton under two tillage systems and nitrogen application following white lupine grain crop. *Journal Cotton Science, 10*, 1–8.

Wright, S. D., Munk, D., Munier, D., Vargas, R., Weir, B., Roberts, B., & Jimenez, Jr., M. (1995). Effect of aminofol/Boll-Set Plus calcium zinc on California cotton. In *Proceedings Beltwide Cotton Conferences, San Antonio, TX, USA, January 4–7* (Vol. 2). Memphis, TN: National Cotton Council.

Wu, F.-B., Wu, L.-H., & Xu, F.-H. (1998). Chlorophyll meter to predict nitrogen sidedress requirement for short-season cotton (*Gossypium hirsutum* L.). *Field Crops Research, 56*, 309–314.

Zeng, Q.-F. (1996). Experimental study on the efficiency of K fertilizer applied to cotton in areas with cinnamon soil or aquic soil. *China Cottons, 23*, 12.

Zhao, D.-L., & Oosterhuis, D. (1999). Physiological, growth and yield responses of cotton to Mepplus and mepiquat chloride. In *Proceedings Beltwide Cotton Conferences, Orlando, Florida, USA, 3–7 January* (pp. 599–602). Memphis, TN: National Cotton Council.

Zhao, D.-L., & Oosterhuis, D. M. (2000). Pix plus and mepiquat chloride effects on physiology, growth, and yield of field-grown cotton. *Journal of Plant Growth Regulation, 19*, 415–422. http://dx.doi.org/10.1007/s003440000018

11

Plant density; plant growth retardants: Its direct and residual effects on cotton yield and fiber properties

Zakaria M. Sawan[1]*

*Corresponding author: Zakaria M. Sawan, Cotton Research Institute, Agricultural Research Center, Ministry of Agriculture and Land Reclamation, 9 Gamaa street, 12619, Giza, Egypt

E-mail: zmsawan@hotmail.com

Reviewing editor: Raffaele Dello Ioio, Universita degli Studi di Roma La Sapienza, UK

Additional information is available at the end of the article

Abstract: Foliar sprays of (PGR's) Cycocel and Alar were applied at concentrations of 250, 500, and 750 ppm after 105 days after plantation (square and boll setting stage) to Egyptian cotton cultivar planted at three plant densities (166.000, 222.000, and 333.000 plant ha^{-1}). Number of opened bolls plant^{-1}, seed-cotton yield plant^{-1}, and earliness increased as plant density decreased in both years, as did seed-cotton and lint yield ha^{-1} in the second season. In the first year, the intermediate plant density gave highest yields. Plant density had no significant effect on lint percentage or fiber properties. Both Cycocel and Alar increased the number of opened bolls plant^{-1}, boll weight, seed and lint indices, seed-cotton yield plant^{-1} and both seed-cotton and lint yield ha^{-1}, but effects were not always significant and response varied for different traits. Neither Cycocel nor Alar affected lint percentage, yield earliness or fiber properties at any plant density.

Subjects: Bioscience; Environment & Agriculture; Environmental Studies and Management

Keywords: alar; boll weight; cycocel; lint yield; micronaire readings; seed and lint indices; yield earliness

ABOUT THE AUTHOR

Effect of different rates and application systems of nitrogen fertilization and indole-3-butyric acid on Egyptian cotton growth and yield. Effect of organic fertilizer and micro-nutrient treatments on cotton agronomy and fiber characteristics of Giza 69. Effect of concentration and time of application of the defoliant Harvade on the lint, seed, protein and oil yields, and oil properties of cottonseed. Effect of 1-naphthalene acetic acid and Kinetin on yield and fiber properties, seed, protein, oil, and fatty acids of Egyptian cotton. Plant growth retardants, plant nutrients, and cotton production. Plant density; plant growth retardants: its direct and residual effects on cotton yield and fiber properties. Plant nutrition and plant growth retardants: their effects on cottonseed, protein, oil yields, and oil properties. Direct and Residual affects of Plant Nutrition's and Plant Growth Retardants, on Cottonseed. Studying the nature relationships between climatic factors and cotton production by different applied statistical methods.

PUBLIC INTEREST STATEMENT

This work confirmed the applicability of some other reports on PGR under Egyptian conditions and indicated that yield components and yield could be improved without affecting fiber properties by applying Cycocel at 500 or 750 ppm or Alar at 250 ppm to a plant density of 166,000 plants ha^{-1}. Yields at higher plant densities could be enhanced by either treatment, but were less than those observed at a plant density of 166,000 plants ha^{-1}. There was a definite correlation between plant density and growth and growth retardants, which suggested that cotton plants produced more when each plant had optimum growing space, that maximum yield depended on an optimum balance of space plant^{-1} vs. number of plants ha^{-1}, and that the yield effect of wider spacing can be enhanced by treatment with growth retardants.

1. Introduction

Chemical may be used to reduce plant size in cotton (*Gossypium barbadense*, L.), which can increase cotton yield by allowing an increased number of plants per unit area. Mondino, Peterlin, and Garay (1999) indicated that to optimize yield, it is necessary to establish a balance between biomass production and harvest index. Short cotton plants necessitate the use of higher plant densities per area unit. Plant size may be reduced genetically or chemically. Plant Growth Regulators, which affects physiological processes using hormones in the plant, can be used to modify plant size. Also, an important objective for using PGR's in cotton is to balance vegetative and reproductive growth as well as to improve lint yield and fiber quality (Zhao & Oosterhuis, 2000). Application of Cycocel and Alar, when plants had at least four fruiting branches, reduced plant height and length of lateral branches (Wang, Chem, & Yu, 1985) They have also been shown to enhance yield-related physiological functions by increasing gross plant photosynthesis or by increasing the retention of bolls by enhanced partitioning of photosynthates to fruiting forms (Guinn, 1984). Treated plants are compact, conical in form (Wang et al., 1985; Zhao & Oosterhuis, 2000) and can be spaced closer to achieve higher plant populations. Also, short, compact, open-canopy plants resulting from such treatments conceivably could improve energy distribution through better light penetration and improve insect control through better insecticide coverage thereby increasing yield.

Koraddi, Modak, Guggari, and Kamath (1993) found that application of 60 ml Cycocel ha^{-1} at 90, 105, and 120 days after sowing increased mean yield of cotton plants. Pipolo, Athayde, Pipolo, and Parducci (1993) found that single and double applications of 25 g ha^{-1} of Cycocel resulted in yield increases of 11.5% and 11.6%, respectively. These treatments also enhanced earliness and seed weight and micronaire. More, Waykar, and Choulwar (1993) found plant height, number of branches, number of leaves plant^{-1}, and number of internodes and internodal length to be significantly decreased when plants were treated with 100, 150, and 200 ppm of Cycocel. Singh and Chouhan (1993) reported cotton yield of a control treatment to be 1.06 t ha^{-1} and to have increased to 1.14 t ha^{-1} when 80 ppm of Cycocel was sprayed once at flower initiation and again 20 days later. Cycocel decreased the percentage of boll shedding and increased net economic return Sawan (2013). Mahmoud, Bondok, and Abdel-halim (1994) found that Cycocel and Alar decreased plant height with application rates of 500 and 5,000 ppm, respectively, when applied at-early growth stages, while late application increased plant height and leaf abscission, but decreased the number of nodes plant^{-1} and number of leaves plant^{-1}.

Bednarz, Bridges, and Brown (2000) indicated that lower cotton population densities resulted in plants with more main-stem nodes and monopodial branches with increased fruit retention, resulting in greater fruit production per plant. They added that mean net assimilation rate from first flower to peak bloom was inversely related to population density. Mohamed, El-Din, and Ragab (1991) and Sawan (2013) found, when cotton was grown at 2, 3, or 4 plants hill^{-1} (166.000, 222.000, and 333.000 plants ha^{-1}, respectively), that increasing plant density decreased number of bolls plant^{-1}, and seed-cotton yield plant^{-1}, but increased yield ha^{-1}. Fiber quality was not significantly affected by plant density. Gannaway, Hake, and Harrington (1995) found that when cotton was grown at 6, 12, 18, and 24 plants m^{-1} of a row, lint gin turnout and boll size decreased, as population increased. Plant population had essentially no effect on fiber length, strength and elongation, but micronaire reading decreased as the population increased. Campanella and Hood (2000) indicated that plots sown at a rate of nine seeds m^{-1} (90.000 ha^{-1}) produced 2–10% more yield, saved 31–66% in sowing costs, and increased profit margins by 7–13%, when compared to sowing rates of 12 and 15 seeds m^{-1}.

Considerable research with Cycocel effects on cotton has been widely reported, but little work has been carried out with Alar. Inadequate information is available on cotton's response to these chemicals under Egyptian growing conditions. Little or no literature was found on interactions between plant density and growth retardant treatments. To fill this gap and confirm the applicability of other work, this study was designed to evaluate the effects of Cycocel and Alar (PGR's) on cotton yield and fiber properties as inter-related to plant density of an Egyptian variety of *G. barbadense* under Egyptian field conditions.

2. Materials and methods

Two field experiments in two consecutive years, I and II) were conducted at Giza Agricultural Research Station (30°N, 31°: 28'E and 19 m altitude), Agricultural Research Center, Egypt, with the Egyptian cotton cv. Giza 75 (*Gossypium barbadense*, L.). The soil texture in both seasons was a clay loam, with an alluvial substratum, (pH = 8.10, 43.0% clay, 28.40% silt, 19.33% fine sand, 4.31% coarse sand, 3.00% calcium carbonate, and 1.83% organic matter). Treatments were arranged in a randomized complete-block design with four replications. Each experiment included 21 treatments, i.e. 3 plant densities × 6 growth retardant concentrations + 1 (0 concentrations as control). The plot size was 3 rows, 4 m long × 0.6 m wide. Plots were planted on April 8 in season I and on April 15 in season II. Seeds were planted in hills 20, 15, and 10 cm apart within the row and thinned to 2 plants hill^{-1} six weeks after planting to achieve plant densities of 166.000, 222.000, and 333.000 plants ha^{-1}, respectively. Plots were foliar-sprayed with 250, 500, and 750 ppm of either 2-chloroethyl trimethyl ammonium chloride (chlormequat chloride, Cycocel, or CCC) or succinic acid 2,2-dimethyl hydrazide (daminozide, SADH, B-Nine, Kylar or Alar). Water alone was sprayed as a control and 166.000 plants ha^{-1} was the control for plant density. This concept is based on the fact that the plant density of 166.000 plants ha^{-1} is the recommended density in Egypt. Treatments were foliar-sprayed 105 days after planting, which was during the square-and-boll setting stage. The Cycocel and Alar were both applied to the leaves with uniform coverage using a knapsack sprayer. The solution volume was 960 L ha^{-1} for all treatments. The application was carried out between 09.00 and 11.00 h. Total irrigation amount during both growing seasons (surface irrigation) was about 6,000-m^3 ha^{-1}. No rainfall occurred during the two growing seasons. The first irrigation was applied 3 weeks after sowing, and the second one was 3 weeks after that. Thereafter, the plots were irrigated every 2 weeks until the end of the season, thus providing a total of nine irrigations. Phosphorus fertilizer was applied at the rate of 24 kg P ha^{-1} as calcium super-phosphate during land preparation (the recommended level for semi-fertile soil). Potassium fertilizer was applied at the rate of 47 kg K ha^{-1} as potassium sulfate before the first irrigation. Nitrogen fertilizer was applied at the rate of 144 kg N ha^{-1} as ammonium nitrate in two equal doses; the first one was applied after thinning just before the second irrigation and the other one was applied before the third irrigation. Fertilizer application (N, P, and K), along with pest and weed management was carried out according to local practices performed at the Experiment Research Station.

Ten plants were randomly chosen from the central row of each plot to determine plant height and the number of opened bolls per plant and boll weight (g seed cotton boll^{-1}), seed-cotton yield (g plant^{-1}), and plant height (cm). Earliness as percentage of the yield harvested in the first picking was calculated as follows: seed-cotton yield in first picking divided by the total seed cotton yield and multiplied by 100. First hand-pickings took place on September 25th and 30th and the final pickings on October 10th and 15th in seasons I and II, respectively. Total seed cotton yield of each plot (including the 10 plant sub samples) was ginned on a laboratory roller gin stand to determine lint yield (kg lint he^{-1}), lint percent, seed index (g per100 seeds), and lint index (g of lint per100 seeds). Fiber tests were made using single instrumentation at a relative humidity of 65 ± 2% and a temperature of 20 ± 1°C. A digital fibrograph was used to determine fiber length in terms of 2.5 and 50% span length (millimeters) and uniformity ratio. Micronaire readings, including a measurement of fiber fineness and maturity, were done on a Micronaire instrument, on harvested fiber. Flat bundle strength (in g tex^{-1}) was determined using a Pressly tester (according to the method of the American Society for Testing Materials (ASTM), 1979) (Sawan, Mahmoud, & Fahmy, 2011).

Results were statistically analyzed as a factorial experiment in a randomized complete-block design with four replications for the studied characters each year, following the procedure outlined by Snedecor and Cochran (1980). The Least Significant Difference (LSD) test method, at 5% level of significance, was used to verify the significance of differences among treatment means and the interactions.

Table 1. Average values of plant height and yield components as affected by plant densities and plant growth retardants

Character	Plant height (cm)		Number of open bolls m^{-2}		Number of open bolls plant^{-1}		Boll weight (g)	
Season	I	II	I	II	I	II	I	II
Treatments:								
Plant density (plant ha^{-1})								
166	108.4	105.6	121.7	131.2	10.52	10.99	2.232	2.396
222	111.8	107.2	125.2	125.9	8.21	8.43	2.205	2.356
333	114	109.6	119.3	123.8	7.4	7.27	2.162	2.291
LSD (p = 0.05)	0.75	0.87	5.01	4.96	0.36	0.37	0.037	0.039
Cycocel & Alar concentration (ppm)								
Control, 0	117.5	112	113.1	116.4	7.96	8.18	2.164	2.297
Cycocel, 250	113.4	108.5	121.4	128.3	8.57	9	2.191	2.299
Cycocel, 500	110.1	106.9	121	130	8.55	9.24	2.198	2.308
Cycocel, 750	106.2	103.6	128.7	132.3	8.97	9.28	2.208	2.355
Alar, 250	114.1	109.3	129.1	127	9.02	8.96	2.19	2.42
Alar, 500	111.3	107.2	120.9	127.4	8.52	8.86	2.218	2.378
Alar, 750	107.1	104.8	120.3	127.3	8.51	8.78	2.226	2.376
LSD (p = 0.05)	1.15	1.33	7.66	7.57	0.55	0.57	NS	0.06

3. Results and discussion

3.1. Plant height
Both Cycocel and Alar significantly decreased plant height in both years (Table 1), with the most pronounced effect at the highest concentration (750 ppm). Cycocel and Alar, therefore, could be used to control the vegetative development of cotton plants. This corroborates the research of Wang et al. (1985), who report that these two PGRs reduce plant height and length of lateral branches. Plant height significantly increased as plant density increased in both years (Table 1). Similar results were obtained by Alfageih, Baswaid, and Atroosh (2001).

3.2. Yield components

3.2.1. Number of open bolls m^{-2} and plant^{-1}
Number of opened bolls m^{-2} and plant^{-1} were significantly greater on plants treated with Cycocel and Alar in both years, compared with the untreated control (Table 1), but there was no difference between application rates. Increased number of opened bolls m^{-2} and plant^{-1} may be due to increased photosynthetic activity of leaves following application of Cycocel and Alar (Gardner, 1988; Wu, Chen, Yuan, & Hu, 1985). Increased photosynthesis has been shown to greatly increase flowering and boll retention (Kler, Raj, & Dhillon, 1989; Wang et al., 1985). Also, favorable metabolite balance may suppress vegetative growth, which increases penetration of photosynthetically active radiation into the canopy. These results agree with those of Singh and Chouhan (1993), where Cycocel decreased percentage of boll shedding and increased net economic return after its application at 80 ppm at flower initiation and 20 days later. Our findings also agree with those of Sawan, Mahmoud, and Momtaz (1997), when Cycocel and Alar were sprayed at 300 ppm 75 days after plantation, and Sawan (2013).

Plant density was significantly and inversely related to the number of opened bolls m^{-2} and plant^{-1} in both years, except in the first year when the highest number of opened bolls m^{-2} was obtained at the intermediate plant density, followed by the lowest density. As to reduce plant density increasing the number of opened bolls m^{-2} and plant^{-1}, Guinn (1984), studying factors influencing abscission of floral buds and bolls, found that conditions that decrease photosynthesis might delay fruiting and reduce retention of squares and bolls. Bednarz et al. (2000) found lower cotton population densities lead to the production of plants with more main-stem nodes and monopodial branches with increased fruit retention, resulting in greater fruit production per plant. More opened bolls plant^{-1} has been reported at low population levels by Abd-El-Malik and El-Shahawy (1999), and Palomo Gil, Gaytan Mascorro, and Godoy Avila (2000).

3.2.2. Boll weight
Alar increased boll weight above the control in second year at all rates; Cycocel did not show significant effects (Table 1). Similar results were obtained by Sawan et al. (1997) and Sawan (2013) with Cycocel and Alar, and by Karthikeyan and Jayakumar (2001) and Lamas (2001) with Cycocel.

Decreasing plant density was accompanied by an increase in boll weight in both years. The effects of plant density may be due to heavy shading by upper leaves in high plant populations, which could limit photosynthate availability to, and growth of, bolls on the lower portion of the plant. In this connection, Bednarz et al. (2000) stated that mean net assimilation rate from first flower to peak bloom and boll size were inversely related to population density. These results agree with those of Gannaway et al. (1995) and Abd-El-Malik and El-Shahawy (1999).

3.2.3. Lint percentage
Lint percentage was not affected by Cycocel or Alar (Table 2). This result agrees with the finding of Sawan et al. (1997) and Sawan (2013) for Cycocel and Alar, and Lamas (2001) for Cycocel, although a reduction in lint percentage due to Cycocel application was observed by Carvalho et al. (1994). Plant density also had no significant effect in either year, supporting the results of Jones, Snipes, and Tupper (2000).

3.2.4. Seed index
Seed index significantly increased in both years due to both Cycocel and Alar applications compared with the control (Table 2). In the first year, there was no difference between rates of either PGR;

Table 2. Average values of yield components as affected by plant densities and plant growth retardants

Character	Lint (%)		Seed index (g)		Lint index (g)	
Season	I	II	I	II	I	II
Treatments:						
Plant density (plant ha^{-1})						
166.000	35.14	33.61	10.34	11.48	5.61	5.81
222.000	35.37	33.64	10.22	11.37	5.59	5.76
333.000	35.57	33.75	10.10	11.23	5.58	5.72
LSD (p = 0.05)	NS	NS	0.07	0.11	NS	NS
Cycocel & Alar concentration (ppm)						
Control, 0	35.13	33.66	10.09	10.95	5.46	5.56
Cycocel, 250	35.47	34.00	10.26	11.02	5.64	5.67
Cycocel, 500	35.12	33.60	10.23	11.02	5.54	5.58
Cycocel, 750	35.08	33.77	10.27	11.22	5.55	5.73
Alar, 250	35.37	33.40	10.20	11.64	5.59	5.84
Alar, 500	35.64	33.48	10.23	11.82	5.66	5.95
Alar, 750	35.73	33.78	10.26	11.82	5.70	6.03
LSD (p = 0.05)	NS	NS	0.10	0.17	0.12	0.17

however, in the second year, higher rates of both PGR's increased seed index. This indicates that treated cotton bolls had larger carbohydrates and other metabolites than untreated bolls (Gardner, 1988; Wu et al., 1985). This agrees with previous works of Pipolo et al. (1993) for Cycocel, Sawan et al. (1997) and Sawan (2013) for Cycocel and Alar, and Lamas (2001) for Cycocel.

Lower plant density significantly increased seed index in both years probably due to improved light penetration at the lower plant density (166.000 plants ha^{-1}). Similar results were obtained by Singh and Warsi (1985).

3.2.5. Lint index

Compared with the untreated control, lint index was significantly increased on plants treated with Alar in both years (Table 2), with the most pronounced effect being at the highest concentration of 750 ppm. Also, Cycocel significantly increased lint index when applied at 250 ppm in the first year and at 750 ppm in the second year. The significant increase in lint index due to both Cycocel and Alar application was similar to the results of Kler et al. (1989) for Cycocel and Sawan et al. (1997) and Sawan (2013) for both Cycocel and Alar. Lint index was not significantly affected by plant density in either year. Lack of effects due to plant density for lint index agreed with the findings of Singh and Warsi (1985).

3.3. Yield

Both Cycocel and Alar significantly increased seed-cotton yield plant^{-1} and seed-cotton and lint yields ha^{-1} in both years (Table 3), with no difference among application rates of either chemical. Increased seed-cotton yield plant^{-1}and seed-cotton and lint yields ha^{-1} due to Cycocel and Alar may be attributed to the stimulatory effect of these substances on certain physiological processes, leading to improvement of all the yield components. Cycocel and Alar have also been associated with increased photosynthesis (Gardner, 1988; Wu et al., 1985) through increased total chlorophyll concentration of plant leaves. In these studies, increased photosynthesis greatly increased flowering, boll retention, and yield. These results agree with those obtained by Sawan et al. (1997) and Sawan (2013) with Cycocel and Alar, and by Karthikeyan and Jayakumar (2001) and Lamas (2001) with Cycocel.

Table 3. Average values of yield and yield earliness as affected by plant densities and plant growth retardants

Character	Seed cotton yield (g plant^{-1})		Seed cotton yield (kg ha^{-1})		Lint yield (kg ha^{-1})		Yield earliness (%)	
Season	I	II	I	II	I	II	I	II
Treatments:								
Plant density (plant ha^{-1})								
166.000	23.48	26.29	2,717.6	3136.8	955.1	1053.8	83.13	78.89
222.000	18.10	19.86	2,759.8	2963.8	976.7	997.8	81.98	74.01
333.000	15.20	16.66	2,577.7	2836.9	917.7	965.4	75.18	68.94
LSD (p = 0.05)	0.88	0.92	116.3	123.0	42.5	43.8	6.02	4.73
Cycocel & Alar concentration (ppm)								
Control, 0	17.25	18.80	2,447.5	2,672.5	860.3	901.2	81.30	73.85
Cycocel, 250	18.78	20.72	2,660.4	2,948.1	943.7	1,002.8	79.02	72.99
Cycocel, 500	18.83	21.32	2,661.2	2,999.5	934.7	1,023.1	79.44	72.58
Cycocel, 750	19.86	21.94	2,843.6	3,117.9	997.7	1,053.4	81.12	72.53
Alar, 250	19.78	21.76	2,820.7	3,075.6	998.6	1,027.3	80.02	75.08
Alar, 500	18.96	21.13	2,684.2	3,023.9	956.7	1,012.4	79.14	74.86
Alar, 750	19.04	20.89	2,677.8	3,016.5	956.9	1,019.3	80.62	75.68
LSD (p = 0.05)	1.35	1.41	177.7	187.9	64.9	66.9	NS	NS

Seed-cotton yield plant^{-1}, and seed-cotton and lint yields ha^{-1} significantly decreased as plant density increased in both years, with two exceptions in the first year, where maximum seed-cotton and lint yields ha^{-1} were obtained from the intermediate plant density, followed by the lowest density. Increase in plants per unit area with dense planting did not nullify the increase in seed-cotton yield plant^{-1} under wide hill spacing. Decreased seed-cotton yield plant^{-1}, and seed-cotton and lint yields ha^{-1}, as plant density increased, may be due to the fact that the increase in plants per unit area with dense planting could not compensate for the decrease in yield components plant^{-1} under narrow hill spacing. A consequence of high plant densities was that many plants were barren of bolls, which decreased fruiting efficiency; such decrease due to a lower percentage of assimilates being converted into fruits. Chhabra and Bishnoi (1993), when planting two cotton cultivars at spacing of 60×30, 60×22.5 and 60×15 cm, found that with narrow spacing, dry matter accumulation in stems, leaves and reproductive parts at all the crop stages was decreased. Increased seed-cotton yield under low plant density was reported by Abd-El-Malik and El-Shahawy (1999), Campanella and Hood (2000), and Palomo Gil et al. (2000).

3.4. Yield earliness
No significant difference was noted in yield earliness (as percentage of the yield harvested in the first picking) due to Cycocel or Alar in either year (Table 3). Similar lack of influence on yield earliness for Cycocel or Alar was obtained by Sawan et al. (1997) and Sawan (2013), but Abdel Al and Eid (1985) found that in one out of two seasons, Alar increased earliness; however, Pipolo et al. (1993) indicated that application of Cycocel 70 days after emergence with 25–100 g ha^{-1} and given also two extra sprays 15 days apart (85 and 100 days after emergence) increased earliness of yield.

Yield earliness was inversely related to plant density probably due to more light penetration to, and air movement around, plants, thus causing earlier boll opening. Abd-El-Malik and El-Shahawy (1999) and Alfageih et al. (2001) reported a similar inverse relationship of yield earliness to plant density.

3.5. Fiber properties
Neither PGRs (Cycocel and Alar) nor plant density had a significant effect on any of the fiber properties investigated (Table 4). There may be specific hormones for cotton fiber, so failure to improve fiber properties consistently may simply be due to failure to test the right compound; potential chemical improvement of fiber length should not be ignored. These results agree with those of Sawan et al. (1997) and Sawan (2013) who reported that Cycocel and Alar did not affect cotton fiber quality. Karthikeyan and Jayakumar (2001) suggested that fiber quality parameters were unaffected by Cycocel application. Mohamed et al. (1991) found that fiber quality was not affected by plant density. Gannaway et al. (1995) found that plant population had essentially no effect upon fiber length and strength, but micronaire reading decreased as population increased. Pettigrew and Johnson (2005) stated that few fiber quality differences were detected among PGR application rates and seedling rates.

3.6. Interactions
Interactions effects between plant density and Cycocel or Alar relative to number of opened bolls m^{-2} and plant^{-1}, seed-cotton yield plant^{-1}, seed-cotton, and lint yield ha^{-1} in both years were noted (Tables 5 and 6), but not for the other measured properties (Tables 7–10). The highest number of opened bolls m^{-2} and plant^{-1}, seed-cotton yield plant^{-1}, seed-cotton, and lint yield ha^{-1} was obtained at the lowest plant densities when either Cycocel or Alar was applied. The plant density × growth retardants interaction implies that effects on cotton yield depend essentially on the space available to each plant, and that optimum plant spacing exists and cotton yield can be affected by growth retardant treatment. Favorable effects of low plant density on number of opened bolls plant^{-1} and seed-cotton yield plant^{-1}, and low or intermediate plant density on number of opened bolls m^{-2}, seed-cotton and lint yields ha^{-1} were greater when treated with Cycocel or Alar, especially Cycocel at 500 or 750 ppm or Alar at 250 ppm. The interaction of plant density × growth retardants was significant for number of opened bolls m^{-2} and plant^{-1} seed-cotton yield plant^{-1} and ha^{-1} and lint

Table 4. Average values of fiber properties as affected by plant densities and plant growth retardants

Character	2.5% span length (mm)		50% span length (mm)		Uniformity ratio (%)		Micronaire reading		Flat bundle strength (g tex^{-1})	
Season	I	II	I	II	I	II	I	II	I	II
Treatments:										
Plant density (plant ha^{-1})										
166.000	30.99	30.89	14.78	14.35	47.70	46.48	4.38	4.46	10.55	10.72
222.000	31.09	31.01	15.01	14.63	48.27	47.19	4.34	4.45	10.64	10.75
333.000	31.14	31.04	14.86	14.48	47.72	46.60	4.35	4.38	10.47	10.68
LSD (p = 0.05)	NS	NS	NS	NS	NS	NS	NS	NS	NS	NS
Cycocel & Alar concentration (ppm)										
Control, 0	30.96	30.89	14.78	14.38	47.70	46.53	4.32	4.38	10.37	10.62
Cycocel, 250	31.22	30.89	14.86	14.33	47.58	46.40	4.42	4.42	10.42	10.80
Cycocel, 500	31.06	30.99	14.96	14.53	48.16	46.85	4.33	4.38	10.57	10.72
Cycocel, 750	31.16	30.91	14.91	14.43	47.81	46.64	4.38	4.39	10.52	10.84
Alar, 250	31.19	31.01	14.99	14.58	48.05	47.03	4.31	4.46	10.66	10.69
Alar, 500	30.96	31.16	14.91	14.55	48.11	46.74	4.40	4.46	10.65	10.68
Alar, 750	30.96	31.06	14.81	14.66	47.84	47.14	4.36	4.54	10.68	10.63
LSD (p = 0.05)	NS	NS	NS	NS	NS	NS	NS	NS	NS	NS

Table 5. Effect of interaction between plant densities and plant growth retardants on number of open bolls m^{-2} and plant^{-1}

Character	Number of open bolls m^{-2}			Number of open bolls plant^{-1}		
Cycocel and Alar concentration (ppm)	Plant density (plants ha^{-1})					
	166.000	222.000	333.000	166.000	222.000	333.000
Season I						
Control, 0	115.75	109.33	114.08	9.98	7.16	6.74
Cycocel, 250	127.05	113.63	123.58	10.99	7.44	7.26
Cycocel, 500	126.78	114.93	121.33	10.96	7.75	7.16
Cycocel, 750	117.10	144.98	123.93	10.12	9.50	7.30
Alar, 250	118.38	147.10	121.68	10.22	9.66	7.18
Alar, 500	121.80	126.33	114.70	10.52	8.28	6.76
Alar, 750	125.20	120.13	115.58	10.83	7.88	6.83
LSD (p = 0.05)		13.26			0.97	
Season II						
Control, 0	120.08	119.55	109.60	10.08	8.01	6.44
Cycocel, 250	134.70	125.28	124.80	11.28	8.40	7.32
Cycocel, 500	147.63	132.75	109.73	12.36	8.91	6.44
Cycocel, 750	141.40	122.95	132.58	11.85	8.22	7.78
Alar, 250	138.45	122.50	119.98	11.61	8.22	7.04
Alar, 500	120.78	131.20	130.10	10.14	8.79	7.64
Alar, 750	115.15	126.80	140.05	9.63	8.49	8.24
LSD (p = 0.05)		13.11			0.99	

Table 6. Effect of interaction between plant densities and plant growth retardants on cotton yield

Character	Seed cotton yield (g plant⁻¹)			Seed cotton yield (kg ha⁻¹)			Lint yield (kg ha⁻¹)		
Cycocel & Alar concentration (ppm)	Plant density (plants ha⁻¹)								
	166.000	222.000	333.000	166.000	222.000	333.000	166.000	222.000	333.000
Season I									
Control, 0	21.78	15.52	14.45	2525.9	2370.6	2446.0	881.6	840.2	859.1
Cycocel, 250	24.11	16.33	15.89	2785.7	2492.9	2702.4	976.4	887.2	967.4
Cycocel, 500	24.23	16.68	15.57	2803.0	2541.3	2639.2	977.0	891.1	936.1
Cycocel, 750	22.54	21.34	15.70	2608.8	3258.8	2663.2	903.5	1144.4	945.2
Alar, 250	22.90	21.17	15.27	2652.3	3222.7	2587.0	941.2	1144.8	909.7
Alar, 500	23.70	18.44	14.72	2745.2	2810.6	2496.9	967.2	1007.9	895.2
Alar, 750	25.11	17.19	14.82	2902.6	2621.6	2509.3	1038.5	921.1	911.1
LSD ($p = 0.05$)		2.35			307.8			112.4	
Season II									
Control, 0	23.40	18.40	14.60	2788.4	2745.2	2484.1	948.4	925.5	829.8
Cycocel, 250	26.22	19.37	16.58	3130.5	2889.3	2824.4	1056.9	982.9	968.7
Cycocel, 500	28.63	20.58	14.76	3418.9	3066.5	2513.2	1135.2	1037.6	896.7
Cycocel, 750	28.44	19.52	17.85	3393.1	2917.3	3043.3	1135.1	988.4	1036.6
Alar, 250	28.76	19.83	16.68	3429.9	2956.0	2841.0	1145.8	987.6	948.6
Alar, 500	25.07	20.97	17.35	2987.4	3130.7	2953.7	1005.9	1038.3	992.6
Alar, 750	23.50	20.36	18.80	2809.7	3041.4	3198.6	949.0	1024.1	1084.8
LSD ($p = 0.05$)		2.44			325.4			115.9	

Table 7. Mean squares from analysis of variance of plant height and yield components

Source	d.f.	Plant height (cm)		Number of open bolls m⁻²		Number of open bolls plant⁻¹		Boll weight (g)	
		I	II	I	II	I	II	I	II
Replicates	3	0.87	2.70	515.8**	1210.9**	2.93**	7.06**	0.0048	0.0134
PD	2	223.59**	110.67**	249.1	401.8*	87.79**	101.44**	0.0345**	0.0777**
GR	6	192.79**	93.23**	359.1**	302.7**	1.49**	1.62**	0.0051	0.0272**
PD × GR	12	1.73	1.31	365.5**	469.2**	1.74**	4.48**	0.0038	0.0061
Error	60	1.99	2.66	88.0	85.9	0.47	0.49	0.0050	0.0056

*Significant at 0.05 probability level.
**Significant at 0.01 probability level.

Table 8. Mean squares from analysis of variance of yield components

Source	d.f.	Lint percentage (%)		Seed index (g)		Lint index (g)	
		I	II	I	II	I	II
Replicates	3	6.457**	0.798	0.2114**	1.7103**	0.1347**	0.1931*
PD	2	1.309	0.152	0.3904**	0.4275**	0.0050	0.0507
GR	6	0.844	0.489	0.0482*	1.8459**	0.0840**	0.3911**
PD × GR	12	0.419	0.275	0.0080	0.0470	0.0294	0.0274
Error	60	0.474	0.515	0.0174	0.0457	0.0249	0.0469

*Significant at 0.05 probability level.
**Significant at 0.01 probability level.

Table 9. Mean squares from analysis of variance of on yield and yield earliness in cotton

Source	d.f.	Seed cotton yield (g plant⁻¹)		Seed cotton yield (kg ha⁻¹)		Lint yield (kg ha⁻¹)		Yield earliness (%)	
		I	II	I	II	I	II	I	II
Replicates	3	10.76*	29.11**	166076*	497098**	41404**	70743**	548.2**	1875.0**
PD	2	494.27**	673.27**	254390**	634835**	24955*	55925**	517.2*	692.1**
GR	6	8.92**	12.97**	202407**	254719**	26076**	28391**	10.6	19.9
PD × GRP	12	9.64**	11.10**	2022300**	205445**	26396**	18851**	43.7	26.3
Error	60	2.77	2.99	47385	52962	6318	6717	126.9	78.6

*Significant at 0.05 probability level.
**Significant at 0.01 probability level.

Table 10. Mean squares from analysis of variance of fiber properties of cotton

Source	d.f.	2.5% span length (mm)		50% span length (mm)		Uniformity ratio (%)		Micronaire reading		Flat bundle strength (g tex⁻¹)	
		I	II	I	II	I	II	I	II	I	II
Replicates	3	0.0024**	0.0020**	0.0034**	0.0015**	8.15**	4.54*	0.025	0.018	1.322**	0.147
PD	2	0.0002	0.0003	0.0006	0.0009	2.92	4.02	0.016	0.061	0.189	0.033
GR	6	0.0002	0.0002	0.0001	0.0003	0.57	0.84	0.024	0.042	0.180	0.079
PD × GR	12	0.0003	0.0004	0.0002	0.0004	0.87	2.59	0.001	0.019	0.218	0.084
Error	60	0.0002	0.0002	0.0004	0.0003	1.56	1.32	0.012	0.030	0.182	0.154

*Significant at 0.05 probability level.
**Significant at 0.01 probability level.

yield ha^{-1}. The lowest plant density of 166.000 plants ha^{-1} produced higher seed-cotton yield plant^{-1} when either Cycocel or Alar was applied. The highest seed-cotton yield plant^{-1} was actually obtained in the first year when 750 ppm Alar was applied to plant density of 166.000 plants ha^{-1} (Table 6). In the second year, the highest seed-cotton yield plant^{-1} was attained from 250 ppm Alar applied to plant density of 166.000 plants ha^{-1}. Highest seed-cotton yield ha^{-1} in the first year was from Cycocel at 750 ppm applied to plant density of 222.000 plants ha^{-1}. Highest lint yield ha^{-1} in the first year was from Alar at 250 ppm or Cycocel at 750 ppm applied to plant density of 222.000 plants ha^{-1}. In the second year, highest seed-cotton yield ha^{-1} and lint yield ha^{-1} were from applying Alar at 250 ppm and Cycocel at 500 ppm, respectively, to the lowest plant density of 166.000 plants ha^{-1} (Table 6).

Interactions effects between plant density and Cycocel or Alar (highest yields at lowest plant density combined with Cycocel or Alar) on seed-cotton yield plant^{-1} and seed-cotton and lint yields ha^{-1}, could logically be expected, as the growth retardants should have the effect of adding extra space between plants while maintaining the optimum population for best yield.

4. Conclusions

This work confirmed the applicability of some other reports on PGR under Egyptian conditions and indicated that yield components and yield could be improved without affecting fiber properties by applying Cycocel at 500 or 750 ppm or Alar at 250 ppm to a plant density of 166.000 plants ha^{-1}. Yields at higher plant densities could be enhanced by either treatment, but were less than those observed at a plant density of 166.000 plants ha^{-1}. There was a definite correlation between plant density and growth and growth retardants, which suggested that cotton plants produced more when each plant had optimum growing space, that maximum yield depended on an optimum balance of space plant^{-1} vs. number of plants ha^{-1}, and that the yield effect of wider spacing can be enhanced by treatment with growth retardants.

Funding
The authors received no direct funding for this research.

Competing Interests
The authors declare no competing interest.

Author details
Zakaria M. Sawan[1]
E-mail: zmsawan@hotmail.com
[1] Cotton Research Institute, Agricultural Research Center, Ministry of Agriculture and Land Reclamation, 9 Gamaa Street, 12619, Giza, Egypt.

References

Abdel Al, M. H., & Eid, E. T. (1985). Effect of some growth retardants on growth, yield and some chemical constituents of cotton plants. *Annals of Agricultural Science, Moshtohor, Zagazig University, 23*, 41–57.

Abd-El-Malik, R. R., & El-Shahawy, M. I. M. (1999). Impact of plant population density through row and hill spacing under different nitrogen levels on Giza 89 cotton cultivar. *Egyptian Journal of Agricultural Research, 77*, 1287–1300.

Alfageih, F. M., Baswaid, A. S., & Atroosh, K. B. (2001). Effect of plant density on some agronomic characters and growth development of medium-staple cotton. *University of Aden Journal of Natural and Applied Sciences, 5*, 35–44.

Annual Book of ASTM Standards. (1979). *Part 33: Textiles, Fibers, Zippers*. Philadelphia, PA: American Society for Testing and Materials.

Bednarz, C. W., Bridges, D. C., & Brown, S. M. (2000). Analysis of cotton yield stability across population densities. *Agronomy Journal, 92*, 128–135. http://dx.doi.org/10.2134/agronj2000.921128x

Campanella, R., & Hood, K. B. (2000). Patterns among seeding rates, remotely sensed data, and yield on a Mississippi delta cotton farm. In *Proceedings Beltwide Cotton Conferences, January 4–8, San Antonio, TX, USA* (pp. 421–426). Memphis, TN: National Cotton Council.

Carvalho, L. H., Chiavegato, E. J., Cia, E., Kondo, J. I., Sabino, J. C., Pettinelli Júnior, A., ... Gallo, P. B. (1994). Plant growth regulators and pruning in cotton crop. *Bragantia, 53*, 247–254. http://dx.doi.org/10.1590/S0006-87051994000200014

Chhabra, K. L., & Bishnoi, K. C. (1993). Response of American cotton varieties to plant spacing and nitrogen levels on growth characters. *Journal of Cotton Research and Development, 7*, 101–109.

Gannaway, J. R., Hake, K., & Harrington, R. K. (1995). Influence of plant population upon yield and fiber quality. In *Proceedings Beltwide Cotton Conferences, January 4–7, San Antonio, TX, USA* (pp. 551–556). Memphis, TN: National Cotton Council.

Gardner, F. P. (1988). Growth and partitioning in peanut as influenced by gibberellic acid and daminozide. *Agronomy Journal, 80*, 159–163. http://dx.doi.org/10.2134/agronj198 8.00021962008000020004x

Guinn, G. (1984). Boll abscission in cotton. In U. S. Gupta (Ed.), *Crop Physiology: Advancing Frontiers* (pp. 177–225). New Delhi: Mohan Primlani for Oxford& IBH Publishing.

Jones, M. A., Snipes, C. E., & Tupper, G. R. (2000). Management systems for transgenic cotton in ultra-narrow rows. In *Proceedings Beltwide Cotton Conferences, January 4–8, San Antonio, TX, USA* (pp. 714–716). Memphis, TN: National Cotton Council.

Karthikeyan, P. K., & Jayakumar, R. (2001). Nitrogen and chlormequatchloride on cotton cultivar. In W. J. Horst, M. K. Schenk, A. Burkert, N. Claassen, H. Flessa, W. B. Frommer, ... L. Wittenmayer (Eds.), *Plant nutrition: Food security and* sustainability of agro-ecosystems through basic and applied research. Fourteenth International Plant Nutrition colloquium (pp. 806–807). Dordrecht: Kluwer Academic Publishers.

Kler, D. S., Raj, D., & Dhillon, G. S. (1989). Modification of micro-environment with cotton canopy for reduced abscission and increased seed yield. Environment and Ecology, 7, 800–802.

Koraddi, V. R., Modak, S. B., Guggari, A. K., & Kamath, K. S. (1993). Studies on efficient utilization of rainwater and soil moisture in rain fed cotton. Journal of Maharashtra agricultural universities, 18, 27–29.

Lamas, F. M. (2001). Comparative study of mepiquat chloride and chlormequat chloride application in cotton. Pesquisa Agropecuária Brasileira, 36, 265–272. http://dx.doi.org/10.1590/S0100-204X2001000200008

Mahmoud, M. M., Bondok, M. A., & Abdel-halim, M. A. (1994). The control of flowering in cotton plants in relation to induced growth correlations. 1-The use of some growth regulators and N levels on vegetative and reproductive growth. Annals of Agricultural Science, Cairo, 39, 1–19.

Mohamed, M. K., El-Din, G. M. S., & Ragab, M. T. (1991). Effect of plant density and defoliation on yield, yield components and fiber quality of Giza 75 cotton variety. Annals of Agricultural Science, Moshtohor, Zagazig University, 29, 1285–1298.

Mondino, M. H., Peterlin, O., & Garay, F. (1999). Optimization of yield of cotton (Gossypium hirsutum L.) by means of management of growth control with different combinations of densities and regulation. In Anais II Congresso Brasileiro de Algodao: O algodao no seculo XX, perspectivas para o seculo XXI (pp. 100–103). Ribeirao Preto, SP, Brasil, 5–10 Setembro, Campina Grande: Empresa Brasileira de Pesquisa Agropecuaia, Embrapa Algodao.

More, P. R., Waykar, S. K., & Choulwar, S. B. (1993). Effect of Cycocel (CCC) on morphological and yield contributing characters of cotton. Journal of Maharashtra agricultural universities, 18, 294–295.

Palomo Gil, A., Gaytan Mascorro, A., & Godoy Avila, S. (2000). Response of cotton cultivars to plant density. I. Yield and yield components. ITEA Produccion Vegetal, 96, 95–102.

Pettigrew, W. T., & Johnson, J. T. (2005). Effects of different seedlings rates and plant growth regulators on early-planted cotton. The Journal of Cotton Science, 9, 189–195.

Pipolo, A. E., Athayde, M. L., Pipolo, V. C., & Parducci, S. (1993). Comparison of different rates of chlorocholine choloride applied to herbaceous cotton. Pesquisa Agropecuaria Brasileira, 28, 915–923.

Sawan, Z. M. (2013). Plant growth retardants, plant nutrients, and cotton production. Communications in Soil Science and Plant Analysis, 44, 1353–1398. http://dx.doi.org/10.1080/00103624.2012.756509

Sawan, Z. M., Mahmoud, H. M., & Momtaz, O. (1997). Influence of nitrogen fertilization and foliar application of plant growth retardants and zinc on quantitative and qualitative properties of Egyptian cotton (Gossypium barbadense L. Var. Giza 75). Journal of Agricultural and Food Chemistry, 45, 3331–3336. http://dx.doi.org/10.1021/jf950817p

Sawan, Z. M., Mahmoud, H. M., & Fahmy, A. H. (2011). Cotton (Gossypium barbadense, L.) yield and fiber properties as affected by plant growth retardants and plant density. Journal of Crop Improvement, 21, 171–189.

Singh, I., & Chouhan, G. S. (1993). Effect sowing time, cycocel spray and nitrogen fertilization on production potential of upland cotton (G. hirsutum). Indian Journal of Agronomy, 38, 193–196.

Singh, J., & Warsi, A. S. (1985). Effect of sowing dates, row spacings and nitrogen levels on ginning out-turn and its components in hirsutum cotton. Indian Journal of Agronomy, 30, 263–264.

Snedecor, G. W., & Cochran, W. G. (1980). Statistical Methods (7th ed.). Ames, IA: Iowa State University Press.

Wang, J.-X., Chem, W.-H., & Yu, Y.-L. (1985). The yield increasing effect of growth regulators on cotton and their application. China Cottons, 3, 32–33.

Wu, L.-Y., Chen, Z.-H., Yuan, Y.-N., & Hu, M.-Y. (1985). A comparison of physiological effects of DPC and CCC on cotton plants. Plant Physiology Communications, 5, 17–19.

Zhao, D., & Oosterhuis, D. M. (2000). Pix plus and mepiquat chloride effects on physiology, growth, and yield of field-grown cotton. Journal of Plant Growth Regulation, 19, 415–422. http://dx.doi.org/10.1007/s003440000018

12

Whitefly species efficiency in transmitting cassava mosaic and brown streak virus diseases

Moffat K. Njoroge[1,2]*, D.L. Mutisya[3], D.W. Miano[2] and D.C. Kilalo[2]

*Corresponding author: Moffat K. Njoroge, Department of Crops, Ministry of Agriculture and Livestock, P.O. Box 16, Kitui, Kenya

E-mail: njorogemoffat@yahoo.com

Reviewing editor: Bernhard Lieb, Johannes Gutenberg Universitat Mainz, Germany

Additional information is available at the end of the article

Abstract: Whiteflies are vectors of plant viral diseases. The rate of disease transmission by whiteflies on cassava continue to present a complex of efficiency in relation to species diversity. An experiment was carried out to access the period required for viruliferous whitefly to feed on cassava plant for it to transmit both cassava mosaic and brown streak diseases (CMD and CBSD) as well as the number of whiteflies required for transmission of the pathogens. The whitefly species used in study were *Bemisia tabaci*, *Trialeuroides vaporariorum* and *Aleurodicus dispersus* in plant cages. The result showed that a minimum period of 6 h was required for whitefly to feed and transmit the viral diseases. Only *B. tabaci* species was capable of transmitting CMD, whereas for CBSD all the species under experimentation transmitted the disease. Higher density of whitefly led to higher transmission of diseases. The findings here highlight the complex transmission rate of these two diseases of cassava by different whitefly species.

Subjects: Agriculture & Environmental Sciences; Botany; Plant & Animal Ecology; Zoology; Entomology

Keywords: whitefly species; cassava brown streak, cassava mosaic; transmission rate

1. Introduction

The whitefly insect has associations with almost 600 different species of plant which comprise a large number of both cultivated and non-cultivated as well as both annual and perennials crops (Bedford, Briddon, Markham, Brown, & Rosell, 1992; Brown, Frohlich, & Rosell, 1995; Martin, Mifsud, &

ABOUT THE AUTHORS

Mr Moffat K. Njoroge works for Ministry of Agriculture and Livestock in Kitui County. He is currently training on Master of Science in Crop Protection at the University of Nairobi. He has carried out a greenhouse study on rate of viral disease transmission by different whitefly species at KALRO Katumani Research Centre where Dr Daniel Mutisya supervised his research activity. At present is the contact extension person on Crops in Kitui County. The other two co-authors Dr D. Miano and Dr C. Kilalo are his Msc supervisors of his thesis proposal from University of Nairobi.

PUBLIC INTEREST STATEMENT

Whiteflies are not true flies, but are small and white looking insects on most vegetative plants. The whitefly feeds on the plant leaf tissue by sucking cell sap. Besides being minor pests, these insects are major disease vectors on most cultivated crops. In the present work, three whitefly species were evaluated in enclosed cages to find out which one spreads most two major viral diseases namely cassava mosaic and brown streak. The cassava brown streak disease causes severe root rot and huge loss of yield once it fully establishes in a cassava field. On the other hand the cassava mosaic disease causes relatively lower yield loss in comparison to the former depending on field incidence and vector presence. In the present study results, the common whitefly species, *Bemisia tabaci* was found to spread the two diseases on cassava.

Rapisarda, 2000; Naranjo, Cañas, & Ellsworth, 2009). Whiteflies are major vectors of viral diseases, such as cassava mosaic disease (CMD) and cassava brown streak disease which can reduce yield by up to 40% and at times up to 100% (Legg & Fauquet, 2004). *Bemisia tabaci* (Genn.) is one of the major vectors of cassava mosaic *begomoviruses* (CMBs) and cassava brown streak viruses (CBSVs) which are causative agents of CMD and cassava brown streak diseases (CBSD), respectively (Legg et al., 2011; Maruthi et al., 2005). Studies on *B. tabaci* reveal convincing evidence of at least 35 cryptic species with extensive genetic diversity and which show diverse behaviour concerning host plant preference, oviposition, ecological adaptation as well as virus dissemination (Ahmed, De Barro, Ren, Greeff, & Qiu, 2013; De Barro, Trueman, & Frohlich, 2005). Even in the same genetic group such as *B. tabaci*, different subclades can differ in important aspects of biology such as virus transmission, fecundity and mating ability (Habibu et al., 2012). The mode of transmission of plant viruses can either be classified as persistent, semi persistent or non-persistent depending on time required for the vector to acquire the ability to transmit virus and length of time a specific vector retains the ability. For whiteflies, *ipomovirus* are transmitted in non-persistent mode (Hollings, Stone, & Bock, 1976) whereas *criniviruses, carlaviruses* and *closteroviruses* are transmitted in semi-persistent mode and *begomoviruses* in a persistent mode (Duffus, Larsen, & Liu, 1986; Goodman & Bird, 1978; Horn et al., 2011). There is also evidence of transovarial passage of *begomoviruses* to progeny and lateral transmission among the adult whiteflies in sex-related manner (Ghanim & Czosnek, 2000; Ghanim, Morin, Zeidan, & Czosnek, 1998).

The altitude above sea level and abundance of whiteflies on cassava plant has no correlation to incidences of CBSD and CMD (Njoroge, Kilalo, Miano, & Mutisya, 2016). Hence, the need for further studies on species involvement in transmission of the two diseases on cassava. The objective of the present study was to elucidate virus transmission rate by three greenhouse mass reared species, *B. tabaci, Trialeuroides vaporariorum* and *A. dispersus* in staggered feeding for 2, 6, 12 and 24 h on virus-free tissue culture plants.

2. Materials and methods

2.1. Whitefly colony establishment
Tomatoes *Lycopersicon esculentum* Mill, pumpkin (*Cucurbita maxima* (L.) and *Tephrosia purpurea* (L.) are reported hosts of whiteflies (Saraf, Al-Musa, & Batta, 1985; Mutisya and Miano, unpublished). These were used for mass rearing purposes of the whitefly species at KALRO Katumani greenhouse and KALRO Kiboko substation. Ten seedlings of tomatoes, five seedlings of pumpkins and five seedlings of *T. purpurea* were established in plastic containers filled with sandy loam soil and followed careful recommended agronomic practices of plant growth requirements. Nitrogen fertilizer (calcium ammonium nitrate) was also applied as top dressing nutrient to increase leaf growth and enhance egg laying by whiteflies. A temperature of around 28°C and relative humidity of 30–50% was maintained by opening and closing of greenhouse as well as watering to encourage optimum growth of the host plants.

The pupal stage of whiteflies were collected from the field and identified to species level before introduction on plants in the pots. The whiteflies were let to develop to adults on the disease-free plants. The disease-free adult whiteflies were isolated in a cage for eight weeks to have a non-viruliferous colony of whiteflies before start of the experimentation (Lapidot, 2007; Mutisya and Miano, unpublished).

2.2. Diseased cassava plants establishment
Cassava cuttings obtained from field diseased material of cassava brown streak disease and one obtained from KALRO Katumani field infected with CMD were planted in plastic containers and placed in cages for establishment. Ad lib watering was done to prevent plant withering and defoliation. Disease-free cassava cuttings were also added to the cages with brown streak infected plants and whiteflies introduced to ensure there was enough inoculum source on the vectors. Some other disease-free cassava cuttings of variety TM-14 were obtained from KALRO-NARL which was verified

through use of RT-PCR method. The variety TM-14 is known to be highly susceptible to both mosaic and brown streak diseases in Kenya. Some 200 cuttings of TM-14 variety were then established in a greenhouse at KALRO Katumani in isolated cages and free of whitefly infestation where they were planted in plastic containers filled with sterilized soil mixed with sand and compost manure at a ratio of 3:2:1. They were then watered until they sprouted and after 2 weeks, about 10 g of nitrogen fertilizer per plant pot was added to enhance vegetative growth.

2.3. Transmission studies

Adult whiteflies which of 3–4 days old were used in the studies as they are known to be active and able to transmit cassava *gemiviruses*. A total of 162 disease-free plantlets were randomly isolated for the transmission studies under insect rearing cages each measuring 45 × 45 × 45 cm and laid in a completely randomized design in a greenhouse.

The non-viruliferous whiteflies were then introduced to diseased plants to feed for 48–72 h according to Mware et al. for acquisition of disease virus and then trapped in standard plastic Petri dishes and introduced to feed on disease-free cassava. The number of whiteflies introduced in disease-free cassava was varied from 5, 10 and 20 and the feeding period (inoculation) on disease-free cassava was also varied from 2, 6, 12 and 24 h. This was replicated for three times on both mosaic and brown streak disease for each of the three whitefly species. The disease symptoms were then monitored and recorded from 21st to 91st day at 7 days interval. To ensure optimum feeding, the whiteflies were disturbed by gently shaking the plantlets to ensure they did not aggregate on top of the leaves, while feeding. After the feeding period was over, the plants were removed, and whiteflies shaken off and sprayed with imidacloprid to kill all the insects. Thereafter the plantlets were cut back, added nitrogen (10 g/plant pot) and watered until leaves emerged and then monitored for another 42 days for disease symptoms observation.

3. Results

3.1. Species pathogen transmission rate

It was observed that only *B. tabaci* species were capable of transmitting both CBSD and CMD diseases, whereas for the CBSD all species were able to transmit the disease but at different rates and with different numbers of whiteflies introduced (Figure 1). Species *B. tabaci* and *T. vaporariorum* had similar ability to transmit CBSD at 50% plant incidence during the observation time period. *Aleurodicus dispersus* had least incidence of 30% CBSD incidence.

As the results indicated, the transmission of CMVD was significant ($p < 0.05$) by viruliferous *B. tabaci* species when fed on healthy plants for 6 h at 44 ± 16% followed by when fed for 24 h at 22 ± 31 and at 11 ± 16% when same species was fed for 12 h. All other species did not transmit the disease irrespective of time of feeding (Table 1). On the other hand *B. tabaci* had CBSD transmission of 22 ± 16% for all time variations of 6, 12 and 42 h. Species *T. vaporariorum* had CBSD incidence of 22 ± 16, 11 ± 16 and 33 ± 27% for 6, 12 and ±24 h. Similarly *A. dispersus* had CBSD incidence of 11 ± 16% for all time variations of 6, 12 and 24 h.

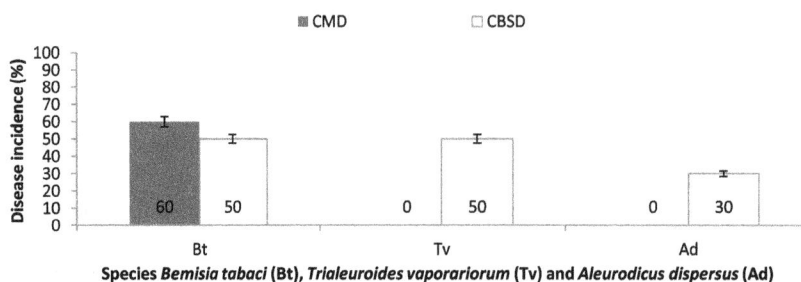

Figure 1. Viral disease transmission incidence of three whitefly species on cassava plants.

Table 1. Mean number (±SD) of CMD and CBSD incidences in relation to feeding period and whitefly species

Feeding time (Hrs)	Whitefly species	CMD (%)	CBSD (%)
2	B. tabaci	0b	0b
2	A. dispersus	0b	0b
2	T. vaporariorum	0b	0b
6	B. tabaci	44 ± 16a	22 ± 16ab
6	A. dispersus	0b	11 ± 16ab
6	T. vaporariorum	0b	22 ± 16ab
12	B. tabaci	11 ± 16b	22 ± 16ab
12	A. dispersus	0b	11 ± 16ab
12	T. vaporariorum	0b	11 ± 16ab
24	B. tabaci	22 ± 31b	22 ± 16ab
24	A. dispersus	0b	11 ± 16ab
24	T. vaporariorum	0b	33 ± 27a

Notes: Similar lower case letters denote insignificant ($p > 0.05$) mean value of CMD and CBSD incidences upon different hours of feeding (SNK at 5% level).

3.2. Species disease transmission efficiency

Species *B. tabaci* showed increases of whitefly numbers leading to higher efficiency of CMD transmission incidence of 8 ± 14, 17 ± 11 and 33 ± 33% scored for 5, 10 and 20 individual insects (Table 2). Closely related incidence of CBSD was observed where *B. tabaci* had 8 ± 14, 17 ± 11 and 25 ± 14% for 5, 10 and 20 respective individuals. On the same CBSD *T. vaporariorum* had 17 ± 16% incidence for both 5 and 10 individuals and 17 ± 29% when increased to 20 individuals. The species *A. dispersus* did not transmit CBSD at 5 and 10 individuals introduced on the plants. When increased to 20 individuals, *A. dispersus* increased transmission efficiency of CBSD to incidence of 17 ± 16%.

3.3. Feeding rate and disease correlation

There was no correlation observable between increased whitefly exposure duration on cassava and rate of disease pathogen transmission of CMD. Highest pathogen transmission occurred at 5 and 24 h exposure periods at 66% disease incidence on the leaves (Figure 2). A transmission rate of 33% was noted for 5 and 12 h exposure feeding periods. The least period of feeding period of two hours showed no CMD symptoms on leaves. The results showed no relationship between exposure period and transmission rate.

Table 2. Mean number (±SD) of CMD and CBSD incidences in relation to number of whitefly and inoculant species

Whitefly numbers	Whitefly species	CMD incidence (%)	CBSD incidence (%)
5	B. tabaci	8 ± 14a	8 ± 14a
5	A. dispersus	0b	0b
5	T. vaporariorum	0b	17 ± 16a
10	B. tabaci	17 ± 11a	17 ± 16a
10	A. dispersus	0b	0b
10	T. vaporariorum	0b	17 ± 16a
20	B. tabaci	33 ± 33a	25 ± 14a
20	A. dispersus	0b	17 ± 16a
20	T. vaporariorum	0b	17 ± 29a

Notes: Similar lower case letters denote insignificant ($p > 0.05$) mean value of CMD and CBSD incidences upon different hours of feeding (SNK at 5% level).

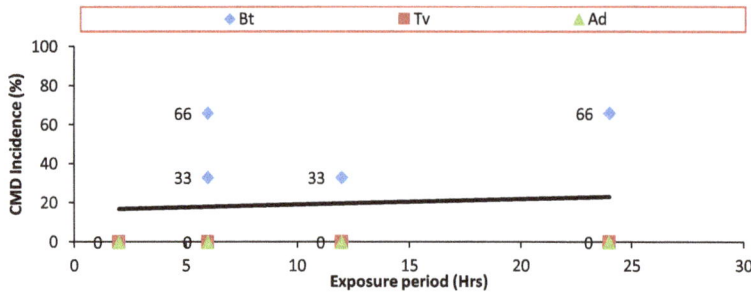

Figure 2. Relationship between whitefly exposure period and CMD incidence for different species.

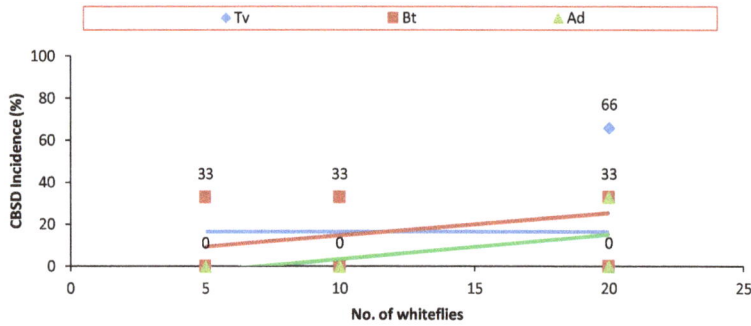

Figure 3. Relationship between number of whitefly and CBSD incidence for different species during transmission studies.

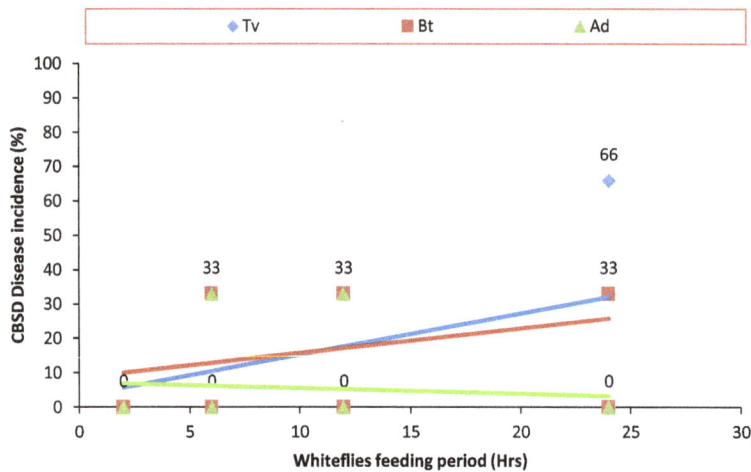

Figure 4. Relationship between whiteflies feeding period and CBSD incidence for different species.

There was similarly no correlation between number of whiteflies introduced of (*T. vaporariorum*, *B. tabaci* and *A. dispersus*) and CBSD percentage incidence (Figure 3). Likewise the analysis showed no positive correlation between period of feeding for whiteflies species of *T. vaporariorum*, *B. tabaci* and *A. dispersus* (Figure 4).

4. Discussion

As the study results showed it is only *B. tabaci* species which was found capable of transmitting the CMD. Moreover, the results were also in contrast to earlier work done by Hillocks (2000) who reported rapid expansion of CMD being closely associated to whitefly population specifically *B. tabaci*. Hence, from the results it was noted that increasing the number of *B. tabaci* did not lead to higher disease incidence in contrast to results from work done by Maruthi et al. (2005) who reported CMD disease incidence being positively correlated to number of whiteflies in an area. This can be clarified that it would depend on the whitefly earlier disease acquisition level and persistence in the insect body. It was also noted that for effective transmission of the virus, the pest would require a minimum of sufficient feeding period. This therefore calls for concerted effort in management of *B. tabaci* as the main vector of CMD disease and ensuring the source of inoculum is eradicated as the pest transmits the disease pathogen.

For CBSD, the results showed all whitefly species were capable of transmitting the viral disease pathogen in agreement with earlier work done by Maruthi et al. (2005) and Alicai et al. (2007). This is also the first time to report *T. vaporariorum* being able to transmit the CBSD disease. Lapidot (2007) has reported *T. vaporariorum* transmitting tomato yellow leaf curl virus in similar rates as *B. tabaci*. The rate of transmission was found to be 17% for *B. tabaci*, 17% for *T. vaporariorum* and 6% for *A. dispersus* in comparatively lower time rates. Mware et al. reported lower transmission rate for *A. dispersus* and high level by *B. tabaci.* In reference to time period required for transmission, 6 h was the minimum time period required for all whitefly species pathogen transfer to plant tissue. For the number of whitefly species required for transmission, it was noted that this varied with species and conditions. All the tested whitefly species did not show difference on transmission rate even as the numbers were increased, indicating that it all depended on whether the insect had disease pathogen or not. The present results therefore calls for good management practices to be put in place for all whitefly species to arrest the spread of the disease to susceptible plants in crop production systems.

The biology of the whitefly is such that it purely relies on the leaf texture and nutrient level on phloem cell tissue of the host plant (Backus, Cline, Ellerseick, & Serrano, 2007). The virus on the leaf tissue is reported to influence more feeding of the whitefly on specific crop species (Fereres & Moreno, 2009). One such classical relationship is the tomato leaf yellow curl virus (TLYCV) and *B. tabaci* species (Moreno-Delafuente, Garzo, Moreno, & Fereres, 2013). As the level of the virus increase on the plant leaf tissue the *B. tabaci* feeds more and moves out to spread the disease to other less infected similar plants within the vicinity (Backus et al., 2007; Czosnek & Rubinstein, 1997; Shah, Zhang, & Liu, 2015). As it would be expected of the increased feeding of the vector more eggs are laid and juvenile development increases with the virus presence on the leaf tissue. The present study objective was mainly on transmission rate of virus diseases by the virulent whitefly adults among the three species. Nevertheless, field spatial occurrence of *B. tabaci* on cassava mosaic infected plants has been reported showing increased disease incidence where the whitefly species is common (Martin et al., 2000; Njoroge et al. 2016).

In conclusion, it has been noted that CMD which is caused by viruses in Begomovirus family and transmitted in persistent manner is spread by *B. tabaci* species only, whereas CBSD caused by virus in Ipomovirus family and transmitted in semi persistent manner is spread by different species of whiteflies (Hull, 2002). This calls for more research on mode of transmission and biochemical analysis for these species.

Acknowledgement
The authors acknowledge the East African Productivity Project (EAAPP) for providing funds for the survey activity and other related studies by which Mr Moffat K. Njoroge undertook his MSc (Crop Protection) study at University of Nairobi.

Competing Interests
The authors declare no competing interest.

Funding
This work was supported by World Bank Group [grant number CONTRACT NO.FED/2012/291-241].

Author details
Moffat K. Njoroge[1,2]
E-mail: njorogemoffat@yahoo.com

D.L. Mutisya[3]
E-mail: dlmutisya@gmail.com
ORCID ID: http://orcid.org/0000-0002-2498-7706
D.W. Miano[2]
E-mail: dwatuku@yahoo.com
D.C. Kilalo[2]
E-mail: ngachalor@gmail.com
[1] Department of Crops, Ministry of Agriculture and Livestock, P.O. Box 16, Kitui, Kenya.
[2] Department of Plant and Crop Protection, University of Nairobi, P.O. Box 30197, Nairobi, Kenya.
[3] Department of Crop Health, KALRO-Katumani, P.O. Box 340, Machakos, Kenya.

References

Ahmed, M. Z., De Barro, P. J., Ren, S. X., Greeff, J. M., & Qiu, B. L. (2013). Evidence for horizontal transmission of secondary endosymbionts in the *Bemisia tabaci* cryptic species complex. *PLoS ONE, 8*, e53084.

Alicai, T., Omongo, C. A., Maruthi, M., Hillocks, R. J., Baguma, Y., Kawuki, R., ... Colvin, J. (2007). Re-emergence of cassava brown streak disease in Uganda. *Plant Disease, 91*, 24–29.

Backus, E. A., Cline, A. R., Ellerseick, M. R., & Serrano, M. S. (2007). *Lygus Hesperus* (Hemiptera: Miridae) feeding on cotton: New methods and parameters for analysis of non-sequential electrical penetration graph data. *Annals of the Entomological Society of America, 100*, 296–310. https://doi.org/10.1603/0013-8746(2007)100[296:LHHMFO]2.0.CO;2

Bedford, I. D., Briddon, R. W., Markham, P. G., Brown, J. K., & Rosell, R. C. (1992). *Bemisia tabaci*-biotype characterization and the threat of this whitefly species to agriculture. In *Proceedings-Brighton Crop Protection Conference Pests and Diseases* (pp. 1235–1240). Oxfordshire: British Crop Protection Council.

Brown, J. K., Frohlich, D. R., & Rosell, R. C. (1995). The sweet potato or silver leaf whiteflies: Biotypes of *Bemisia tabaci* or a species complex? *Annual Review of Entomology, 40*, 511–534. https://doi.org/10.1146/annurev.en.40.010195.002455

Czosnek, H., & Rubinstein, G. (1997). Long-term association of tomato yellow leaf curl virus with its whitefly vector *Bemisia tabaci*: Effect on the insect transmission capacity, longevity and fecundity. *Journal of General Virology, 78*, 2683–2689. https://doi.org/10.1099/0022-1317-78-10-2683

De Barro, P. J., Trueman, J. W. H., & Frohlich, D. R. (2005). *Bemisia argentifolii* is a race of *B. tabaci* (Hemiptera: Aleyrodidae): The molecular genetic differentiation of *B. tabaci* populations around the world. *Bulletin of Entomological Research, 95*, 193–203. https://doi.org/10.1079/BER2004351

Duffus, J. E., Larsen, R. L., & Liu, H. Y. (1986). Lettuce infectious yellow virus—A new type of whitefly transmitted virus. *Phytopathology, 76*, 97–100. https://doi.org/10.1094/Phyto-76-97

Fereres, A., & Moreno, A. (2009). Behavioural aspects influencing plant virus transmission by homopteran insects. *Virus Research, 141*, 158–168. https://doi.org/10.1016/j.virusres.2008.10.020

Ghanim, M., & Czosnek, H. (2000). Tomato yellow leaf curl geminivirus (TYLCV-Is) is transmitted among whiteflies (*Bemisia tabaci*) in sex related manner. *Journal of Virology, 74*, 4738–4745. https://doi.org/10.1128/JVI.74.10.4738-4745.2000

Ghanim, M., Morin, S., Zeidan, M., & Czosnek, H. (1998). Evidence for transovarial transmission of tomato yellow leaf curl virus by its vector, the whitefly *Bemisia tabaci*. *Virology, 240*, 295–303. https://doi.org/10.1006/viro.1997.8937

Goodman, R. M., & Bird, J. (1978). Bean golden mosaic virus. *CMI/AAB Description of Plant Viruses* (p. 192).

Habibu, M., Rey, M. E. C., Alicai, T., Ateka, E., Atuncha, H., Ndunguru, J., & Sseruwagi, P. (2012). Genetic diversity and geographic distribution of *Bemisi tabaci* (Gennadius) (Hemiptera: Aleyrodidae) genotypes associated with cassava in East Africa. *Ecology & Evolution Journal, 2*, 2749–2762.

Hillocks, R. J. (2000). Integrtaed crop management for smallholder farmers in Africa with special reference to coffee in Malawi. *Pest Management Science, 56*, 963–968.

Hollings, M., Stone, O. M., & Bock, K. R. (1976). Sweet potato mild mottle virus. CMI/AAB descriptions of plant viruses. (Vol. 162, 4 p.).

Horn, T., Sandmann, T., Fischer, B., Axelsson, E., Huber, W., & Boutros, M. (2011). Mapping of signaling networks through synthetic genetic interaction analysis by RNAi. *Nature Methods, 8*, 341–346.

Hull, R. (2002). *Matthews' plant virology* (4th ed.). New York, NY: Academic Press.

Lapidot, M. (2007). Screening for TYLCV-resistant plants using whitefly-mediated inoculation. In H. Czosnek (Ed.), *Tomato yellow leaf curl disease* (pp. 329–342). New York, NY: Springer Online Books. https://doi.org/10.1007/978-1-4020-4769-5

Legg, J. P., & Fauquet, C. M. (2004). Cassava mosaic geminiviruses in Africa. *Plant Molecular Biology, 56*, 585–599. https://doi.org/10.1007/s11103-004-1651-7

Legg, J. P., Obiero, H. M., Maruthi, M. N., Ndyetabula, I., Okao-Okuja, G., Bouwmeester, H., ... Lava, K. P. (2011). Comparing the regional epidemiology of the cassava mosaic and cassava brown streak pandemics in Africa. *Virus Research, 159*, 161–170. doi:10.1016/j.virusres.2011.04.018

Martin, J. H., Mifsud, D., & Rapisarda, C. (2000). The whiteflies (Hemiptera: Aleyrodidae) of Europe and Mediterranean Basin. *Bulletin of Entomological Research, 90*, 407–448.

Maruthi, M. N., Hillocks, R. J., Mtunda, K., Raya, M. D., Muhanna, M., Kiozia, H., ... Thresh, J. M. (2005). Transmission of Cassava brown streak virus by *Bemisia tabaci* (Gennadius). *Journal of Phytopathology, 153*, 307–312. https://doi.org/10.1111/jph.2005.153.issue-5

Moreno-Delafuente, A., Garzo, E., Moreno, A., & Fereres, A. (2013). A plant virus manipulate the behavior of its whitefly vector to enhance its transmission efficacy and spread. *PLOS One, 8*(4), 1–10.

Naranjo, S. E., Cañas, L., & Ellsworth, P. C. (2009). Mortality and population dynamics of *Bemisia tabaci* within a multi-crop system. In P. G. Mason, D. R. Gillespie, & C. D. Vincent (Eds.), *Proceedings of the Third International Symposium on Biological Control of Arthropods* (pp. 202–207). Christchurch: USDA Forest Service.

Njoroge, M. K., Kilalo, D. C., Miano, D. W., & Mutisya, D. L. (2016). Whiteflies species distribution and abundance on cassava in different agro-ecological zones of Kenya. *Journal of Entomology and Zoology Studies, 4*, 258–262.

Saraf, N. S., Al-Musa, A. M., & Batta, Y. (1985). Effect of different plants on population development of the sweetpotato whitefly (*Bemisia tabaci* Genn. Homoptera: Aleyrodidae). *Dirasat, 11*, 89–100.

Shah, M. M., Zhang, S., & Liu, T. (2015). Whitefly, host plant and parasitoid: A review on their interactions. *Asian Journal of Applied Science and Engineering, 4*, 47–60.

13

Botanical insecticide as simple extractives for pest control

Wafaa M. Hikal[1,2]*, Rowida S. Baeshen[1] and Hussein A.H. Said-Al Ahl[3]

*Corresponding author: Wafaa M. Hikal, Faculty of Science, Department of Biology, University of Tabuk, P.O. Box 741, Tabuk 71491, Saudi Arabia; Parasitology Laboratory, Water Pollution Researches Department, National Research Center, 33 El-Bohouth St., Dokki, Giza, Egypt, 12622

E-mail: wafaahikal@gmail.com

Reviewing editor: Karol Ujházy, Technical University in Zvolen, Slovakia

Additional information is available at the end of the article

Abstract: One of the most important global problems is protecting crops from insects. For the control of insects, synthetic chemicals are continuously used, and their toxicity endangers health of farm operators, animals and food consumers. The negative effects on human health led to a resurgence of interest in botanical insecticides due to their minimal costs and ecological side effects. In this, we review the use of plant compounds (essential oils, flavonoids, alkaloids, glycosides, esters and fatty acids) having anti-insect effects and their importance as an alternative to the chemical compounds used in the elimination of insects in different ways, namely repellents, feeding deterrents/antifeedants, toxicants, growth retardants, chemosterilants, and attractants. Botanical insecticides affect only target insects, not destroy beneficial natural enemies and provide residue-free food and safe environment. We, therefore, recommend using botanical insecticides as an integrated insect management program which can greatly reduce the use of synthetic insecticides.

Subjects: Botany; Plant & Animal Ecology; Entomology

Keywords: natural products; insecticides; active constituents; crop protection; insect

ABOUT THE AUTHORS

Wafaa M. Hikal, PhD, parasitology. Her research interests focus on parasitology and published peer-reviewed articles in the areas of parasitology and books, and collaborate in water borne parasites projects. She is a member of various scientific societies and editorial board member of some international journals.

Rowida S. Baeshen is an assistant professor at the University of Tabuk, Saudi Arabia. Her research areas include entomology, and this paper is one of the main concerns of volatile oils application and one of the researcher's top priorities. She published peer-reviewed articles and book in the areas of entomology.

A.H. Hussein Said-Al Ahl is a professor at the National research Centre, and collaborates in medicinal and aromatic plants projects. His research interests focus on medicinal and aromatic plants. He is a member of various scientific societies and editorial board member of some international journals. Award-winning of the National Research Center Scientific Encouragement for agricultural sciences and published peer-reviewed articles and books.

PUBLIC INTEREST STATEMENT

Provision of food has always been a challenge facing mankind. A major cornerstone in this challenge is the competition from insects. Insect pests inflict damage to humans, farm animals and crops. They are responsible for destroying one fifth of the world's total crop production annually. Considerable losses of agricultural products add a serious burden to people's daily life. There is an urgent need to move toward natural products as insecticides because it reduces negative impacts to human health and the environment, and we recommend the botanical insecticides to reduce the insect population and increase food production. Therefore, this study is useful for farmers, producers and exporters of crops and recommends the use of botanical insecticides as an alternative to chemicals or synthetic insecticides to eliminate insects and maintain crops and meet the risks of food shortages as well as the preservation of the environment and human and animal health.

1. Introduction

Insects are the most diverse species of animals living on earth and can be found in all habitats. Less than 0.5% of the total number of the known insect species are considered pests, and only a few of these can be a serious menace to people. Insects inflict damage to humans, farm animals and crops. Some insects can constitute a major threat to entire countries or a group of nations. One prominent example is the tsetse fly that puts about 100 million people and 60 million head of cattle at risk in sub-Saharan Africa due to the transmission of trypanosomiasis (ICIPE, 1997; Imms, 1964).

Provision of food has always been a challenge facing mankind. A major cornerstone in this challenge is the competition from insects. Herbivorous insects are said to be responsible for destroying one-fifth of the world's total crop production annually. Particularly in the tropics and sub-tropics, where the climate provides a highly favorable environment for a wide range of insects. In the developing countries, the problem of competition from insects is further complicated with a rapid annual increase in the human population. Taking into consideration sudden problems caused by drought in places such as Africa, considerable losses of agricultural products add a serious burden to people's daily life. The introduction of alien insects into new habitats due to the global increase of trade and transport causes another dilemma. Insects inflict their damage on stored products mainly by direct feeding. Some species feed on the endosperm causing loss of weight and quality, while other species feed on the germ, resulting in poor seed germination and less viability. Thus, due to damage done by insects, grains lose value for marketing, consumption or planting. In addition to direct consumption of the product, insect pests contaminate their feeding media through excretion, molting, dead bodies and their own existence in the product, which is not commercially desirable. Damage done by insects encourages infection with bacterial and fungal diseases through transmission of their spores (Malek & Parveen, 1989; Santos, Maia, & Cruz, 1990).

In this review, we will focus on biological control of insects in crop production using botanical products. We provide an overview of botanical insecticides from the chemical point of view and classify their effects on insects.

2. Types of botanical insecticides

2.1. Essential oils

Plant secondary natural products are natural chemicals extracted from plants and used as an excellent alternative to synthetic or chemical pesticides (Regnault-Roger & Philogène, 2008; Sithisut, Fields, & Chandrapathya, 2011). In addition insecticide resistance to synthetic pesticides, which led to a significant losses of food arising from failure of chemicals in pest and annually caused economic losses of several billion dollars worldwide (Elzen & Hardee, 2003; Pereira, Sanaveerappanavar, & Murthy, 2006; Shelton, Zhao, & Roush, 2002), Furthermore, United States Food and Drug Administration (FDA) recognized botanical pesticides (essential oils) as safe than synthetic pesticides which caused increase in risk of ozone depletion, neurotoxic, carcinogenic, teratogenic and mutagenic effects in non-targets and cross- and multi-resistance in insects (Regnault-Roger, Vincent, & Arnason, 2012).

Essential oils extracted from aromatic plants have increased considerably as insecticides owing to their popularity with organic growers and environmentally conscious consumers. They have repellent, insecticidal, antifeedants, growth inhibitors, oviposition inhibitors, ovicides, and growth-reducing effects on a variety of insects (Don-Perdo, 1996; Elzen & Hardee, 2003; Koshier & Sedy, 2001; Lu, 1995; Pereira et al., 2006; Regnault-Roger et al., 2012; Shelton et al., 2002; Sithisut et al., 2011; Tripathi, Prajapati, Khanuja, & Kumar, 2003). Essential oils possess interesting larvicidal effects on *Limantria dispar* (Lepi-doptera: Lymantridae, gypsy moth) larvae (Moretti, Sanna-Passino, Demontis, & Bazzoni, 2000), insecticidal activity, repellent properties against ants, cockroach, bedbugs, head lices flied and moth and toxic to termites. *Mentha piperita* oil repels ants, flies, lice, moths' and is effective against *Callosobruchus maculatus* and *Tribolium castanum* (Kordali, Cakir, Mavi, Kilic, & Yildirim, 2005). *Trachyspermum* sp. oil has larvicidal against *Aedes aegypti* and southern house mosquito, *Culex quinquefasciatus* (Tripathi et al., 2000). Nepetalactone, the active constituent in catnip (*Nepeta cateria*) essential oil is highly effective for repelling mosquitoes, bees and other flying insects. It repels mosquitoes more than DEET. It is particularly effective against *A. aegypti* mosquito, a vector for yellow fever virus. Oils of *Zingiber officinale* rhizomes and *Piper cubebaberries* were exhibited insecticidal and antifeeding activities against *Tribolium castaneum* and *Sitophilus oryzae* (Chaubey, 2012a, 2012b). Tagetes species oil possesses as anti-insect activity on *Ceratitis capitata* and *Triatoma infestans* (López et al., 2011). *Melaleuca alternifolia* essential oil possesses the fumigant toxicity against *Sitophilus zeamais* (Min et al., 2016). Rosemary, oregano, yarrow, eucalyptus and mint oils used as the safe compounds for surface treating or fumigation in cockroach control. Oregano oil used as a potential repellent against *Supella longipalpa* (Sharififard, Safdari, Siahpoush, Hamid, & Kassiri, 2016). Kanat, Hakk, and Alma (2003) found that essential oils of many plants have insecticidal effects against the larvae of pine processionary moth, *Thaumetopoea pityocampa*. Also, *Laurus nobilis* essential oil was toxic activities against *Rhyzopertha dominica* and *T. castaneum* (Ben Jemba, Tersim, Toudert, & Khouja, 2012). *Lavandula hybrida, Rosmarinus officinalis,* and *Eucalyptus globulus* oils were insecticidal on *Acanthoscelides obtectus* adults (Papachristos, Karamanoli, Stamopoulos, & Menkissoglu-Spiroudi, 2004). Moreover, essential oil of *Tagetes minuta* has toxicity against *Cochliomyia macellaria* (Diptera: Calliphoridae) and as acaricidal, repellent (Chaaban, de Souza, Martins, Bertoldi, & Molento, 2017). Eugenol which the principle compound of the essential oils from basil oil have a strong repellent effect on mosquitoes and linalool also in basil oil has toxic effect to the *Bruchid zabrotes* sub *fasciatus* and other storage pests (Chogo & Crank, 1981; Weaver, Dunkel, Ntezurubanza, Jackson, & Stock, 1991). Essential oil of *Zingiber zerumbet* has repellant against *Lasioderma serricorne* (Wu et al., 2017). *Juniperus procera* essential oil was exhibited significant repellent against the malarial vector *Anopheles arabiensis* (Karunamoorthi, Girmay, & Hayleeyesus, 2014). Terpinen-4-ol; 1,8-cineol; verbenone and camphorin Eucalyptus oil were active against *A. obtectus* adults (Tholl, 2006), antifeedant against biting insects, insecticidal agents and prevent mosquito bite (Bakkali, Averbeck, Averbeck, & Idaomar, 2008; Batish, Singh, Kohli, & Kaur, 2008; Fradin & Day, 2002; Isman & Machial, 2006). Lucia et al. (2007) reported that essential oil from *Eucalyptus globules* is toxic to *A. aegypti* larvae. Seyoum, Killeen, Kabiru, Knols, and Hassanali (2003) reported that burning of leaves of *Eucalyptus citriodora* used as protection against mosquitoes in Africa. Also, CDC (Center for Disease Control and Prevention, USA) recommended the use of lemon eucalyptus oil (with p-menthane-3,8-diol, PMD, as active ingredient) for protection against West Nile virus that causes neurological disease or even death and is spread by mosquitoes (CDC, 2005).

Toloza et al. (2006) concluded the fumigant toxicity/repellent activity of essential oil from *Eucalyptus cinerea, Eucalyptus viminalis,* and *Eucalyptus saligna,* against permethrin-resistant humanhead lice. The pesticidal and antifeedant activities of eucalyptus oils has been due to 1,8-cineole, citronellal, citronellol, citronellyl acetate, p-cymene, eucamalol, limonene, linalool, α-pinene, Υ-terpinene, α-terpineol, alloocimene, and aromadendrene components (Batish, Singh, Setia, Kaur, & Kohli, 2006; Cimanga et al., 2002; Li, Madden, & Potts, 1996; Liu, Chen, Wang, Xie, & Xu, 2008; Watanabe et al., 1993). Eucalyptus oils rich in cineole have affective against varroa mite, *Varroa jacobsoni*-an important parasite of honey bee (Calderone & Spivak, 1995), *Tetranychus urticae, Phytoseiulus persimilis* and *Dermatophagoides pteronyssinus* (Choi, Lee, Park, & Ahn, 2004; El-Zemity, Hussien, Saher, & Ahmed, 2006). Eucalyptus essential oil significantly reduced the number of tick bites in humans and concluded that it could be used to reduce tick-borne infections (Chagas et al., 2002; Gardulf, Wohlfart, & Gustafson, 2004). Pujiarti and Fentiyanti (2017) reported that *Eucalyptus deglupta* essential oils have repellent activity against *C. quinquefasciatus* mosquito. Eugenol, isoeugenol, methyleugenol, methyl isoeugenol, coumarin, coniferyl aldehyde, diniconazole, ethyl cinnamate, and rosmarinic acid against red palm weevil posses as antifeedant activity against adults of red palm weevil (*Rhynchophorus ferrugineus*) (AlJabr, Hussain, Rizwan-ul-Haq, & Al-Ayedh, 2017; Aref, Valizadegan, & Farashiani, 2015; Shukla, Vidyasgar, Aldosari, & Abdel-Azim, 2012).

2.2. Alkaloid

Alkaloids are the most important group of natural substances playing an important role in insecticidal (Balandrin, Klocke, Wurtele, & Bollinger, 1985; Rattan, 2010). Wachira et al. (2014) concluded that pyridine alkaloids extracted from *Ricinus communis* against the malaria vector *Anopheles gambiae*. Furocoumarin and quinolone alkaloids extracted from *Ruta chalepensis* leaves showed larvicidal and antifeedant activities against the larvae *Spodoptera littoralis* (Emam, Swelam, & Megally, 2009). Acheuk and Doumandji-Mitiche (2013) found that alkaloids extract of *Pergularia tomentosa* caused antifeeding and larvicidal effects. Lee (2000) concluded that pipernonaline and piperidine alkaloids have mosquito larvicidal activity. Alkaloids from *Arachis hypogaea* extract have larvicidal activity against chikungunya and malarial vectors (Velu et al., 2015).

2.3. Flavonoids

Flavonoids could be useful in a pest-management strategy. Flavonoids play an important role in the protection of plants against plant feeding insects' and herbivores (Acheuk & Doumandji-Mitiche, 2013). Both flavonoids and isoflavonoids protect the plant against insect pests by influencing their behavior, growth, and development (Simmonds, 2003; Simmonds & Stevenson, 2001). Rutin and quercetin-3-glucoside in *Pinus banksiana* inhibit the development and increase the mortality of *L. dispar* (Gould & Lister, 2006). Quercetin and rutin glycosides in peanuts caused increased mortality of the tobacco armyworm (*Spodoptera litura*) (Mallikarjuna, Kranthi, Jadhav, Kranthi, & Chandra, 2004). In rice, three flavone glucosides inhibit digestion in insects and function as deterrent agents in *Nilaparvata lugens* and herbivores (Acheuk & Doumandji-Mitiche, 2013). Diwan and Saxena (2010) found that flavinoid glycosides isolated from *Tephrosia purpuria* showed insecticidal property on *C. maculatus* grubs. Isoflavonoids and proanthocyanidins are other classes of flavonoids responsible for plant protection against insects. Forexample, naringenin procyanidin inhibits the development of *Aphis craccivora* andherbivores (Acheuk & Doumandji-Mitiche, 2013). Kumar, Bhadauria, and Mishra (2015) concluded that quercetin/azadirachtin insecticide can be a safe and efficient insecticide which improved the activity of *Euphaedra orientalis* and is non toxic to it, and also is environmentally less harmful as it is easily biodegradable. Goławska, Sprawka, Łukasik, and Goławski (2014) found that two polyphenolic flavonoids (flavanone naringenin and flavonol quercetin) used as insecticides of the pea aphid, *Acyrthosiphon pisum* (Hemiptera: Aphididae). Santos et al. (2016) conclude that *Tagetes erecta* and *Tagetes patula* have phytotoxic compounds (flavonoids) that can promote and expand its use as a natural insecticide. Golawska, Kapusta, Łukasik, and Wojcicka (2008) suggests that quercetin, kaempferol + RCO-, kaempferol, tricin, apigenin + RCO-, and apigenin are good for control of the insect pests.

Goławska and Łukasik (2012) showed the effects of isoflavone genistein and flavone luteolin on the feeding behavior of the pea aphid. Flavonoid glycosides in alfalfa affect feeding behavior of pea aphid (Goławska, Łukasik, Goławski, Kapusta, & Janda, 2010; Goławska, Łukasik, Kapusta, & Janda, 2012). Morimoto, Kumeda, and Komai (2000) showed that flavonoids can act as feeding deterrents. Flavonoids from *A. hypogaea* extract have larvicidal activity against chikungunya and malarial vectors (Velu et al., 2015).

2.4. Glycosides

Cyanogenic glucosides presented in plant species and considered to have an important role in plant defense against herbivores (Zagrobelny et al., 2004). Al-Rajhy, Alahmed, Hussein, and Kheir (2003) concluded that the cardiac glycoside, digitoxin, from *Digitalis purpurea*, a cardiac glycosidal (cardenolide) extract from *Calotropis procera*, azadirachtin and neem oil from *Azadirachta indica* posses against larvae and adult stages of the camel tick, *Hyalomma dromedarii*. Also, Kubo and Kim (1989) found that flavan glycosides, Viscutin-1, -2 and -3, which exhibit insect growth inhibitory activity against the cotton pest insect; *Pectinophora gossypiella*. However, iridoid glycosides posses as insect antifeeding on gypsy moths (*L. dispar*, Lymantriidae) and buckeyes (*Junonia coenia*, Nymphalidae) (Bowers & Puttick, 1989). Also, cyanogenic glycosides are known as plant defense chemicals and found in cassava, bamboo, flax, and other plants. They are effective against stored-product insects as fumigants. Due to their insecticidal activity to insects, cyanohydrins can be used as an alternative fumigant and also as soil fumigants (Park & Coats, 2002). Dave and Lediwane (2012) found that anthraquinones isolated from *Cassia* species posses as antimalarial and insecticidal activity. Glycosides from *A. hypogaea* extract have larvicidal activity against chikungunya and malarial vectors (Velu et al., 2015). Juvenogens have a potential application in insect pest control (Wimmer et al., 2007).

2.5. Esters and fatty acids

Allyl cinnamate caused rapid toxic effects in *S. littoralis* larvae at low concentrations potential for use in pest control (Giner, Avilla, Balcells, Caccia, & Smagghe, 2012). Schmidt, Tomasi, Pasqualini, and Ioriatti (2008) demonstrated that ethyl (*E,Z*)-2,4-decadienoate (pear ester) have insecticide against *Cydia pomonella*.

Fatty acids methyl esters were isolated from *Solanum lycocarpum* have larvicidal activity against the vector *C. quinquefasciatus* (Silva, Ribeiro Neto, Alves, & Li, 2015). Mullens, Reifenrath, and Butler (2009) showed that saturated fatty acids (particularly C8, C9 and C10) used as repellents or antifeedants against houseflies, horn flies and stable flies. Samuel, Oliver, Wood, Coetzee, and Brooke (2015) concluded that fatty acids mixture (C8910) has toxicity and repellence against insecticide susceptible and resistant strains of the major malaria vector *Anopheles funestus*. Yousef, EL-Lakwah, and EL Sayed (2013) reported that toxicity and reduction in larval body weight of linoleic acid against the larvae of *S. littoralis*.

3. Insecticide effects on insects

Botanical insecticides affect various insects in different ways depending on the physiological characteristics of the insect species as well as the type of the insecticidal plant. The components of various botanical insecticidal can be classified into six groups namely; repellents, feeding deterrents/antifeedants, toxicants, growth retardants, chemosterilants, and attractants (Rajashekar, Bakthavatsalam, & Shivanandappa, 2012).

3.1. Repellents

A botanical pesticide have a repellent property, where keeps away the insect pest, and protect the crops (Isman, 2006) with minimal impact on the ecosystem, as they drive away the insect pest from the treated materials by stimulating olfactory or other receptors (Talukder, 2006; Talukder, Islam, Hossain, Rahman, & Alam, 2004). Botanical pesticides are considered safe in pest control because they have low or none pesticide residue making them safe to the people, environment and ecosystem (Talukder et al., 2004). Ghavami, Poorrastgoo, Taghiloo, and Mohammadi (2017) found that essential oils of *Ziziphora tenuiore, Myrtus communis, Achillea wilhelmsii* and *M. piperita* have repellent activities against human fleas. Rahdari and Hamzei (2017) demonstrated the efficacy of *M. piperita, R. officinalis,* and *Coriandrum sativum* oils for applying in organic food protection due to repellent activity of essential oils on *Tribolium confusum*. Zhang et al. (2017) reported the repellent activities of six *Zanthoxylum* species including *Z. armatum, Z. dimorphophyllum, Z. dimorphophyllum* var. *spinifolium, Z. piasezkii, Z. stenophyllum,* and *Z. dissitum* essential oils against two storage pests including *T. castaneum* and *L. serricorne* adults and the essential oils of these six *Zanthoxylum* species essential oils possessed significant repellent activities against *T. castaneum* and *L. serricorne* adults. The different repellent activities on two insects might be attributed to the different anti-insect mechanism and different non persistent volatility of essential oil sample. Kimutai et al. (2017) demonstrated that the essential oils of *Cymbopogon citratus* and *T. minuta* on the sandfly, *Phlebotomus duboscqi*. However, the effectiveness of the repellents depends on multiple factors including the type of repellents (active ingredients), formulation, mode of application, environmental factors (temperature, humidity, and wind), the attractiveness of individual people to insects, loss due to removal by perspiration and abrasion, the sensitivity of the insects to repellents, and the biting density. In addition, *Origanum onites* essential oil had repellent activity against *Amblyomma americanum* (L.) and *A. aegypti* (L.). Carvacrol and thymol were strongly repellent to *A. aegypti* and *A. americanum*. Thus, carvacrol-rich *O. onites* essential oil, carvacrol, and possibly thymol appear to have potential for use to protect humans and domestic animals against mosquitoes and ticks.

Various natural fatty acids have insecticidal properties, some involving action on acetylcholinesterase and octopaminergic receptors (Perumalsamy, Jang, Kim, Kadarkarai, & Ahn, 2015). A saturated fatty acid mixture composed of octanoic acid (also called caprylic acid, nonaoic acid and decanoic acid (also called capric acid), collectively called "C8910 acids" (C8, C9, and C10 mixture), repelled horn flies. C8910 acids deterred horn flies from feeding by > 85% and the pest was also strongly repelled (Zhu, Brewer, Boxler, Friesen, & Taylor, 2015). Feeding deterrence caused by C8910 acids and > 50% antifeedancy was observed (Zhu et al., 2015). The naturally occurring oleic and linoleic acids, and methyl oleate synergized the repellency of DEET and the monoterpenoids cuminyl alcohol, cuminaldehyde, and α-phellandrene (Hieu, Choi, Kim, Wang, & Ahn, 2015; Showler, 2017).

3.2. Feeding deterrents/antifeedants

Botanical pesticides that inhibit feeding or disrupt insect feeding by rendering the treated materials unattractive or unpalatable (Rajashekar et al., 2012; Talukder, 2006). The insects remain on the treated material indefinitely and eventually starve to death. Liao et al. (2017) demonstrated that oil of *M. alternifolia* and their chemical constituents possessed obvious antifeedant activities against *Helicoverpa armigera* Hubner. The phytoconstituents found in the leaf extract of *Khaya senegalensis* include tannins, saponins, flavonoids, steroids, and alkaloids may have been responsible for the mortality of *Dinoderus porcellus* (Loko et al., 2017). Chaudhary et al. (2017) and Ghoneim and Hamadah (2017) pointed that azadirachtin which is prominent constituent of neem established as a pivotal insecticidal ingredient. It acts as an antifeedant, repellent, and repugnant agent and induces sterility in insects by preventing oviposition and interrupting sperm production in males. The same result obtained by Abdullah et al. (2017) who found that, 1,8-cineol found in Galangal essential oil exhibited antifeedant activity, repellent activity, and toxicity effect toward the termites. Jose and Sujatha (2017) revealed that terpenoids, coumarin and phenols, present in the methanol extracts of *Gliricidium sepium* exhibited significant antifeedant activity. This indicated that the active compounds present in the plant inhibit the larval feeding behavior while others disrupt hormonal balance or make the food unpalatable. These active substances may directly act on the chemosensilla of the larvae resulting in feeding deterrence.

3.3. Toxicity

Some botanical pesticides are toxic cause death to stored product insects (Padin, Fuse, Urrutia, & DalBello, 2013). Rotenone is considered as a toxic compound since it is a mitochondrial poison which blocks the electron transport chain and prevents energy production (Hollingworth, Ahammadsahib, Gadelhak, & McLaughlin, 1994). As an insecticide, it is a stomach poison because it must be ingested to be effective (Isman, 2006). Essential oil of *Lavandula angustifolia* exhibited good fumigant and contact toxicity against granary weevil adults. In addition, a strong repellent activity is able to disrupt granary weevil orientation to an attractive host substrate (Germinara et al., 2017). Trivedi, Nayak, and Kumar (2017) demonstrated fumigant toxicity against the stored grain pest *Callosobruchus chinensis*. The essential oils of cinnamon, clove, rosemary, bergamot, and Japanese mint showed potential to be developed as possible natural fumigants or repellents for control of the pulse beetle. Lucia, Toloza, Guzmán, Ortega, and Rubio (2017) found that the mortality of adults and eggs for head lice associated with the use of (geraniol, citronellol, 1,8-cineole, linalool, α-terpineol, nonyl alcohol, thymol, menthol, carvacrol, and eugenol) essential oils. Bouguerra, Djebbar, and Soltani (2017) showed that *Thymus vulgaris* essential oil exhibited significant activity and could be considered as potent natural larvicidal agent against *Culex pipiens*. Zhao, Liu, Liu, and Liu (2017) indicated that the essential oil of *Echinops grijsii* roots and the isolated thiophenes have an excellent potential for use in the control of *Aedes albopictus, Anopheles sinensis,* and *C. pipiens* pallens larvae and could be used in the search for new, safer and more effective natural compounds as larvicides. Wu et al. (2017) observed the toxicity and repellent activities of the rhizomes of *Z. zerumbet* (L.) Smith (Zingiberaceae) essential oil contains the component α-caryophyllene against cigarette beetles (*L. serricorne*). Alkan, Gokce, and Kara (2017) indicate that *Heracleum platytaenium* and *Humulus lupulus* extracts have great potentials as insecticides in the management of larvae of *Leptinotarsa decemlineata*. Papanastasiou et al. (2017) showed the toxicity of limonene, linalool and α-pinene on adult Mediterranean fruit flies. Qari, Nilly, Abdel-Fattah, and Shehawy (2017) showed DNA damage due to alterations in enzymatic system (acetylcholinesterase, acid phosphatase, alkaline phosphatase, lactate dehydrogenase and phenol oxidase), total protein and DNA concentration after treatment with essential oils of *Citrus aurantium, Eruca sativa, Z. officinale* and *Origanum majorana* against *R. dominica*. Park, Jeon, Lee, Chung, and Lee (2017) demonstrated that *T. vulgaris* oil had the highest insecticidal toxicity followed by *R. graveolens, C. aurantium, L. petersonii,* and *A. millefolium* oils. The insecticidal toxicity of *T. vulgaris* oil against *P. shantungensis* nymphs was about 1.3-fold more than that against *P. shantungensis* adults. Differences in the insecticidal toxicities of plant-derived oils may be explained on the basis of species-specific responses to plant species, phytochemicals, and the weight and size of *P. shantungensis* adults and nymphs.

3.4. Growth retardants and development Inhibitors

Botanical pesticides showed deleterious effects on the growth and development of insects, reducing the weight of larva, pupa and adult stages and lengthening the development stages (Talukder, 2006). Plant derivatives also reduce the survival rates of larvae and pupae as well as adult emergence (Koul, Waliai, & Dhaliwal, 2008). It has been reported that both azadirachtin and neem seed oil increased aphid nymphal mortality significantly at 80 and 77%, respectively, and at the same time increasing development time of those surviving to adulthood (Kraiss & Cullen, 2008). Many botanical pesticides have been reported to have a pronounced effect on the developmental period, growth, and adult emergence (Shaalan, Canyon, Younes, Abdel-Wahab, & Mansour, 2005).

3.5. Sterility/reproduction inhibitors

Sterility can be induced by sterile insect technique (SIT) or a chemosterilant, a chemical compound that interferes with the reproductive potential of sexually reproducing organism (Morrison et al., 2010). Chemosterilants are used to control economically destructive or disease-causing pests (usually insects) by causing temporary or permanent sterility of one or both of the sexes or preventing maturation of the young to a sexually functional adult stage (Navarro-Llopis, Vacas, Sanchis, Primo, & Alfaro, 2011; Wilke et al., 2009). It has been reported that plant parts, oil, extracts, and powder mixed with grain reduced insect oviposition, egg hatchability, postembryonic, and progeny development (Asawalam & Adesiyan, 2001; Shaalan et al., 2005). Hexane extracts of *Andrographis lineat, A. paniculata,* and *T. erecta* showed 100% ovicidal activity against *Anopheles subpictus* (Elango et al., 2009). Some botanical insecticides are used as chemosterilants, for example, at the physiological level azadirachtin blocks the synthesis and release of molting hormones from the prothoracic gland, leading to incomplete ecdysis in immature insects and in adult insects it leads to sterility (Isman, 2006).

Garlic essential oil and its constituents, diallyl sulfide and diallyl disulfide have been highly toxic to *S. zeamais* and *T. castaneum* (Ho, Koh, Ma, Huang, & Sim, 1996; Huang, Lam, & Ho, 2000) at different developmental stages. Plata-Rueda et al. (2017) showed that *Tenebrio molitor* was more susceptible in the pupal stage followed by larvae and adults exposed to diallyl sulfide and diallyl disulfide. One possible explanation for the developmental stages difference is that efficacy may be affected by the penetration of the garlic compounds into the body and the ability of the insect to metabolize these compounds. When insects exposed to the garlic essential oil displayed altered locomotion activity, and muscle contractions and paralysis were observed. Paralysis and muscle contractions can be explained by the toxic effect in the nervous system. The toxicity of essential oils in insects indicates neurotoxic action with hyperactivity, hyperextension of the legs and abdomen and rapid knockdown effect or immobilization (Prowse, Galloway, & Foggo, 2006; Zhao et al., 2013). Acetylcholinesterase is an enzyme that has been shown to be inhibited by garlic compounds and can act only or in synergism as diallyl disulfide, diallyl trisulfide, and allicin (Bhatnagar-Thomas & Pal, 1974; Singh & Singh, 1996). Diallyl sulfide in garlic compounds have toxic effect in *T. molitor* and may cause inhibition by cross-linking with essential thiol compounds in enzyme structures, altering the functional shape of the protein and denaturalization (Halliwell & Gutteridge, 1999). Also, diallyl disulfide as main volatile compound in garlic essential oil has repellent properties to *S. zeamais* and *T. castaneum* (Huang et al., 2000). Diallyl disulfide, and diallyl sulfide have high activities of behavioral deterrence against *T. molitor*, as evaluated by the behavioral responses of larvae and adults to different odor sources and the number of insects repelled, indicating their potential to the pest control in stored products. As well as garlic essential oil compromised the respiration rate of *T. molitor*. Thus, low respiration rate is an indicator of physiological stress, and essential oils can compromise insect respiration by impairing muscle activity, leading to paralysis (Correa, Faroni, Haddi, Oliveira, & Pereira, 2015; de Araújo et al., 2017; Guedes, Oliveira, Guedes, Ribeiro, & Serrão, 2006).

Dehghani-Samani, Madreseh-Ghahfarokhi, Dehghani-Samani, and Pirali-Kheirabadi (2015) showed that essential oil of *E. globulus* had repellent activity against *Dermanyssus gallinae* due to the essential oil components such as 1, 8-cineole, citronellal, citronellol, citronellyl acetate, p-cymene, eucamalol, limonene, linalool, α-pinene, g-terpinene, α-terpineol, alloocimene, and aromadendrene. Among the various components of eucalyptus oil, 1, 8-cineole is the most important one which is largely responsible for a variety of its pesticidal properties and insecticide effects. Jayakumar, Arivoli, Raveen, and Tennyson (2017) recorded the repellent activity of camphor, citronella, eucalyptus, lemon and wintergreen oil essential oils against stored product pests; the adult rice weevil *S. oryzae* Linnaeus 1763 (Coleoptera: Curculionidae) due to essential oils components repellent nor attractant activity. The protection of stored products and the phytochemical constituents from essential oils can work synergistically, improving their effectiveness.

Ho, Ma, Goh, and Sim (1995) concluding that adult mortality might be attributed to the contact toxicity or to the abrasive effect on the pest cuticle (Mathur, Shankar, & Ram, 1985), which might also interfere with the respiratory mechanism of insect (Agarwal, Lal, & Gupta, 1988; Kim, Roh, Kim, Lee, & Ahn, 2003; Schoonhoven, 1978). Fumigation studies showed that the essential oils had a "knock down effect" on the test insect. Essential oils act by inhibiting insect acetylcholinesterase (AChE) and thus, ultimately blocking the nerve functions. Also, Obeng-Ofori and Amitaye (2005) who observed signs of immobilization with flexed legs and clinging to the grain outstretched meta thoracic wings from the elytra and paralysis of the dead or dying insects. The enzyme AChE is also the target site of inhibition by organophosphates and carbamate insecticides (Matsumura, 1985). The observed rapid action of essential oils could be attributed to their property of acting in the vapor phase, hence gaining entry into the insect's internal systems with ease through the spiracles; whereas, in topical application procedures, the insect is protected by its exoskeleton against external influences.

Essential oils possess acute contact and fumigant toxicity to insects (Abdelgaleil, Mohamed, Badawy, & El-arami, 2009), repellent activity (Nerio, Olivero-Verbel, & Stashenko, 2009), antifeedant activity (Huang et al., 2000), as well as development and growth inhibitory activity (Waliwitiya, Kennedy, & Lowenberger, 2008). Repellent activity has been linked to the presence of essential oils that cause death of insects by inhibiting AChE activity in the nervous system (Houghton, Ren, & Howes, 2006). Essential oils being more useful as insect fumigants (Regnault-Roger & Hamraoui, 1993; Weaver et al., 1991) and have strong toxicity to insects due to high volatility and lipophilic properties can penetrate into insects rapidly and interfere in physiological functions (Lee et al., 2002; Negahban, Moharramipour, & Sefidkon, 2007). Due to their high volatility, they have fumigant and gaseous action on stored product insects. Carvacrol component has broad insecticidal and acaricidal activity against agricultural, stored product, medical pests and acts as a fumigant being highly toxic to adults of *S. oryzae* (Ahn, Lee, Lee, & Kim, 1998). Besides, menthol, methonene, limonene, β-pipene, α-pipene, pulegone, linalool and linalyl acetate exhibited fumigant toxicity in *S. oryzae* and inhibited AChE activity (Koul et al., 2008; Lee, Choi, Lee, & Park, 2001; Lee, Lee et al., 2001; Singh, Siddiqui, & Sharma, 1989). Caryophyllene, a volatile compound, was reported to be a strong fumigant and toxic to *S. zemais* (Chu, Liu, Jiang, & Liu, 2010).

The insecticidal activity varies with plant-derived material, insect species, and exposure time. The presence of volatile compounds having strong odor would have blocked the tracheal respiration of the insects leading to their death. Similar observation was made by Liu and Ho (1999) against *S. zeamais* and *T. castaneum*. Brown (1951) however, pointed out that the amount of fumigant absorbed depends on whether the insect's initial contact with the fumigant resulted in supplication or stimulation of the tracheal opening. Moreover, the ability of the insect to exclude vapor from its cuticle and prevent dehydration of body fluid plays a vital role in susceptibility or tolerance to fumigants of various life stages of insects particularly beetles and weevils infesting stored products (El-Nahal, Schmidt, & Risha, 1989). The toxic effect of essential oils, apart from the variability of phytochemical patterns, involves several other factors. The point of entry of the toxin is one of them where essential oils can be inhaled, ingested or skin absorbed by insects (Regnault-Roger, 1997).

The presence of volatile compounds is responsible for strong odor that could block the tracheal respiration of the insects leading to their death (Pugazhvendan, Ross, & Elumalai, 2012). The mode of action of oils was partially attributed to interference in normal respiration, resulting in suffocation (Schoonhoven, 1978). Most insects breathe through the trachea which usually leads to the opening of the spiracle.

These spiracles might have been blocked thereby leading to suffocation (Adedire, Obembe, Akinkurolere, & Oduleye, 2011; Ileke & Olotuah, 2012) Essential oils are presumed to interfere with basic metabolic, biochemical, physiological, and behavioral functions of insects (Mann & Kaufman, 2012). Essential oils block the spiracles, resulting in blockage of respiratory siphons (asphyxiation) and death (Kaufmann & Briegel, 2004; Rotimi, Chris, Olusola, Joshua, & Josiah, 2011). Further, Rattan (2010) reviewed the mechanism of action of essential oils on the body of insects and documented several physiological disruptions, such as inhibition of AChE, disruption of the molecular events of morphogenesis, and alteration in the behavior and memory of cholinergic system. Of these, the most important activity is the inhibition of AChE activity as it is a key enzyme responsible for terminating the nerve impulse transmission through synaptic pathway. Plant oils affect AChE and have an action on the nervous system (Mikhaiel, 2011). Recent research has demonstrated the interference of Monoterpenes with AChE activity in insects (Chaubey, 2012a, 2012b). Essential oils are lipophilic in nature and can be inhaled or ingested. The rapid action against insect pests is indicative of a neurotoxic mode of action and interference with the neuromodulator octopamine (Enan, 2005) or GABA-gated chloride channels (Priestley, Burgess, & Williamson, 2006).

Several essential oil components act on the octopaminergic system of insects. Octopamine is a neurotransmitter, neurohormone, and circulating neurohormone-neuromodulator, and its disruption results in total breakdown of the nervous system (Hollingworth, Johnstone, & Wright, 1984). Thus, the octopaminergic system of insects represents a target for insect control. Low molecular weight terpenoids are too lipophilic to be soluble in the hemolymph after crossing the cuticle, and the proposed route of entry is tracheae (Veal, 1996) and may also bind to target sites on receptors that modulate nervous activity (Hollingworth et al., 1984) and interrupt normal neurotransmission leading to paralysis and death.

3.6. Attractants

Botanicals chemicals that cause insects to make oriented movements toward their source are called insect attractants. They influence both gustatory (taste) and olfactory (smell) receptor or sensilla. Iso-thiocyantes from seeds of *Crucifera*, sugar and molasses and terpenes from bark in with pheromones are natural attractants for various insects of Cruciferaea and bark beetles. In onion propylmercapton from *Umbelliferae* and phenylacetaldehyde from flowers of *Araujia serisofera* are attracted carrot fly (*Psila rosae*) and Lepidoptera, respectively. Insect attractants can be used in three ways for the control of insects. In sampling or monitoring insect populations to assess the extent of infestation and decides the measure of control to be adapted in lusting insect to insecticide-coated traps or poison baits and in distracting insects from normal mating, aggregation feeding or ovipostion. They do not kill the insects therefore, do not disturb ecosystem. They can be used to misguide the insects to wrong oviposition sites whereby their number will go down by starvation or by producing unfertilized eggs. They cannot be relied as sole control measure used only in integrated control program (Arora, Singh, & Dhawan, 2012).

4. Conclusion

Botanical insecticides are natural chemicals extracted from plants with insecticidal properties and used as an excellent alternative to synthetic or chemical pesticides for crop protection to avoid negative or side effects of synthetic insecticides. Botanical pesticides (essential oils, flavonoids, alkaloids, glycosides, esters and fatty acids) have various chemical properties and modes of action and affect on insects in different ways namely; repellents, feeding deterrents/antifeedants, toxicants, growth retardants, chemosterilants, and attractants. So it is preferable to use the botanical insecticides instead of synthetic insecticide and these botanical insecticides are recognized by organic crop producers in industrialized countries. So, we recommended using botanical insecticidal and being promoted and research is being conducted to find new sources of botanical insecticides.

Funding
The authors received no direct funding for this research.

Competing Interests
The authors declare no competing interest.

Author details
Wafaa M. Hikal[1,2]
E-mail: wafaahikal@gmail.com
ORCID ID: http://orcid.org/0000-0002-7443-072X
Rowida S. Baeshen[1]
E-mail: rbaeben@ut.edu.sa
Hussein A.H. Said-Al Ahl[3]
E-mail: shussein272@yahoo.com
[1] Faculty of Science, Department of Biology, University of Tabuk, P.O. Box 741, Tabuk 71491, Saudi Arabia.
[2] Parasitology Laboratory, Water Pollution Researches Department, National Research Center, 33 El-Bohouth St., Dokki, Giza 12622, Egypt.
[3] Medicinal and Aromatic Plants Researches Department, National Research Centre, 33 El-Bohouth St., Dokki, Giza 12622, Egypt.

References

Abdelgaleil, S., Mohamed, M., Badawy, M., & El-arami, S. (2009). Fumigant and contact toxicities of monoterpenes to *Sitophilus oryzae* (L.) and *Tribolium castaneum* (Herbst) and their inhibitory effects on acetylcholinesterase activity. *Journal of Chemical Ecology, 35,* 518–525. https://doi.org/10.1007/s10886-009-9635-3

Abdullah, F., Subramanian, P., Ibrahim, H., Abdul Malek, S. N., Lee, G. S., & Hong, S. L. (2017). Chemical composition, antifeedant, repellent, and toxicity activities of the rhizomes of galangal, *Alpinia galanga* Against Asian Subterranean Termites, *Coptotermes gestroi* and *Coptotermes curvignathus* (Isoptera: Rhinotermitidae). *Journal of Insect Science, 15*(7), 2015.

Acheuk, F., & Doumandji-Mitiche, B. (2013). Insecticidal activity of alkaloids extract of *Pergularia tomentosa* (Asclepiadaceae) against fifth instar larvae of *Locusta migratoria* cinerascens (Fabricius 1781) (Orthoptera: Acrididae). *International Journal of Science and Advanced Technology, 3*(6), 8–13.

Adedire, C. O., Obembe, O. M., Akinkurolere, R. O., & Oduleye, S. O. (2011). Response of *Callosobruchus maculatus* Fabricius (Coleoptera: Chrysomelidae: Bruchinae) to extracts of cashew kernels. *Journal of Plant Diseases Protection, 118*(2), 75–79. https://doi.org/10.1007/BF03356385

Agarwal, A., Lal, S., & Gupta, K. C. (1988). Natural products as protectants of pulse betles. *Bulletin of Grain Technology, 26,* 154–164.

Ahn, Y. J., Lee, S. B., Lee, H. S., & Kim, G. H. (1998). Insecticidal and acaricidal activity of carvacrol and b-thujaplicine derived from *Thujopsis dolabrata* var. hondai sawdust. *Journal of Chemical Ecology, 24,* 1–90.

AlJabr, A. M., Hussain, A., Rizwan-ul-Haq, M., & Al-Ayedh, H. (2017). Toxicity of plant secondary metabolites modulating detoxification genes expression for natural red palm weevil pesticide development. *Molecules, 22,* 169. https://doi.org/10.3390/molecules22010169

Alkan, M., Gokce, A., & Kara, K. (2017). Contact toxicity of six plant extracts to different larval stages of Colorado potato beetle (*Leptinotarsa decemlineata* SAY (Col: Chrysomelidae)). *Journal of Agricultural Science, 23,* 309–316.

Al-Rajhy, D. H., Alahmed, A. M., Hussein, H. I., & Kheir, S. M. (2003). Acaricidal effects of cardiac glycosides, azadirachtin and neem oil against the camel tick, *Hyalomma dromedarii* (Acari: Ixodidae). *Pest Management Science, 59*(11), 1250–1254. https://doi.org/10.1002/(ISSN)1526-4998

de Araújo, A. M. N., Faroni, D. L. R., de Oliveira, J. V., Navarro, D. M. F., Barbosa, D. R. S., Breda, M. O., & de França, S. M. (2017). Lethal and sublethal responses of *Sitophilus zeamais* populations to essential oils. *Journal of Pest Science, 90*(2), 589–600. https://doi.org/10.1007/s10340-016-0822-z

Aref, S. P., Valizadegan, O., & Farashiani, M. E. (2015). *Eucalyptus dundasii* Maiden essential oil, chemical composition and insecticidal values against *Rhyzopertha dominica* (F.) and *Oryzaephilus surinamensis* (L.). *Journal of Plant Protection Research, 55,* 35–41.

Arora, R., Singh, B., & Dhawan, A. K. (2012). *Theory and Practice of Integrated Pest Management.* Jodhpur: Scientific Publishers.

Asawalam, E., & Adesiyan, S. (2001). Potential of *Ocimum basilicum* (Linn) for the control of maize weevil *Sitophilus zeamais* (Motsch). *Nigeria Agricultural Journal, 32*(1), 195–201.

Bakkali, F., Averbeck, S., Averbeck, D., & Idaomar, M. (2008). Biological effects of essential oils – A review. *Food and Chemical Toxicology, 46,* 446–475. https://doi.org/10.1016/j.fct.2007.09.106

Balandrin, M. F., Klocke, J. A., Wurtele, E. S., & Bollinger, W. H. (1985). Natural plant chemicals: Sources of industrial and medicinal materials. *Science, 7; 228*(4704), 1154–1160. https://doi.org/10.1126/science.3890182

Batish, D. R., Singh, H. P., Setia, N., Kaur, S., & Kohli, R. K. (2006). Chemical composition and phytotoxicity of volatile essential oils from intact and fallen leaves of *Eucalyptus citriodora. Zeitschrift für Naturforschung C, 61,* 465–471.

Batish, D. R., Singh, H. P., Kohli, R. K., & Kaur, S. (2008). Eucalyptus essential oil as a natural pesticide. *Forest Ecology and Management, 256,* 2166–2174. https://doi.org/10.1016/j.foreco.2008.08.008

Ben Jemaa, J. M., Tersim, N., Toudert, K. T., & Khouja, M. L. (2012). Insecticidal activities of essential oils from leaves of *Laurus nobilis* L. from Tunisia, Algeria and Morocco, and comparative chemical composition. *Journal of Stored Products Research, 48,* 97–104.

Bhatnagar-Thomas, P. L., & Pal, A. K. (1974). Studies on the insecticidal activity of garlic oil 2. Mode of action of the oil as a pesticide in *Musca domestica* nebulo Fabr and *Trogoderma granarium* Everts. *Journal of Food Science and Technology, 11*, 153–158.

Bouguerra, N., Djebbar, F. T., & Soltani, N. (2017). *Algerian Thymus vulgaris essential oil: Chemical composition and larvicidal activity against the mosquito Culex pipiens International Journal of Mosquito Research, 4*(1), 37–42.

Bowers, M. D., & Puttick, G. M. (1989). Iridoid glycosides and insect feeding preferences: gypsy moths (*Lymantria dispar*, Lymantriidae) and buckeyes (*Junonia coenia*, Nymphalidae). *Ecological Entomology, 14*, 247–256. https://doi.org/10.1111/j.1365-2311.1989.tb00953.x

Brown, A. W. A. (1951). *Insect control by chemicals* (p. 817). New York, NY; London: Wiley; Chapman & Hall.

Calderone, N. W., & Spivak, M. (1995). Plant extracts for control of the parasitic mite *Varroa jacobsoni* (Acari: Varroidae) in colonies of the western honey bee (Hymenoptera: Apidae). *Journal of Economic Entomology, 88*, 1211–1215. https://doi.org/10.1093/jee/88.5.1211

CDC (Center for Disease Control and Prevention, USA). (2005). *CDC adopts new repellent guidance for upcoming mosquito season.* Retrieved from http://www.cdc.gov/ncidod/dvbid/westnile/RepellentUpdates.htm

Chaaban, A., de Souza, A. L. F., Martins, C. E. N., Bertoldi, F. C., & Molento, M. B. (2017). Chemical composition of the essential oil of *Tagetes minuta* and its activity against *Cochliomyia macellaria* (Diptera: Calliphoridae). *European Journal of Medicinal Plants, 18*(1), 1–10.

Chagas, A. C. S., Passos, W. M., Prates, H. T., Leitem, R. C., Furlong, J., & Fortes, I. C. P. (2002). Acaricide effect of *Eucalyptus* spp. essential oils and concentrated emulsion on *Boophilus microplus*. *Brazilian Journal of Veterinary Research and Animal Science, 39*, 247–253.

Chaubey, M. K. (2012a). Responses of *Tribolium castaneum* (Coleoptera: Tenebrionidae) and *Sitophilus oryzae* (Coleoptera: Curculionidae) against essential oils and pure compounds. *Herba Polonica, 58*(3), 33–45.

Chaubey, M. K. (2012b). Biological effects of essential oils against rice weevil *Sitophilus oryzae* L. (Coleoptera: Curculionidae). *Journal of Essential Oil Bearing Plants, 15*, 809–815. https://doi.org/10.1080/0972060X.2012.10644124

Chaudhary, S., Kanwar, R. K., Sehgal, A., Cahill, D. M., Barrow, C. J., Sehgal, R., & Kanwar, J. R. (2017). Progress on *Azadirachta indica* based biopesticides in replacing synthetic toxic pesticides. *Frontiers in Plant Science, 8*, 610.

Chogo, J. B., & Crank, G. (1981). Chemical composition and biological activity of the tanzanian plant *Ocimum suave*. *Journal of Natural Products, 44*, 308–311. https://doi.org/10.1021/np50015a012

Choi, W., Lee, S. G., Park, H. M., & Ahn, Y. J. (2004). Toxicity of plant essential oils to *Tetranychus urticae* (Acari: Tetranychidae) and *Phytoseiulus persimilis* (Acari: Phytoseiidae). *Journal of Economic Entomology, 97*, 553–558. https://doi.org/10.1093/jee/97.2.553

Chu, S. S., Liu, S. L., Jiang, G. H., & Liu, Z. L. (2010). Composition and toxicity of essential oil of *Illicium simonsii* Maxim (Illiciaceae) fruit against the maize weevils. *Records of Natural Products, 4*, 205–210.

Cimanga, K., Kambu, K., Tona, L., Apers, S., De Bruyne, T., Hermans, N., ... Vlietinck, A. J. (2002). Correlation between chemical composition and antibacterial activity of essential oils of some aromatic medicinal plants growing in the Democratic Republic of Congo. *Journal of Ethnopharmacology,, 79*, 213–220. https://doi.org/10.1016/S0378-8741(01)00384-1

Correa, Y. D. C. G., Faroni, L. R., Haddi, K., Oliveira, E. E., & Pereira, E. J. G. (2015). Locomotory and physiological responses induced by clove and cinnamon essential oils in the maize weevil *Sitophilus zeamais. Pesticide Biochemistry and Physiology, 125*, 31–37. https://doi.org/10.1016/j.pestbp.2015.06.005

Dave, H., & Lediwane, L. (2012). A review on anthraquinones isolated from *Cassia* species and their applications. *Indian Journal of Natural Products and Resources, 3*, 291–319.

Dehghani-Samani, A., Madreseh-Ghahfarokhi, S., Dehghani-Samani, A., & Pirali-Kheirabadi, K. (2015). Acaricidal and repellent activities of essential oil of *Eucalyptus globulus* against *Dermanyssus gallinae* (Acari: Mesostigmata). *Journal of HerbMed Pharmacology, 4*(3), 81–84.

Diwan, R. K., & Saxena, R. C. (2010). Insecticidal property of flavinoid isolated from *Tephrosia purpuria*. *International Journal of Chemical Sciences, 8*(2), 777–782.

Don-Perdo, K. M. (1996). Investigation of single and joint fumigant insecticidal action of citrus peel oil components. *Journal of Pest Science, 46*, 79–84.

Elango, G., Rahuman, A. A., Bagavan, A., Kamaraj, C., Zahir, A. A., & Venkatesan, C. (2009). Laboratory study on larvicidal activity of indigenous plant extracts against *Anopheles subpictus* and *Culex tritaeniorhynchus*. *Parasitology Research, 104*(6), 1381–1388. https://doi.org/10.1007/s00436-009-1339-7

El-Nahal, A. K. M., Schmidt, G. H., & Risha, E. M. (1989). Vapours of *Acorus calamus* oil – A space treatment for stored-product insects. *Journal of Stored Products Research, 25*, 211–216. https://doi.org/10.1016/0022-474X(89)90026-X

El-Zemity, S., Hussien, R., Saher, F., & Ahmed, Z. (2006). Acaricidal activities of some essential oils and their monoterpenoidal constituents against house dust mite, *Dermatophagoides pteronyssinus* (Acari: Pyroglyphidae). *Journal of Zheijang University Science B, 7*, 957–962.

Elzen, G. W., & Hardee, D. D. (2003). United state department of agricultural-agricultural research on managing insect resistance to insecticides. *Pest Management Science, 59*, 770–776. https://doi.org/10.1002/(ISSN)1526-4998

Emam, A. M., Swelam, E. S., & Megally, N. Y. (2009). Furocoumarin and quinolone alkaloid with larvicidal and antifeedant activities isolated from *Ruta chalepensis* leaves. *Journal of Natural Products, 2*, 10–22.

Enan, E. E. (2005). Molecular and pharmacological analysis of an octopamine receptor from american cockroach and fruit fly in response to plant essential oils. *Archives of Insect Biochemistry and Physiology, 59*, 161–171. https://doi.org/10.1002/(ISSN)1520-6327

Fradin, M. S., & Day, J. F. (2002). Comparative efficacy of insect repellents against mosquito bites. *The New England Journal of Medicine, 347*, 13–18. https://doi.org/10.1056/NEJMoo011699

Gardulf, A., Wohlfart, I., & Gustafson, R. (2004). A prospective cross-over field trial shows protection of lemon eucalyptus extract against tick bites. *Journal of Medical Entomology, 41*, 1064–1067. https://doi.org/10.1603/0022-2585-41.6.1064

Germinara, G. S., Distefano, M. G., Acutis, L. D., Pati, S., Delfne, S., Cristofaro, A. D., & Rotundo, G. (2017). Bioactivities of *Lavandula angustifolia* essential oil against the stored grain pest *Sitophilus granaries*. *Bulletin of Insectology, 70*(1), 129–138.

Ghavami, M. B., Poorrastgoo, F., Taghiloo, B., & Mohammadi, J. (2017). Repellency effect of essential oils of some native plants and synthetic repellents against human flea, *Pulex irritans* (Siphonaptera: Pulicidae). *Journal of Arthropod-Borne Diseases, 11*(1), 105–115.

Ghoneim, K., & Hamadah, K. (2017). Antifeedant activity and detrimental effect of Nimbecidine (0.03% Azadirachtin) on the nutritional performance of Egyptian cotton leafworm *Spodoptera littoralis* Boisd. (Noctuidae: Lepidoptera). *Bio Bulletin, 31*, 39–55.

Giner, M., Avilla, J., Balcells, M., Caccia, S., & Smagghe, G. (2012). Toxicity of allyl esters in insect cell lines and in *Spodoptera littoralis* larvae. *Archives of Insect Biochemistry and Physiology, 79*(1), 18–30. https://doi.org/10.1002/arch.2012.79.issue-1

Goławska, S., & Łukasik, I. (2012). Antifeedant activity of luteolin and genistein against the pea aphid. *Journal of Pest Science, 85*, 443–450. https://doi.org/10.1007/s10340-012-0452-z

Golawska, S., Kapusta, I., Łukasik, I., & Wojcicka, A. (2008). Effect of phenolics on the pea aphid, *Acyrthosiphon pisum* (Harris) population on *Pisum sativum* L. (Fabaceae). *Pestycydy/Pesticides, 3–4*, 71–77.

Goławska, S., Łukasik, I., Goławski, A., Kapusta, I., & Janda, B. (2010). Alfalfa (*Medicago sativa* L.) apigenin glycosides and their effect on the pea aphid (*Acyrthosiphon pisum*). *Polish Journal of Environmental Studies, 19*, 913–920.

Goławska, S., Łukasik, I., Kapusta, I., & Janda, B. (2012). Do the contents of luteolin, tricin, and chrysoeriol glycosides in alfalfa (*Medicago sativa* L.) affect the behavior of pea aphid (*Acyrthosiphon pisum*)? *Polish Journal of Environmental Studies, 21*, 1613–1619.

Goławska, S., Sprawka, I., Łukasik, I., & Goławski, A. (2014). Are naringenin and quercetin useful chemicals in pest-management strategies? *Journal of Pest Science, 87*, 173–180. https://doi.org/10.1007/s10340-013-0535-5

Gould, K. S., & Lister, C. (2006). Flavonoid functions in plants. In *Flavonoids: Chemistry, biochemistry,and applications* (pp. 397–443). Boca Raton, FL: CRC Press LLC.

Guedes, R. N. C., Oliveira, E. E., Guedes, N. M. P., Ribeiro, B., & Serrão, J. E. (2006). Cost and mitigation of insecticide resistance in the maize weevil, *Sitophilus zeamais*. *Physiological Entomology, 31*, 30–38. https://doi.org/10.1111/pen.2006.31.issue-1

Qari, S. H., Nilly, A. H., Abdel-Fattah, A. H., & Shehawy, A. A. (2017). Assessment of DNA damage and biochemical responses in *Rhyzopertha dominica* exposed to some plant volatile oils. *Journal of Pharmacology and Toxicology, 12*, 87–96. https://doi.org/10.3923/jpt.2017.87.96

Halliwell, B., & Gutteridge, J. M. C. (1999). *Free radicals in biology and medicine* (3rd ed.). Oxford: Oxford University Press.

Hieu, T. T., Choi, W. S., Kim, S. I., Wang, M., & Ahn, Y. J. (2015). Enhanced repellency of binary mixtures of *Calophyllum inophyllum* nut oil fatty acids or their esters and three terpenoids to *Stomoxys calcitrans*. *Pest Management Science, 71*, 1213–1218. https://doi.org/10.1002/ps.2015.71.issue-9

Ho, S. H., Ma, Y., Goh, P. M., & Sim, K. Y. (1995). Star anise, *Illicium verum* Hook F., as a potential grain protectant against *Tribolium castaneum* (Herbst) and *Sitophilus zeamais* (Motsch.). *Postharvest Biology and Technology, 6*, 341–347. https://doi.org/10.1016/0925-5214(95)00015-X

Ho, S. H., Koh, L., Ma, Y., Huang, Y., & Sim, K. Y. (1996). The oil of garlic, *Allium sativum* L. (Amaryllidaceae), as apotential grain protectant against *Tribolium castaneum* (Herbst) and *Sitophilus zeamais* Motsch. *Postharvest Biology and Technology, 9*, 41–48. https://doi.org/10.1016/0925-5214(96)00018-X

Hollingworth, R. M., Johnstone, E. M., & Wright, N. (1984). Pesticide synthesis through rational approaches. In P. S. Magee, G. K. Kohn, & J. J. Menn (Eds.), *ACS symposium series No. 255* (pp. 103–125). Washington, DC: American Chemical Society.

Hollingworth, R., Ahammadsahib, K., Gadelhak, G., & McLaughlin, J. (1994). New inhibitors of complex I of the mitochondrial electron transport chain with activity as pesticides. *Biochemical Society Transactions, 22*(1), 230–233. https://doi.org/10.1042/bst0220230

Houghton, P. J., Ren, Y., & Howes, M. J. (2006). Acetylcholinesterase inhibitors from plants and fungi. *Natural Product Reports, 23*(2), 181–199. https://doi.org/10.1039/b508966 m

Huang, Y., Lam, S. L., & Ho, S. H. (2000). Bioactivities of essential oils from *Elletaria cardamomum* (L.) Maton. to *Sitophilus zeamais* Motschulsky and *Tribolium castaneum* (Herbst). *Journal of Stored Products Research, 36*, 107–117. https://doi.org/10.1016/S0022-474X(99)00040-5

ICIPE (International Centre of Insect Physiology and Ecology) (1997). *Vision and strategic framework towards 2020*. Nairobi: ICIPE Science Press.

Ileke, K. D., & Olotuah, O. F. (2012). Bioactivity of *Anacardium occidentals* and *Allium sativum* powders and oils extracts against cowpea bruchid, *Callosobruchus maculatus* (Fab) (Coleoptera: Bruchidae). *International Journal of Biological Science, 4*(1), 96–103.

Imms, A. D. (1964). *Outlines of entomology* (5th ed., p. 224). London: Methuen.

Isman, M. B. (2006). Botanical insecticides, deterrents, and repellents in modern agriculture and an increasingly regulated world. *Annual Review of Entomology, 51*, 45–66. https://doi.org/10.1146/annurev.ento.51.110104.151146

Isman, M. B., & Machial, C. M. (2006). Pesticides based on plant essential oils: From traditional practice to commercialization. In: Rai, M., Carpinella, M.C. (Eds.), naturally occurring bioactive compounds. *Advances in Phytomedicine, 3*, 29–44. https://doi.org/10.1016/S1572-557X(06)03002-9

Jayakumar, M., Arivoli, S., Raveen, R., & Tennyson, S. (2017). Repellent activity and fumigant toxicity of a few plant oils against the adult rice weevil *Sitophilus oryzae* Linnaeus 1763 (Coleoptera: Curculionidae). *Journal of Entomology and Zoology Studies, 5*(2), 324–335.

Jose, S., & Sujatha, K. (2017). Antifeedant activity of different solvent extracts of *Gliricidia sepium* against third in star larvae of *Helicoverpa armigera* (Hubner) (Lepidoptera: Noctuidae). *International Journal of Advanced Research in Biological Sciences (IJARBS), 4*(4), 201–204. https://doi.org/10.22192/ijarbs

Karunamoorthi, K., Girmay, A., & Hayleeyesus, S. F. (2014). Mosquito repellent activity of essential oil of Ethiopian ethnomedicinal plant against Afro-tropical malarial vector *Anopheles arabiensis*. *Journal of King Saud University of Science, 26*, 305–310. https://doi.org/10.1016/j.jksus.2014.01.001

Kanat, M., Hakk, M., & Alma, M. (2003). Insecticidal effects of essential oils from various plants against larvae of pine processionary moth (*Thaumetopoea pityocampa* Schiff) (Lepidoptera: Thaumetopoeidae). *Pest Management Science, 60*, 173–177.

Kaufmann, C., & Briegel, H. (2004). Flight performance of the malaria vectors *Anopheles gambiae* and *Anopheles atroparous*. *Journal of Vector Ecology, 29*(1), 140–153.

Kim, S. I., Roh, J. Y., Kim, D. H., Lee, H. S., & Ahn, Y. J. (2003). Insecticidal activities of aromatic plant extracts and essential oils against *Sitophilus oryzae* and *Callosobruchus chinensis*. *Journal of Stored Products Research, 39*, 293–303. https://doi.org/10.1016/S0022-474X(02)00017-6

Kimutai, A., Ngeiywa, M., Mulaa, M., Njagi, P. G. N., Ingonga, J., Nyamwamu, L. B., ... Ngumbi, P. (2017). Repellent effects of the essential oils of *Cymbopogon citratus* and *Tagetes minuta* on the sandfly, *Phlebotomus duboscqi*. *BMC Research Notes, 10*, 98. https://doi.org/10.1186/s13104-017-2396-0

Kordali, S., Cakir, A., Mavi, A., Kilic, H., & Yildirim, A. (2005). Screening of chemical composition and antifungal and antioxidant activities of the essential oils from three Turkish artemisia species. *Journal of Agricultural and Food Chemistry, 53*, 1408–1416. https://doi.org/10.1021/jf048429n

Koshier, E. L., & Sedy, K. A. (2001). Effect of plant volatiles on the feeding and oviposition of Thripstabaci. In R. Marullo, & L. Kound (Eds.), *Thrips and Tospoviruses* (pp. 185–187). Australia: CSIRO.

Koul, O., Waliai, S., & Dhaliwal, G. S. (2008). Essential oils as green pesticides: Potential and constraints. *Biopesticides International, 4*(1), 63–84.

Kraiss, H., & Cullen, E. M. (2008). Insect growth regulator effects of azadirachtin and neem oil on survivorship, development and fecundity of *Aphis glycines* (Homoptera: Aphididae) and its predator, *Harmonia axyridis* (Coleoptera: Coccinellidae). *Pest Management Science, 64*(6), 660–668. https://doi.org/10.1002/(ISSN)1526-4998

Kubo, I., & Kim, M. (1989). New insect growth inhibitory flavan glycosides from *Viscum tuberculatum* Tetrahedron. *Letters, 28*(9), 921–924.

Kumar, P., Bhadauria, T., & Mishra, J. (2015). Impact of application of insecticide quercetin/azadirachtin and chlorpyrifos on earthworm activities in experimental soils in Uttar Pradesh India. *Science Postprint, 1*(2), e00044.

Lee, S. E. (2000). Mosquito larvicidal activity of pipernonaline, a piperidine alkaloid derived from long pepper, *Piper longum*. *Journal of the American Mosquito Control Association, 16*(3), 245–247.

Lee, B. H., Choi, W. S., Lee, S. E., & Park, B. S. (2001). Fumigant toxicity of essential oils and their constituent compounds towards the rice weevil, *Sitophilus oryzae* (L.). *Crop Protection, 20*, 317–320. https://doi.org/10.1016/S0261-2194(00)00158-7

Lee, S. E., Lee, B. H., Choi, W. S., Park, B. S., Kim, J. G., & Campbell, B. C. (2001). Fumigant toxicity of volatile natural products from Korean spices and medicinal plants towards the rice weevil, *Sitophilus oryzae* (L.). *Pest Management Science, 57*, 548–553. https://doi.org/10.1002/(ISSN)1526-4998

Lee, B. H., Lee, S. E., Annis, P. C., Pratt, S. J., Park, B. S., & Tumaalii, F. (2002). Fumigant toxicity of essential oils and Monoterpenes against the red flour beetle, *Tribolium castaneum* Herbst. *Journal of Asia-Pacific Entomology, 5*(2), 237–240. https://doi.org/10.1016/S1226-8615(08)60158-2

Li, H., Madden, J. L., & Potts, B. M. (1996). Variation in volatile leaf oils of the Tasmanian *Eucalyptus* species II. Subgenus Symphyomyrtus. *Biochemical Systematics and Ecology, 24*, 547–569. https://doi.org/10.1016/0305-1978(96)00040-3

Liao, M., Xiao, J. J., Zhou, L. J., Yao, X., Tang, F., Hua, R. M., ... Cao, H. Q. (2017). Chemical composition, insecticidal and biochemical effects of *Melaleuca alternifolia* essential oil on the *Helicoverpa armigera*. *Journal of Applied Entomology, 2017*, 1–8.

Liu, Z. H., & Ho, S. H. (1999). Bioactivity of essential oil extracted from *Evodia rutracurpa* (Hook F. et Thomas) against the grain storage insects, *Sitophilus zeamais* (Motsch.) and Tribolium castaneum (Herbst.). *Journal of Stored Products Research, 35*, 317–328. https://doi.org/10.1016/S0022-474X(99)00015-6

Liu, X., Chen, Q., Wang, Z., Xie, L., & Xu, Z. (2008). Allelopathic effects of essential oil from *Eucalyptus grandis*, E. urophylla on pathogenic fungi and pest insects. *Frontiers of Forestry in China, 3*, 232–236. https://doi.org/10.1007/s11461-008-0036-5

Loko, L. Y., Alagbe, O., Dannon, E.A., Datinon, B., Orobiyi, A., Thomas-Odjo, A., ... Tamò, M. (2017). Repellent effect and insecticidal activities of *Bridelia ferruginea*, *Blighia sapida*, and *Khaya senegalensis* leaves powders and extracts against *Dinoderus porcellus* in infested dried yam chips. *Psyche*, 2017 Article ID 5468202, 18 pages.

López, S. B., López, M. L., Aragón, L. M., Tereschuk, M. L., Slanis, A. C., Feresin, G. E., ... Tapia, A. A. (2011). Composition and anti-insect activity of essential oils from *Tagetes* Species (Asteraceae, Helenieae) on *Ceratitis capitata* Wiedemann and *Triatoma infestans* Klug. *Journal of Agricultural and Food Chemistry, 59*(10), 5286–5292. https://doi.org/10.1021/jf104966b

Lu, F. C. (1995). A review of the acceptable daily intakes of pesticides assessed by the world health organization. *Regulatory Toxicology and Pharmacology, 21*, 351–364.

Lucia, A., Audino, P. G., Seccacini, E., Licastro, S., Zerba, E., & Masuh, H. (2007). Larvicidal effect of *Eucalyptus grandis* essential oil and turpentine and their major components on *Aedes aegypti* larvae. *Journal of the American Mosquito Control Association, 23*, 299–303. https://doi.org/10.2987/8756-971X(2007)23[299:LEOEGE]2.0.CO;2

Lucia, A., Toloza, A. C., Guzmán, E., Ortega, F., & Rubio, R. G. (2017). Novel polymeric micelles for insect pest control: Encapsulation of essential oil monoterpenes inside a triblock copolymer shell for head lice control. *Peer-Reviewed Journal, 5*, e3171. https://doi.org/10.7717/peerj.3171

Malek, M., & Parveen, B. (1989). Effect of insects infestation on the weight loss and viability of stored BE paddy. *Bangladesh Journal of Zoology, 17*(1), 83–85.

Mallikarjuna, N., Kranthi, K. R., Jadhav, D. R., Kranthi, S., & Chandra, S. (2004). Influence of foliar chemical compounds on the development of *Spodoptera litura* in interspecific derivatives of groundnut. *Journal of Applied Entomology, 128*, 321–328. https://doi.org/10.1111/jen.2004.128.issue-5

Mann, R. S., & Kaufman, P. E. (2012). Natural product pesticides: Their development, delivery and use against insect vectors. *Mini-Reviews in Organic Chemistry, 9*, 185–202. https://doi.org/10.2174/157019312800604733

Mathur, Y. K., Shankar, K., & Ram, S. (1985). Evaluation of some grain protectants against *Callosobruchus chinensis* (L.) on black gram. *Bulletin of Grain Technology, 23*, 253–259.

Matsumura, F. (1985). *Toxicology of insecticides* (2nd ed., pp. 11–43). New York, NY: Plenum Press. https://doi.org/10.1007/978-1-4613-2491-1

Mikhaiel, A. A. (2011). Potential of some volatile oils in protecting packages of irradiated wheat flour against *Ephestia kuheniella* and *Tribolium castaneum*. *Journal of Stored Products Research, 47*(4), 357–364. https://doi.org/10.1016/j.jspr.2011.06.002

Min, L., Jin-Jing, X., Li-Jun, Z., Liu, Y., Xiang-Wei, W., Ri-Mao, H., ... Hai-Qun, C. (2016). Insecticidal activity of *Melaleuca alternifolia* essential oil and RNA-seq analysis of *Sitophilus zeamais* transcriptome in response to oil fumigation. *PLoS One, 11*(12), e0167748.

Moretti, M. D. L., Sanna-Passino, G., Demontis, S., & Bazzoni, E. (2000). Essential oil formulations useful as a new tool for insect pest control. *AAPS Pharmaceutical Science and Technology, 3*(2), 13.

Morimoto, M., Kumeda, S., & Komai, K. (2000). Insect antifeedant flavonoids from *Gnaphalium affine*. *Journal of Agricultural and Food Chemistry, 48*, 1888–1891. https://doi.org/10.1021/jf990282q

Morrison, N. I., Franz, G., Koukidou, M., Miller, T. A., Saccone, G., Alphey, L. S., ... Polito, L. C. (2010). Genetic improvements to the sterile insect technique for agricultural pests. *Asia-Pacific Journal of Molecular Biology and Biotechnology, 18*(2), 275–295.

Mullens, B. A., Reifenrath, W. G., & Butler, S. M. (2009). Laboratory trials of fatty acids as repellents or antifeedants against houseflies, horn flies and stable flies (Diptera: Muscidae). *Pest Management Science, 65*(12), 1360–1366. https://doi.org/10.1002/ps.v65:12

Navarro-Llopis, V., Vacas, S., Sanchis, J., Primo, J., & Alfaro, C. (2011). Chemosterilant bait stations coupled with sterile insect technique: An integrated strategy to control the mediterranean fruit fly (Diptera: Tephritidae). *Journal of Economic Entomology, 104*(5), 1647–1655. https://doi.org/10.1603/EC10448

Negahban, M., Moharramipour, S., & Sefidkon, F. (2007). Fumigant toxicity of essential oil from *Artemisia sieberi* Besser against three stored product insects. *Journal of Stored Products Research, 43*(2), 123–128. https://doi.org/10.1016/j.jspr.2006.02.002

Nerio, L. S., Olivero-Verbel, J., & Stashenko, E. (2009). Repellency activity of essential oils from seven aromatic plants grown in Colombia against *Sitophilus zeamais* Motschulsky (Coleoptera). *Journal of Stored Products Research*, 45, 212–214. https://doi.org/10.1016/j.jspr.2009.01.002

Obeng-Ofori, D., & Amitaye, S. (2005). Efficacy of mixing vegetable oils with pirimiphos-methyl against the maize weevil, *Sitophilus zeamias* Motchulsky, in stored maize. *Journal of Stored Products Research*, 41(1), 57–66. https://doi.org/10.1016/j.jspr.2003.11.001

Padin, S. B., Fuse, C., Urrutia, M. I., & DalBello, G. M. (2013). Toxicity and repellency of nine medicinal plants against *Tribolium castaneum* in stored wheat. *Bulletin of Insectology*, 66(1), 45–49.

Papachristos, D. P., Karamanoli, K., Stamopoulos, D. C., & Menkissoglu-Spiroudi, U. (2004). The relationship between the chemical composition of three essentialoils and their insecticidal activity against *Acanthoscelides obtectus* (Say). *Pest Management Science*, 60, 514–520. https://doi.org/10.1002/(ISSN)1526-4998

Papanastasiou, S. A., Bali, E. M. D., Ioannou, C. S., Papachristos, D. P., Zarpas, K. D., & Papadopoulos, N. T. (2017). Toxic and hormetic-like effects of three components of citrus essential oils on adult Mediterranean fruit flies (*Ceratitis capitata*). *PLoS One*, 12(5), e0177837. https://doi.org/10.1371/journal.pone.0177837

Park, D. S., & Coats, J. R. (2002). Cyanogenic glycosides: Alternative insecticides? *The Korean Journal of Pesticide Science*, 6(2), 51–57.

Park, J., Jeon, Y., Lee, C., Chung, N., & Lee, H. (2017). Insecticidal toxicities of carvacrol and thymol derived from *Thymus vulgaris* Lin. against *Pochazia shantungensis* Chou & Lu., newly recorded pest *Scientific Reports*, 7, 40902. https://doi.org/10.1038/srep40902

Pereira, S. G., Sanaveerappanavar, V. T., & Murthy, M. S. (2006). Geographical variation in the susceptibility of the diamond back moth *Ptlutella xylostella* L. to *Bacillus thuringiensis* products and acylurea compounds. *Pest Management*, 15, 26–26.

Perumalsamy, H., Jang, M. J., Kim, J. R., Kadarkarai, M., & Ahn, Y. J. (2015). Larvicidal activity and possible mode of action of four flavonoids and two fatty acids identified in *Millettia pinnata* seed toward three mosquito species. *Parasites & Vectors*, 8, 237–244. https://doi.org/10.1186/s13071-015-0848-8

Plata-Rueda, A., Martínez, L. C., Santos, M. H. D., Fernandes, F. L., Wilcken, C. F., Soares, M. A., ... Zanuncio, J. C. (2017). Insecticidal activity of garlic essential oil and their constituents against the mealworm beetle, *Tenebrio molitor* Linnaeus (Coleoptera: Tenebrionidae). *Scientific Reports*, 7, 46406. https://doi.org/10.1038/srep46406

Priestley, C. M., Burgess, I. F., & Williamson, E. M. (2006). Lethality of essential oil constituents towards the human louse, *Pediculus humanus* and its eggs. *Fitoterapia*, 77, 303–309. https://doi.org/10.1016/j.fitote.2006.04.005

Prowse, M. G., Galloway, T. S., & Foggo, A. (2006). Insecticidal activity of garlic juice in two dipteran pests. *Agricultural and Forest Entomology*, 8, 1–6. https://doi.org/10.1111/afe.2006.8.issue-1

Pugazhvendan, S. R., Ross, P. R., & Elumalai, K. (2012). Insecticidal and repellent activities of four indigenous medicinal plants against stored grain pest, *Tribolium castaneum* (Herbst) (Coleoptera: Tenebrionidae). *Asian Pacific Journal of Tropical Disease*, 2, S16–S20. https://doi.org/10.1016/S2222-1808(12)60116-9

Pujiarti, R., & Fentiyanti, P. K. (2017). Chemical compositions and repellent activity of Eucalyptus tereticornis and *Eucalyptus deglupta* essential oils against *Culex quinquefasciatus* mosquito. *Thai Journal of Pharmaceutical Sciences*, 41(1), 19–24.

Rahdari, T., & Hamzei, M. (2017). Repellency effect of essential oils of *Mentha piperita, Rosmarinus officinalis* and *Coriandrum sativum* on *Tribolium confusum* duval (Coleoptera: Tenebrionidae). *Chemistry Research Journal*, 2(2), 107–112.

Rajashekar, Y., Bakthavatsalam, N., & Shivanandappa, T. (2012). Botanicals as grain protectants. *Psyche*, 2012, 1–13.

Rattan, R. S. (2010). Mechanism of action of insecticidal secondary metabolites of plant origin. *Crop Protection*, 29, 913–920. https://doi.org/10.1016/j.cropro.2010.05.008

Regnault-Roger, C. (1997). The potential of botanical essential oils for insect pest control. *Integrated Pest Management Reviews*, 2, 25–34. https://doi.org/10.1023/A:1018472227889

Regnault-Roger, C., & Hamraoui, A. (1993). Influence d'huiles essentielles sur *Acanthoscelides obtectus* Say, bruche du haricot. *Acta Botanica Gallica*, 140, 217–222. https://doi.org/10.1080/12538078.1993.10515584

Regnault-Roger, C., & Philogène, B. J. R. (2008). Past and current prospects for the use of botanicals and plant allelochemicals in integrated pest management. *Pharmaceutical Biology*, 46, 41–52. https://doi.org/10.1080/13880200701729794

Regnault-Roger, C., Vincent, C., & Arnason, J. T. (2012). Essential oils in insect control: Low-risk products in a high-stakes world. *Annual Review of Entomology*, 57, 405–424. https://doi.org/10.1146/annurev-ento-120710-100554

Rotimi, O. A., Chris, O. A., Olusola, O. O., Joshua, R., & Josiah, A. O. (2011). Bioefficacy of extracts of some indigenous Nigerian plants on the developmental stages of mosquito (*Anopheles gambiae*). *Jordan Journal of Biological Sciences*, 4(4), 237–242.

Samuel, M., Oliver, S. V., Wood, O. R., Coetzee, M., & Brooke, B. D. (2015). Evaluation of the toxicity and repellence of an organic fatty acids mixture (C8910) against insecticide susceptible and resistant strains of the major malaria vector *Anopheles funestus* Giles (Diptera: Culicidae). *Parasites & Vectors*, 8, 321. https://doi.org/10.1186/s13071-015-0930-2

Santos, J. P., Maia, J. D. G., & Cruz, I. (1990). Damage to germination of seed corn caused by maize weevil (*Sitophilus zeamais*) and Angoumois grain moth (*Sitotroga cerealella*). *Pesquisa Agropecuaria Brasileira*, 25(12), 1687–1692.

Santos, P. C., Santos, V. H. M., Mecina, G. F., Andrade, A. R., Fegueiredo, P. A., Moraes, V. M. O., ... Silva, R. M. G. (2016). Insecticidal activity of *Tagetes* sp. on *Sitophilus zeamais* Mots. *International Journal of Environmental & Agriculture Research*, 2(4), 31–38.

Schmidt, S., Tomasi, C., Pasqualini, E., & Ioriatti, C. (2008). The biological efficacy of pear ester on the activity of *Granulosis virus* for codling moth. *Journal of Pest Science*, 81, 29. https://doi.org/10.1007/s10340-007-0181-x

Schoonhoven, A. V. (1978). The use of vegetable oils to protect stored beans from bruchid attack. *Journal of Economic Entomology*, 71(2), 254–256. https://doi.org/10.1093/jee/71.2.254

Seyoum, A., Killeen, G. F., Kabiru, E. W., Knols, B. G. I., & Hassanali, A. (2003). Field efficacy of thermally expelled or live potted repellent plants against African malaria vectors in western Kenya. *Tropical Medicine & International Health*, 8, 1005–1011. https://doi.org/10.1046/j.1360-2276.2003.01125.x

Shaalan, E. A. S., Canyon, D., Younes, M. W. F., Abdel-Wahab, H., & Mansour, A. H. (2005). A review of botanical phytochemicals with mosquitocidal potential. *Environment International*, 31(8), 1149–1166. https://doi.org/10.1016/j.envint.2005.03.003

Sharififard, M., Safdari, F., Siahpoush, A., Hamid, H., & Kassiri, H. (2016). Evaluation of some plant essential oils against the brown-banded cockroach, *Supella longipalpa* (Blattaria: Ectobiidae): A mechanical vector of human pathogens. *Journal of Arthropod-Borne Diseases*, 10(4), 528–537.

Shelton, A. M., Zhao, J. Z., & Roush, R. T. (2002). Economic, ecological, food safety, and social consequences of the deployment of B-transgenic plants. *Annual Review of Entomology, 47*, 845–881. https://doi.org/10.1146/annurev.ento.47.091201.145309

Showler, A. T. (2017). Botanically based repellent and insecticidal effects against horn flies and stable flies (Diptera: Muscidae). *Journal of Integrated Pest Management, 8,1*(15), 1–11.

Shukla, P., Vidyasgar, P. S. P. V., Aldosari, S. A., & Abdel-Azim, M. (2012). Antifeedant activity of three essential oils against the red palm weevil, *Rhynchophorus ferrugineus*. *Bulletin of Insectology, 65*(1), 71–76.

Silva, V. C. B., Ribeiro Neto, J. A., Alves, S. N., & Li, L. A. R. S. (2015). Larvicidal activity of oils, fatty acids, and methyl esters from ripe and unripe fruit of *Solanum lycocarpum* (Solanaceae) against the vector *Culex quinquefasciatus* (Diptera: Culicidae). *Revista da Sociedade Brasileira de Medicina Tropical, 48*(5), 610–613. https://doi.org/10.1590/0037-8682-0049-2015

Simmonds, M. S. (2003). Flavonoid–insect interactions: Recent advances in our knowledge. *Phytochemistry, 64*, 21–30. https://doi.org/10.1016/S0031-9422(03)00293-0

Simmonds, M. S., & Stevenson, P. C. (2001). Effects of isoflavonoids from cicer on larvae of *Heliocover paarmigera*. *Journal of Chemical Ecology, 27*, 965–977. https://doi.org/10.1023/A:1010339104206

Singh, V. K., & Singh, D. K. (1996). Enzyme inhibition by allicin, the molluscicidal agent of *Allium sativum* L. (garlic). *Phytotherapy Research, 10*, 383–386. https://doi.org/10.1002/(ISSN)1099-1573

Singh, D., Siddiqui, M. S., & Sharma, S. (1989). Reproduction retardant and fumigant properties in essential oils against rice weevil (Coleptera: Curculionidae) in stored wheat. *Journal of Economic Entomology, 82*, 727–733. https://doi.org/10.1093/jee/82.3.727

Sithisut, D., Fields, P. G., & Chandrapathya, A. (2011). Contact toxicity, feeding reduction and repellency of essential oils from three plants from the ginger family (Zingiberaceae) and their major components against *Sitophiluszeamais* and *Tribolium castaneum*. *The Journal of Stored Products, 104*, 1445–1454.

Talukder, F. A. (2006). Plant products as potential stored-product insect management agents-A mini review. *Emirates Journal of Food and Agriculture, 18*(1), 17–32. https://doi.org/10.9755/ejfa.

Talukder, F., Islam, M., Hossain, M., Rahman, M., & Alam, M. (2004). Toxicity effects of botanicals and synthetic insecticides on *Tribolium castaneum* (Herbst) and *Rhyzopertha dominica* (F.). *Bangladesh. Journal of Environmental Sciences, 10*(2), 365–371.

Tholl, D. (2006). Terpene synthases and the regulation, diversity and biological rolesof terpene metabolism. *Current Opinion in Plant Biology, 9*, 297–304. https://doi.org/10.1016/j.pbi.2006.03.014

Toloza, A. C., Zygadlo, J., Cueto, G. M., Biurrun, F., Zerba, E., & Piccolo, M. S. (2006). Fumigant and repellent properties of essential oils and component compounds against permethrin-resistant *Pediculus humanus* capitis (Anoplura: Pediculidae) from Argentina. *Journal of Medical Entomology, 43*(5), 889–895. https://doi.org/10.1093/jmedent/43.5.889

Tripathi, A. K., Prajapati, V., Aggarwal, K. K., Kumar, S., Kukreja, A. K., Dwivedi, S., & Singh, A. K. (2000). Effects of volatile oil constituents of *Mentha* species against stored grain pests, *Callosobrunchus maculatus* and *Tribolium castaneum*. *Journal of Medicinal and Aromatic Plant Sciences, 22*, 549–556.

Tripathi, A. K., Prajapati, V., Khanuja, S. P. S., & Kumar, S. (2003). Effect of d-limonene on three stored-product beetles. *Journal of Economic Entomology, 96*, 990–995. https://doi.org/10.1093/jee/96.3.990

Trivedi, A., Nayak, N., & Kumar, J. (2017). Fumigant toxicity study of different essential oils against stored grain pest *Callosobruchus chinensis*. *Journal of Pharmacognosy and Phytochemistry, 6*(4), 1708–1711.

Veal, L. (1996). The potential effectiveness of essential oils as a treatment for headlice, *Pediculus humanus* capitis. *Complementary Therapies in Nursing and Midwifery, 2*, 97–101. https://doi.org/10.1016/S1353-6117(96)80083-7

Velu, K., Elumalai, D., Hemalatha, P., Babu, M., Janaki, A., & Kaleena, P. K. (2015). Phytochemical screening and larvicidal activity of peel extracts of *Arachis hypogaea* against chikungunya and malarial vectors. *International Journal of Mosquito Research, 2*(1), 01–08.

Wachira, S. W., Omar, S., Jacob, J. W., Wahome, M., Alborn, H. T., Spring, D. R., ... Torto, B. (2014). Toxicity of six plant extracts and two pyridine alkaloids from *Ricinus communis* against the malaria vector *Anopheles gambiae*. *Parasites & Vectors, 7*, 312. https://doi.org/10.1186/1756-3305-7-312

Waliwitiya, R., Kennedy, C., & Lowenberger, C. (2008). Larvicidal and oviposition altering activity of monoterpenoids, trans-anethole and rosemary oil to the yellow fever mosquito *Aedes aegypti* (Diptera: Culicidae). *Pest Management Science, 65*(3), 241–248.

Watanabe, K., Shono, Y., Kakimizu, A., Okada, A., Matsuo, N., Satoh, A., & Nishimura, H. (1993). New mosquito repellent from *Eucalyptus camaldulensis*. *Journal of Agricultural and Food Chemistry, 41*, 2164–2166. https://doi.org/10.1021/jf00035a065

Weaver, D. K., Dunkel, F. V., Ntezurubanza, L., Jackson, L. L., & Stock, D. T. (1991). The efficacy of linalool, a major component of freshly-milled *Ocimum canum* Sims (Lamiaceae), for protection against postharvest damage by certain stored product Coleoptera. *Journal of Stored Products Research, 27*, 213–220. https://doi.org/10.1016/0022-474X(91)90003-U

Wilke, A. B. B., Nimmo, D. D., John, O., Kojin, B. B., Capurro, M. L., & Marrelli, M. T. (2009). Mini-review: Genetic enhancements to the sterile insect technique to control mosquito populations. *Asia-Pacific Journal of Molecular Biology and Biotechnology, 17*(3), 65–74.

Wimmer, Z., Alexandra, J. F. D. M. Floro, Zarevúcka, M., Wimmerová, M., Sello, G., & Orsini, F. (2007). Insect pest control agents: Novel chiral butanoate esters (juvenogens). *Bioorganic & Medicinal Chemistry, 15*(18), 6037–6042.

Wu, Y., Guo, S., Huang, D., Wang, C., Wei, J., Li, Z., ... Du, S. (2017). Contact and repellent activities of zerumbone and its analogues from the essential oil of *Zingiber zerumbet* (L.) Smith against *Lasioderma serricorne*. *Journal of Oleo Science, 66*(4), 399–405. https://doi.org/10.5650/jos.ess16166

Yousef, H., EL-Lakwah, S. F., & EL Sayed, Y. A. (2013). Insecticidal activity of linoleic acid against *Spodoptera littoralis* (BOISD.). *Egyptian Journal of Agricultural Research, 91*(2), 573.

Zagrobelny, M., Bak, S., Rasmussen, A. V., Jørgensen, B., Naumann, C. M., & Møller, B. L. (2004). Cyanogenic glucosides and plant-insect interactions. *Phytochemistry, 65*(3), 293–306. https://doi.org/10.1016/j.phytochem.2003.10.016

Zhang, W., Zhang, Z., Chen, Z., Liang, J., Geng, Z., Guo, S., ... Deng, Z. (2017). Chemical composition of essential oils from six *Zanthoxylum* species and their Repellent activities against two stored-product insects. *Journal of Chemistry*, Article ID 1287362, 7 pages.

Zhao, N. N., Zhang, H., Zhang, X. C., Luan, X. B., Zhou, C., Liu, Q. Z., ... Liu, Z. L. (2013). Evaluation of acute toxicity of essential oil of garlic (*Allium sativum*) and its selected major constituent compounds against overwintering *Cacopsylla chinensis* (Hemiptera: Psyllidae). *Journal of Economic Entomology, 106*, 1349–1354. https://doi.org/10.1603/EC12191

Zhao, M. P., Liu, Q. Z., Liu, Q., & Liu, Z. L. (2017). Identification of larvicidal constituents of the essential oil of *Echinops grijsii* roots against the three species of mosquitoes. *Molecules, 22*, 205. https://doi.org/10.3390/molecules22020205

Zhu, J. J., Brewer, G. J., Boxler, D. J., Friesen, K., & Taylor, D. B. (2015). Comparisons of antifeedancy and spatial repellency of three natural product repellents against horn flies, *Haematobia irritans* (Diptera: Muscidae). *Pest Management Science, 71*, 1553–1560. https://doi.org/10.1002/ps.3960

14

Onosma bracteatum Wall and *Commiphora stocksiana* Engl extracts generate oxidative stress in *Brassica napus*: An allelopathic perspective

Joham Sarfraz Ali[1], Ihsan ul Haq[2], Attarad Ali[1], Madiha Ahmed[2] and Muhammad Zia[1]*

*Corresponding author: Muhammad Zia, Department of Biotechnology, Quaid-i-Azam University, Islamabad 45320, Pakistan
E-mail: ziachaudhary@gmail.com

Reviewing editor: Yoselin Benitez-Alfonso, University of Leeds, UK
Additional information is available at the end of the article

Abstract: The use of synthetic chemicals as herbicides for crop protection is a big threat due to toxicity, non-degradability, and negative impact on environment. *Onosma bracteatum* leaves and flowers, and *Commiphora stocksiana* Engl bark ethanolic extracts are evaluated for allelopathic potential against *Brassica napus*, a model plant. Complete allelopathic tendency was depicted by crude extract of *O. bracteatum* leaves and partial trend by flower and *C. stocksiana* extracts. *B. napus* seed germination efficiency and plant architecture is adversely influenced by the presence of plants extracts. The antioxidative analysis of *Brassica* plants depicts that extracts in the growth environment produces oxidative stress that eventually increased free radical scavenging activity, total antioxidative potential, and reducing power capability. Though *Brassica* plants produced phenolics and flavonoids to combat the oxidative stress but at insufficient concentration. Based on the findings, it can be concluded that the plants extracts produce oxidative stress to the seedlings and plants that eventually results in toxicity and allelopathic effect. Furthermore *O. bracteatum* can be a good candidate for natural herbicide either in form of extracts or the allelopathic compounds isolated from this plant species, which can be used as replacement of expensive and harmful synthetic herbicide.

Subjects: Plant Biology; Environmental & Ecological Toxicology; Agriculture

Keywords: allelopathy; allelochemicals; germination; weight associated parameters; phytochemical analysis and antioxidant evaluation

1. Introduction
There is an alarming increase in weed distribution as they possess faster colonization and survive under adverse situations (Colautti, Grigorovich, & MacIsaac, 2006; Hamilton et al., 2005). They are also a major threat to plant species due to their abundance mainly caused by low regulation of

ABOUT THE AUTHOR
Muhammad Zia and his group is working on different aspects of plant biotechnology. They are interested in phytochemistry of plants and their possible use/role in different aspects of life including agriculture, food, medicine, and environment.

PUBLIC INTEREST STATEMENT
The use of herbicides in crop cultivation is cause of toxicology to environment and food chain. Plants or their extract may act as agent that inhibit germination of weeds or kill them. This manuscript describes allelopathic potential of two plant, which shows significant inhibition of target specie. Furthermore, how the plants extract inhibited seed germination or growth of target species is also proposed to evaluate the mechanism.

competing plant species and natural predators (Callaway, DeLuca, & Belliveau, 1999). Consequently, in the areas under cultivation synthetic herbicides are used due to their affectivity at low concentration and cost (Dayan, Cantrell, & Duke, 2009). However, extensive use of synthetic herbicides leads to an increased risk of herbicide resistant weeds (Heap, 2014) and have toxic effect to other plants (Macías, Oliveros-Bastidas, & Marín, 2007). The indiscriminate use of synthetic herbicides is also hazardous to human and other biological communities (Peres & Moreira, 2007), environment (Akhtar, Sengupta, & Chowdhury, 2009) and even deplete nutrients from soil (Batish et al., 2007). Therefore an alternate source of natural herbicide apart from biodegradability, affectivity, low contamination (Souza Filho et al., 2006), and ecofriendly is a glittering option. Medicinal plants can be a promising tool to overcome the problems caused by synthetic herbicides (Gilani et al., 2010) as they possess versatile properties due to presence of bioactive compounds (Fujii, Parvez, Parvez, Ohmae, & Iida, 2003). In phytotoxicity, phytochemicals affect the growth, development, and functioning of the target specie (Mallik, 2008; Rice, 1984) as cover crop (Mallik, 2008) or the chemical compounds released (Macías, López, Varela, Torres, & Molinillo, 2008; Macías et al., 2001). Pelargonic acid (Copping & Duke, 2007), glufosinate (Duke, Dayan, Romagni, & Rimando, 2000), sorgoleone (Einhellig & Souza, 1992), and arteether (Bagchi, Jain, & Kumar, 1997) are few examples of herbicides isolated from plant species. The natural herbicides possess numerous advantages i.e. water solubility, less halogenated, ecofriendly, and novel mode of action (Duke et al., 2000; Macías et al., 2008). These compounds function by interrupting the basic procedures of respiration, enzyme activities, phytohormone level, water availability, mineral relation, permeability of cell wall, and cell division (Cruz-Ortega, Anaya, Hernández-Bautista, & Laguna-Hernández, 1998; Ferreira & Aquila, 2000). The phytotoxic ability of medicinal plant extract can be easily explored by analyzing the effect on seed germination, growth, and developmental parameters (Duke et al., 2000; Vyvyan, 2002).

The plants *Onosma bracteatum* Wall and *Commiphora stocksiana* Engl are traditionally well known in ayurvedic medicines for medicinal properties. *O. bracteatum* Wall locally known as Gaozaban is used as an antibacterial (Ahmad et al., 2009), anti-dandruff (Khan et al., 2013), and anti-inflammatory agent (Binzet & Akcin, 2012). It is also effective against respiratory (Kirtikar & Basu, 1999) and gastrointestinal problems (Ata, Faiza, & Shabnam, 2011). *C. stocksiana* Engl also known as *Balsamodendron pubescens* Stock and locally known by its gums Guggal famous for its medicinal properties that dates back to its ancient folkloric use. "Guggal" is related with the cure of laryngitis, rheumatism, inflammations, pneumonia, chronic dyspepsia, diarrhea, gum complaints, chronic endometritis, bronchitis, indolent ulcers, tonsillitis, leucorrhea piles, and whooping cough (Ishnava, & Mohan, 2008; Sharma & Kumar, 2012). The purpose of this study was to evaluate the allelopathic potential of *O. bracteatum* and *C. stocksiana* collected from natural habitat. The target plant species (*Brassica*) is also evaluated for antioxidative response and non-enzymatic antioxidative molecules to define possible mechanism of phytotoxicity by extracts.

2. Material and methods

2.1. Plant materials and extraction procedures
Fresh plant materials of *O. bracteatum* Wall was collected from Chaman and *C. stocksiana* Engl of Turbat region Baluchistan, Pakistan during July 2013 and were identified by Division of Plant Protection, Baluchistan Agricultural College, Quetta. The leaves and flowers of *O. Bracteatum* and the woody part of *C. stocksiana* were washed with tap water to remove dirt and debris. The plants material was shade dried at room temperature for two weeks and then dried in an incubator at 61°C.

2.2. Extraction procedure
Plant materials were cut into small pieces and pulverized in grinder and 100 g was dipped in ethanol (3:1). After three days, the extract was filtered using muslin cloth and Whatmann filter paper. The residue was dipped again in ethanol for additional three days and filtered thereafter. The filtrates were combined and concentrated by rotary evaporator (Rotovapor R 200 Buchi, Flawil Switzerland) at 40°C. The extracts were kept at 4°C until further analysis.

2.3. Phytotoxicity assay

The phytotoxic activity of the extracts is evaluated following the method of Attarad, Phull, Zia, Shah, and Haq (2015), with minor modifications. *Brassica napus*, a dicotyledonous specie is used as a target specie due to its sensitivity to allelopathic extracts and growth pattern. The extracts were diluted at three concentrations i.e. 10,000, 1,000, and 100 ppm and poured in 28 mm Petri plates containing doubled layered Whatmann filter paper. The plates were placed in draft chamber till the solvent evaporated. Under aseptic conditions, *B. napus* seeds were sterilized by sodium hypochlorite solution for three min and rinsed with autoclaved distilled water. Twenty seeds were placed in each Petri plate. 2,4-D (2,4 Dichlorophenoxy acetic acid; an auxin homology well-known herbicide) was used as positive control while distilled water was used as negative control. The experiment was performed in triplicate. All the plates were rapped with parafilm to avoid the loss of moisture and kept in dark at room temperature i.e. 25 ± 1°C. Seed germination efficiency was monitored on daily basis until germination prevalence i.e. the emergence of radical by rupturing the seed coat. The germination indices to evaluate the phytotoxic potential are;

(1) Final germination (FG) percentage is calculated as;

$$FG\% = \text{No of germinated seeds/No of total planted seeds} \times 100$$

(2) Rate of Germination (RG) = $\Sigma\, Ni/Di$

(3) Mean period of final germination (MPFG) = $\Sigma\, (NiDi)/S$

where N is the daily increase in the seedling number, D is the number of days from seed placement, and S is the total number of seeds germinated.

(4) Germination index (GI) = (% Germination against sample)/(% Germination in the control)
(5) Percentage inhibition (%) = 1 − [FG in extract (%)/FG in control (%)] × 100

Fifteen days old plantlets were examined for fresh and dry weight. The percentage seedling weight loss and relative dry weight percent is observed using following formula;

Percentage seedling weight loss (%Wt. loss) = [(Fresh weight of negative control − Fresh weight of sample) /Fresh weight of negative control] × 100

Relative dry weight percent (RDW %) = D.W of sample/D.W of control × 100

Germination percentage depicts the percent germination of a seeds after a specific period of time until it became constant as delayed germination cannot be explained using germination percentage. For this purpose RG and germination index is measured. Seedling weight loss (%Wt. loss) and relative dry weight percent (RDW %) are used to monitor the weight associated parameters.

2.4. Phytochemical and biochemical evaluation

The plantlets obtained after phytotoxic assay are analyzed for phytochemical and antioxidant potential to evaluate stress on germinated plantlets.

2.5. Sample preparation

The dried powdered plants were suspended in DMSO at 4 mg/ml. After 24 h the suspension was centrifuged at 4,000 rpm for 5 min and the supernatant is used for biochemical and antioxidative response. Total flavonoid content was determined by the aluminum chloride colorimetric method described by Almajano, Carbó, Jiménez, and Gordon (2008). The absorbance was measured at 450 nm using an Agilent Spectrophotometer (DAD, 8453, Agilent Technologies, Waldbronn Analytical Division, Germany). The contents are determined as Quercetin equivalents (µg QE/mg D.W).

Total phenolics were determined according to the method of Astill, Birch, Dacombe, Humphrey, and Martin (2001). The absorbance was measured at 630 nm. The total phenolic contents are determined as Gallic acid equivalents (µg GAE/mg D.W).

Antioxidant activities of the test samples is calculated using three different methods i.e. DDPH-based free radical scavenging activity, total antioxidant activity, and total reducing power.

The free radical quenching activity of the test samples against 2,2-diphenyl-1-picrylhydrazyl reagent (DPPH) is determined using procedure modified by Fatima et al. (2015). The absorbance was measured at 515 nm. Percent inhibition is calculated for each sample using formula:

Percent inhibition of the test sample = % scavenging activity = $(1 - Abs/Abc) \times 100$

where Abs is the absorbance of DPPH solution with sample and Abc indicates the absorbance of negative control (containing the reagent except the sample).

Ascorbic acid is used as a standard and values above 50% are considered significant.

Total antioxidant activity of extracts is evaluated by method described by Clarke, Heather, Gregory, and Daniel (2013). The absorbance was measured at 630 nm and Ascorbic acid is used as standard at final concentration of 50 µg/ml. The total antioxidant activity is determined as Ascorbic acid equivalents (µg AAE/mg D.W).

The reduction potential of the extracts is investigated according to the procedure described by Jafri, Saleem, Haq, Ullah, and Mirza (2014). The absorbance was measured at 630 nm and Ascorbic acid is used as standard at final concentration of 100 µg/ml. The reducing power capacity is determined as Ascorbic acid equivalents (µg AAE/mg D.W).

2.6. Statistical analysis
All the experiments were conducted in triplicates. For allelopathy assay three Petri plates were seeded with 60 seeds (20 seeds in each) against each concentration. The selection of plants extract concentration for assay was completely randomized. The plant parameters were studied after 15 days of seed germination. Each plant was considered as unit. The data is presented as mean ± standard deviation. The means are further analyzed for least significant difference (LSD) at $p \leq 0.05$ after analysis of variance.

3. Results
Phytotoxic (allelopathic) effect of crude extracts of *C. stocksiana* and *O. bracteatum* evaluated against seeds of *B. napus* depicts complete to partial allelopathic potential (Tables 1 and 2). Application of low concentration (100 ppm) of all extracts did not show any significant effect on *B. napus* seed germination. FG, RG, germination index, fresh weight, relative dry weight percent, and dry weight are found to decrease with increase in concentration of the extracts. While MPFG, percent inhibition and percent weight loss increase with increase in concentration of the extracts (Tables 1 and 2). The results are in correspondence to the positive control and constituents followed a similar pattern as that of 2,4-D, in fact few parameters (FG, MPFG, RG, GI, and RDW %) showed much better results than 2,4-D.

The phytochemical and biochemical evaluation of the *Brassica* plantlets is performed to observe the impact of stress on phytochemical profile. For 2,4-D at 10,000 ppm there were not sufficient plantlets, due to its potent allelopathic capacity, to conduct any of the assay as it completely prevented germination of the target plant species.

The presence of extracts in growth media caused oxidative stress on *Brassica* plants confirmed by variation in different activities. DPPH-based free radical scavenging activity decreased in plants due

Table 1. Phytotoxic effect of crude extracts of *O. bracteatum* Wall and *C. stocksiana* Engl with germination associated parameters

Extract	Conc. (ppm)	Allelopathy	FG (%)	RG	MPFG	PI (%)	GI
C. stocksiana							
Bark	10,000	Partial	65.4 ± 1	2.4 ± 0.2	1.3 ± 0.3	34.6 ± 1	0.5 ± 0.1
	1,000	Nil	91.3 ± 1.5	3.2 ± 0.1	1.15 ± 0.2	8.7 ± 0.3	0.9 ± 0.2
	100	Nil	95.5 ± 1.3	4.2 ± 0.3	1.1 ± 0.2	4.5 ± 0.6	1.0 ± 0.2
O. bracteatum							
Leaf	10,000	Complete	11.1 ± 0.2	0.6 ± 0.2	1.35 ± 0.2	88.9 ± 1	0.1 ± 0
	1,000	Partial	90.9 ± 0.9	4 ± 0.3	1.15 ± 0.2	9.1 ± 0.3	0.9 ± 0.1
	100	Nil	91.3 ± 0.6	4.2 ± 0.4	1.1 ± 0.3	8.7 ± 0.5	0.9 ± 0.3
Flower	10,000	Partial	50.0 ± 0.4	3.2 ± 0.3	1.6 ± 0.3	50.0 ± 1	0.5 ± 0.1
	1,000	Nil	76.0 ± 0.8	3.8 ± 0.3	1.35 ± 0.2	24.0 ± 1	0.8 ± 0.1
	100	Nil	81.5 ± 0.6	4.4 ± 0.3	1.25 ± 0.1	18.5 ± 0.5	0.8 ± 0.2
Standard							
2,4-D	10,000	Complete	6.3 ± 0.3	0.2 ± 0.1	1 ± 0.2	100.0	0.1 ± 0
	1,000	Partial	50.0 ± 1	2 ± 0.3	0.95 ± 0.1	93.8 ± 1	0.5 ± 0.1
	100	Nil	78.9 ± 0.8	3 ± 0.4	0.8 ± 0.1	21.1 ± 0.5	0.8 ± 0.2
dH$_2$O		Nil	100.0	5.2 ± 0.3	1.3 ± 0.2	0.0 ± 0	1.0 ± 0.1

Notes: FG: final germination (%), RG: rate of germination, MPFG: mean period of final germination, PI: percent inhibition (%), GI: germination index, F.W: fresh weight, D.W: dry weight, %Wt. L: percentage seedling weight loss, and RDW %: relative dry weight percent.

Table 2. Phytotoxic effect of crude extracts of *O. bracteatum* Wall and *C. stocksiana* Engl with weight associated parameters

Extract	Conc. (ppm)	Allelopathy	F.W (mg)	%Wt. L	D.W (mg)	RDW (%)
C. stocksiana						
Bark	10,000	Partial	73.8 ± 1[i]	75.0 ± 1	15.4 ± 0.7[ef]	51.3 ± 1
	1,000	Nil	227.9 ± 1.5[c]	22.8 ± 0.5	16.7 ± 0.8[e]	55.7 ± 1
	100	Nil	233.3 ± 1[d]	20.1 ± 0.8	19.5 ± 0.5[cd]	65.0 ± 1.5
O. bracteatum						
Leaf	10,000	Complete	67.9 ± 1[j]	77.0 ± 1	16.9 ± 0.5[e]	56.3 ± 1
	1,000	Partial	139.4 ± 1.5[g]	52.8 ± 0.5	21.0 ± 1[c]	70.0 ± 1
	100	Nil	152.0 ± 1[f]	48.5 ± 1	21.2 ± 1[c]	70.7 ± 1
Flower	10,000	Partial	7.4 ± 0.4[i]	97.5 ± 0.8	3.2 ± 0.5[h]	10.7 ± 1
	1,000	Nil	204.0 ± 1[e]	30.9 ± 1	18.5 ± 1[d]	61.7 ± 1
	100	Nil	273.6 ± 1[b]	7.3 ± 0.6	27.2 ± 1[b]	90.7 ± 1
Standard						
2,4-D	10,000	Complete	0.0 ± 0	100.0	0.0 ± 0	0.0 ± 0
	1,000	Partial	47.9 ± 1[k]	83.8 ± 0.6	9.2 ± 1[g]	30.7 ± 1.5
	100	Nil	82.1 ± 1[h]	72.2 ± 0.5	12.5 ± 0.5[f]	41.7 ± 1
dH$_2$O		Nil	295.2 ± 1[a]	0.0 ± 0	30.0 ± 1[a]	100.0

Notes: F.W: fresh weight, D.W: dry weight, %Wt. L: percentage seedling weight loss, and RDW %: relative dry weight percent.

Mean values (±standard errors) followed by the same letter in the column are not significantly different after LSD test ($p \leq 0.05$).

to the presence of 2,4-D in the environment. Likewise, extracts also induced oxidative stress at different concentrations making the free radical scavenging activity variable. The samples, however, show no significant DPPH scavenging activity (Figures 1 and 2). The free radical scavenging activity is linked with TAC and TRP. Variation in TAC and TRP in *Brassica* plants show that the metabolites in extracts make the plants under stress resulting in increase in these activities. However, the extent of stress makes the activities variable. The highest antioxidant capacity (88.5 µg AAE/mg D.W) is observed for bark extract of *C. stocksiana* at 1,000 ppm (Figure 1) and the flower extract of *O. bracteatum* at 10,000 ppm i.e. 101.8 µg AAE/mg D.W (Figure 2). Highest value of reducing power was observed among the leaf extract of *O. bracteatum* (148.2 µg AAE/mg D.W) at 1,000 ppm and for *C. stocksiana* (115.7 µg AAE/mg D.W) at 10,000 ppm of bark extracts (Figures 1 and 2). Although extracts constituents may stress the plants leading to death but the natural mechanism to combat the stress may assist the plant to survive. Enzymatic and non-enzymatic systems are present for this purpose and non-enzymatic system is more powerful because during stress the excessive reactive oxygen species (ROS) activates the natural defense mechanism. Secondary metabolites, meantime, produce to overcome the cells oxidative stress preventing alarming deadly accumulation of ROS. For this reason phenolics and flavonoids concentrations are analyzed but the varying trends of phenolics and flavonoids may be due to stress caused by the phytotoxic extracts that masked the production of these antioxidative components. The highest phenolic content are also observed in plantlets exposed to bark extracts of *C. stocksiana* at 100 ppm i.e. 55.1 µg GAE/mg D.W (Figure 3) while highest flavonoid content (22.4 µg QE/mg D.W) was observed at 10,000 ppm. In case of *O. bracteatum*, highest phenolic content was observed by the ones exposed to leaf extracts at 1,000 ppm i.e. 52.4 µg

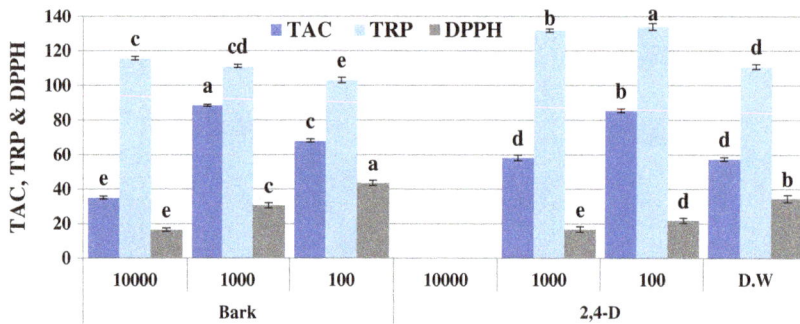

Figure 1. DPPH radical scavenging activity (% inhibition), total antioxidant capacity (TAC µg AAE/mg D.W), and reducing power (TRP µg AAE/mg D.W) of allelopathic plantlets exposed to *C. stocksiana* Engl bark at varying concentration along with 2,4-D and dH$_2$O.

Note: Mean values (±standard errors) followed by the same letter are not significantly different after LSD test ($p \leq 0.05$).

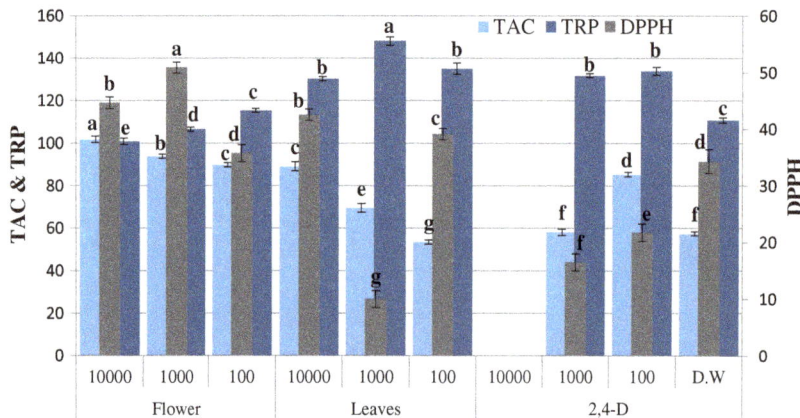

Figure 2. DPPH radical scavenging activity (% inhibition), total antioxidant capacity (TAC µg AAE/mg D.W), and reducing power (TRP µg AAE/mg D.W) of allelopathic plantlets exposed to *O. bracteatum* Wall leaves and flowers at varying concentration along with 2,4-D and dH$_2$O.

Note: Mean values (±standard errors) followed by the same letter are not significantly different after LSD test ($p \leq 0.05$).

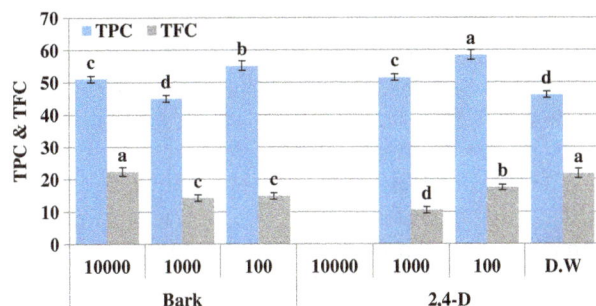

Figure 3. Total phenolic content (TPC μg GAE/mg D.W) and total flavonoid content (TFC μg QE/mg D.W) of allelopathic plantlets exposed to *C. stocksiana* Engl bark extract at varying concentration along with 2,4-D and dH₂O.

Note: Mean values (±standard errors) followed by the same letter are not significantly different after LSD test ($p ≤ 0.05$).

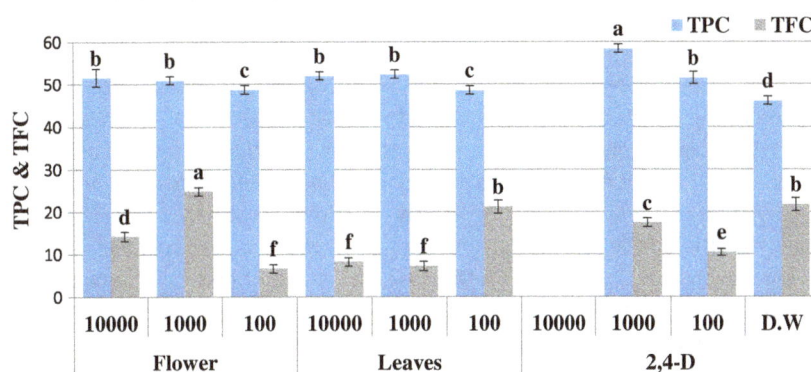

Figure 4. Total phenolic content (TPC μg GAE/mg D.W) and total flavonoid content (TFC μg QE/mg D.W) of allelopathic plantlets exposed to *O. bracteatum* Wall leaves and flowers extracts at varying concentration along with 2,4-D and dH₂O.

Note: Mean values (±standard errors) followed by the same letter are not significantly different after LSD test ($p ≤ 0.05$).

GAE/mg D.W (Figure 4) and highest flavonoid content (24.8 μg QE/mg D.W) was observed in flower extracts of *O. bracteatum* at 1,000 ppm. The results depict oxidative stress in the model plants that was higher as compared to positive and negative control with few exceptions. The phenolic and flavonoid content are affected by the oxidative stress caused by the extracts as lower values of phenolic content and higher values of flavonoid content as compared to 2,4-D suggest the unusual inhibition and enhancement of these natural antioxidative molecules.

4. Discussion

Phytomedicines are well known for numerous advantages and are considered highly valuable in terms of affectivity against different diseases (Jiang et al., 2006). Along with these characteristics, they also tend to be effective allelopathic agents. The current study proposed a complete phytotoxic tendency of leaves extract of *O. bracteatum* and partial allelopathic trend of *C. stocksiana*. The seed germination inhibition potential might be associated with the presence of allelochemicals that affect germination and growth of plants (Evenari, 1949; Gressel & Holm, 1964). Khanh, Chung, Xuan, and Tawata (2005) reported that maximum allelopathic suppression may occur when allelochemicals interfere with early growth stages of plant. Although details of the biochemical mechanism through which allelochemicals exert a toxic effect on germination and growth of the respective plants is still unknown (Zhou & Yu, 2006) but dos Santos et al. (2008) suggested that allelochemicals might affect enzymatic activity or cause depolarization of the root cell membrane or lipid peroxidation blocking. The extracts affected germination and weight associated parameters in a concentration dependent manner similar to one observed by Inderjit and Keating (1999). FG, RG, relative dry weight percent, germination index, fresh weight, and dry weight decreased with increase in concentration of the sample. The variations if fresh weight and dry weight depict that the plants extracts varied water holding capacity of *Brassica* plantlets. Furthermore it indicates that presence of different concentrations of extracts stress the plants to produce biomass. Previous studies also suggest that the effect of phytotoxins on the plants is species dependent both in biochemical and physiological characteristics (Kobayashi, 2004; Prati & Bossdorf, 2004; Xuan, Shinkichi, Khanh, & Chung, 2005). Germination inhibition due to allelochemicals is known to be associated to lipid mobilization and low levels of isothiocynates (Baleroni, Ferrarese, Souza, & Ferrarese-Filho, 2000; Petersen, Belz, Walker, & Hurle, 2001). The identification of compounds responsible for causing allelopathy is of great interest as it can be used as an effective alternative method for weed control. Allelochemicals mainly function by affecting cell division and elongation or by suppressing enzymes responsible for the movement of nutrients, important for germination (Batlang & Shushu, 2007). The current focus of agriculture industry is to find a safe and biological solution to reduce foreseen hazardous impacts of herbicides and insecticides (Khanh & Chung, 2004).

The significant increase in phenolic and flavonoid contents shows varying trend that might be linked to the mediocre point where the effect of phytotoxins get masked. The high phytochemical content can be associated to the stress and to the allelopathic treatment as previous studies depict a strong relation of phenolics and flavonoids with the allelopathic potential (Macías et al., 2008). The antioxidative evaluation of the plantlets shows that stress increases production of secondary metabolites suggesting that not only medicinal plants can be used as an herbicide but it might also help in the enhancement of plants antioxidative potential. Along with that the phytochemical and antioxidant assays help to identify the nature of allelochemicals involved in causing a specific effect. These effects may be associated to the phenolic, saponins, tannins, alkaloids, or terpenoids present in the extract being used (Einhellig, 1995; Inderjit & Nilsen, 2003) and later help out in isolation of compounds that can be used as potent herbicide with lesser side effects and more effective mode of action in comparison to synthetic ones.

5. Conclusion

The allelopathic tendency depicted by the crude extract of two plants and their effect on different parameters suggests the isolation of an important compound that can act as potent herbicide for future with lesser side effects to the environment. The phytochemical and antioxidant evaluation of the plantlets also depict that the extract cause oxidative stress leading to cell/plant death. Furthermore, the ROS and oxidative stress are one of the biochemical concerns that govern allelopathic effect.

Funding
The authors received no direct funding for this research.

Competing Interests
The authors declare no competing interest.

Author details
Joham Sarfraz Ali[1]
E-mail: jsa_09@yahoo.com
Ihsan ul Haq[2]
E-mail: ihsn99@yahoo.com
Attarad Ali[1]
E-mail: attarad.ali@kiu.edu.pk
Madiha Ahmed[2]
E-mail: pharmacyst_madiha@yahoo.com
Muhammad Zia[1]
E-mail: ziachaudhary@gmail.com
[1] Department of Biotechnology, Quaid-i-Azam University Islamabad, Islamabad 45320, Pakistan.
[2] Department of Pharmacy, Quaid-i-Azam University Islamabad, Islamabad 45320, Pakistan.

References
Ahmad, B., Ali, N., Bashir, S., Choudhary, M. I., Azam, S., & Khan, I. (2009). Parasiticidal, antifungal and antibacterial activities of *Onsomagriffithi*. *African Journal of Botany, 8*, 5084–5087.
Akhtar, W., Sengupta, D., & Chowdhury, A. (2009). Impact of pesticides use in agriculture: Their benefits and hazards. *Interdisciplinary Toxicology, 2*(1), 1–12. http://dx.doi.org/10.2478/v10102-009-0001-7
Almajano, M. P., Carbó, R., Jiménez, J., & Gordon, M. H. (2008). Antioxidant and antimicrobial activities of tea infusions. *Food Chemistry, 108*, 55–63. http://dx.doi.org/10.1016/j.foodchem.2007.10.040
Astill, C., Birch, M. R., Dacombe, C., Humphrey, P. G., & Martin, P. T. (2001). Factors affecting the caffeine and polyphenol contents of black and green tea infusions. *Journal of Agricultural Food Chemistry, 49*, 5340–5347. http://dx.doi.org/10.1021/jf010759+
Ata, S., Faiza, F., & Shabnam, J. (2011). Elemental profile of 24 common medicinal plants of Pakistan and its direct link with traditional uses. *Journal of Medicinal Plant Research, 5*, 6164–6168.

Attarad, A., Phull, A. R., Zia, M., Shah, A. M. A., & Haq, I. U. (2015). Phytotoxicity of River Chenab sediments: In vitro morphological and biochemical response of Brassica napus L. Environmental Nanotechnology, Monitoring & Management, 4, 74–84.

Bagchi, G. D., Jain, D. C., & Kumar, S. (1997). Arteether: A potent plant growth inhibitor from Artemisia annua. Phytochemistry, 45, 1131–1133. http://dx.doi.org/10.1016/S0031-9422(97)00126-X

Baleroni, C. R. S., Ferrarese, M. L. L., Souza, N. E., & Ferrarese-Filho, O. (2000). Lipid accumulation during canola seed germination in response to cinnamic acid derivatives. Biologia Plantarum, 43, 313–316. http://dx.doi.org/10.1023/A:1002789218415

Batish, D., Singh, H., Kohli, R., Kaur, S., Saxena, D., & Yadav, S. (2007). Assessment of parthenin against some weeds. ZeitschriftfürNaturforschung, 62, 367–372.

Batlang, U., & Shushu, D. D. (2007). Allelopathic activity of sunflower (Helianthus annuus L.) on growth and nodulation of bambara groundnut (Vigna subterranean (L.) Verdc.). Journal of Agronomy, 6, 541–547.

Binzet, R., & Akcin, O. E. (2012). The anatomical properties of two Onosma L. (Boraginaceae) species from Turkey. Journal of Medicinal Plant Research, 6, 3288–3294.

Callaway, R. M., DeLuca, T. H., & Belliveau, W. M. (1999). Biological-control herbivores may increase competitive ability of the noxious weed Centaurea maculosa. Ecology, 80, 1196–1201. http://dx.doi.org/10.1890/0012-9658(1999)080[1196:BCHMIC]2.0.CO;2

Clarke, D., Heather, W., Gregory, S., & Daniel, R. (2013). Detection and learning of floral electric fields by bumblebees. Journal of Inorganic Biochemistry, 53, 573–583.

Colautti, R. I., Grigorovich, I. A., & MacIsaac, H. J. (2006). Propagule pressure: A null model for biological invasions. Biological Invasions, 8, 1023–1037. http://dx.doi.org/10.1007/s10530-005-3735-y

Copping, L. G., & Duke, S. O. (2007). Natural products that have been used commercially as crop protection agents. Pest Management Science, 63, 524–554. http://dx.doi.org/10.1002/(ISSN)1526-4998

Cruz-Ortega, R., Anaya, A. L., Hernández-Bautista, B. E., & Laguna-Hernández, G. (1998). Effescts of allelochemical stress produced by Sicyos deppei on seedling root ultraestrucutre of Phaseolus vulgaris and curcubita ficifolia. Journal of Chemical Ecology, 24, 2039–2057. http://dx.doi.org/10.1023/A:1020733625727

Dayan, F. E., Cantrell, C. L., & Duke, S. O. (2009). Natural products in crop protection. Bioorganic and Medicinal Chemistry, 17, 4022–4034. http://dx.doi.org/10.1016/j.bmc.2009.01.046

dos Santos, W. D., Ferrarese, M. L., Nakamura, C. V., Mourão, K. S., Mangolin, C. A., & Ferrarese-Filho, O. (2008). Soybean (Glycine max) root lignification induced by ferulic acid. The possible mode of action. Journal of Chemical Ecology, 34, 1230–1241. http://dx.doi.org/10.1007/s10886-008-9522-3

Duke, S. O., Dayan, F. E., Romagni, J. G., & Rimando, A. M. (2000). Natural products as sources of herbicides: Current status and future trends. Weed Research, 40, 99–111. http://dx.doi.org/10.1046/j.1365-3180.2000.00161.x

Einhellig, F. A. (1995). Allelopathy: Organisms, processes and applications. In Inderjit, K. M. M. Dakshini, F. A. Einhellig, (Eds). Journal of American Chemical Society (Vol. 582, pp. 96–116). Washington, DC: American Chemical Society.

Einhellig, F. A., & Souza, I. F. (1992). Phytotoxicity of sorgoleone found in grain Sorghum root exudates. Journal of Chemical Ecology, 18, 1–11. http://dx.doi.org/10.1007/BF00997160

Evenari, M. (1949). Germination inhibitors. The Botanical Review, 15, 153–194. http://dx.doi.org/10.1007/BF02861721

Fatima, H., Khan, K., Zia, M., Ur-Rehman, T., Mirza, B., & Haq, I.-U. (2015). Extraction optimization of medicinally important metabolites from Daturainnoxia Mill: An in vitro biological and phytochemical investigation. BMC Complementary and Alternative Medicine, 15, 1.

Ferreira, A. G., & Aquila, M. E. A. (2000). Alelopatia: Umaáreaemergente de ecofisiologia [Alellopathy: An emerging topic in ecophysiology]. RevistaBrasileira de Fisiologia Vegetal/Brazilian Journal of Plant Physiology, 12, 175–204.

Fujii, Y., Parvez, S. S., Parvez, M. M., Ohmae, Y., & Iida, O. (2003). Screening of 239 medicinal plant species for allelopathic activity using the sandwich method. Weed Biology and Management, 3, 233–241. http://dx.doi.org/10.1046/j.1444-6162.2003.00111.x

Gilani, S. A., Fujii, Y., Shinwari, Z. K., Adnan, M., Kikuchi, A., & Watanabe, K. N. (2010). Phytotoxic studies of medicinal plant species of Pakistan. Pakistan Journal of Botany, 42, 987–996.

Gressel, J. B., & Holm, L. G. (1964). Chemical inhibition of crop germination by weed seeds and the nature of inhibition by Abutilon theophrasti. Weed Research, 4, 44–53. http://dx.doi.org/10.1111/wre.1964.4.issue-1

Hamilton, M. A., Murray, B. R., Cadotte, M. W., Hose, G. C., Baker, A. C., Harris, C. J., & Licari, D. (2005). Life-history correlates of plant invasiveness at regional and continental scales. Ecology Letters, 8, 1066–1074. http://dx.doi.org/10.1111/j.1461-0248.2005.00809.x

Heap, I. (2014). The international survey of herbicide resistant weeds. Retrieved fromhttp://www.weedscience.org

Inderjit, & Keating, K. I. (1999). Allelopathy: Principles, procedures, processes, and promises for biological control. Advances in Agronomy, 67, 141–231. http://dx.doi.org/10.1016/S0065-2113(08)60515-5

Inderjit, & Nilsen, E. (2003). Bioassays and field studies for allelopathy in terrestrial plants: Progress and problems. Critical Reviews in Plant Sciences, 22, 221–238. http://dx.doi.org/10.1080/713610857

Jafri, L., Saleem, S., Haq, I.U., Ullah, N., & Mirza, B. (2014). In vitro assessment of antioxidant potential and determination of polyphenolic compounds of Hedera nepalensis K. Koch. Arab Journal of Chemistry. doi:10.1016/j.arabjc.2014.05.002

Jiang, C., Chang, M., Wen, C., Lin, Y., Hsu, F., & Lee, M. (2006). Natural products of cosmetics: Analysis of extracts of plants endemic to Taiwan for the presence of tyrosinese inhibitory, melanin reducing and free radical scavenging activities. Journal of Food and Drug Analysis, 14, 346–352.

Ishnava, K. B., & Mohan, J. S. S. (2008). Intraspecific isozymes variation in Commiphora wightii (Arn.) Bhandari: A traditional hypocholesteremic medicinal shrub from Gujarat, India. Journal of Herbs, Spices and Medicinal Plants, 13, 25–40.

Khan, S. M., Page, S., Ahmad, H., Shaheen, H., Ullah, Zahid, Ahmad, M., & Harper, D. M. (2013). Medicinal flora and ethnoecological knowledge in the Naran Valley, Western Himalaya, Pakistan. Journal of Ethnobiology and Ethnomedicine, 9, 4. http://dx.doi.org/10.1186/1746-4269-9-4

Khanh, T., & Chung, I. (2004). Weed-suppressing potential of dodder (Cuscuta hygrophilae) and its phytotoxic constituents. Weed Sciences, 56, 119–127.

Khanh, T. D., Chung, M. I., Xuan, T. D., & Tawata, S. (2005). The exploitation of crop allelopathy in sustainable agricultural production. Journal of Agronomy and Crop Science, 191, 172–184. http://dx.doi.org/10.1111/j.1439-037X.2005.00172.x

Kirtikar, K. R., & Basu, B. D. (1999). *Indian medicinal plants* (Vol. 3, 2nd ed., p. 1699). New Delhi: International Book Distributors.

Kobayashi, K. (2004). Factors affecting phytotoxic activity of allelochemicals in soil. *Weed Biology Management, 4*, 1–7. http://dx.doi.org/10.1111/wbm.2004.4.issue-1

Macías, F. A., López, A., Varela, R. M., Torres, A., & Molinillo, J. M. G. (2008). Helikauranoside A, a new bioactive diterpene. *Journal of Chemical Ecology, 34*, 65–69. http://dx.doi.org/10.1007/s10886-007-9400-4

Macías, F. A., Molinillo, J. M. G., Galindo, J. C. G., Varela, R. M., Simonet, A. M., & Castellano, D. (2001). The use of allelopathic studies in the search for natural herbicides. *Journal of Crop Production, 4*, 237–255. http://dx.doi.org/10.1300/J144v04n02_08

Macías, F. A., Oliveros-Bastidas, A., & Marín, D. (2007). Plant biocommunicators: Their phytotoxicity, degradation studies and potential use as herbicide models. *Phytochemistry Reviews, 7*, 179–194. http://dx.doi.org/10.1007/s11101-007-9062-4

Mallik, A. U. (2008). Allelopathy: Advances, challenges and opportunities. In R. S. Zeng, A. U. Mallik, & S. M. Luo (Eds.), *Allelopathy in Sustainable Agriculture and Forestry* (pp. 25–38). New York, NY: Springer Science, Business Midia. http://dx.doi.org/10.1007/978-0-387-77337-7

Peres, F., & Moreira, J. C. (2007). *Saúde e ambiente e sua relação com o consumo de agrotóxicos em um pólo agrícola do estado do Rio de Janeiro, Brasil* [Health and environment and its relation with the consumption of agrochemicals in an agricultural center in the state of Rio de Janeiro, Brazil]. Rio deJaneiro: Caderno de Saúde Pública.

Petersen, J., Belz, R., Walker, F., & Hurle, K. (2001). Weed suppression by release of isothiocyanates from turnip-rape mulch. *Agronomy Journal, 93*, 37–43. http://dx.doi.org/10.2134/agronj2001.93137x

Prati, D., & Bossdorf, O. (2004). Allelopathic inhibition of germination by *Alliaria petiolata* (Brassicaceae). *American Journal of Botany, 91*, 285–288. http://dx.doi.org/10.3732/ajb.91.2.285

Rice, E. L. (1984). *Allelopathy*. Orlando, FL: Academic Press.

Sharma, S., & Kumar, A. (2012). Traditional uses of herbal medicinal plants of Rajashthan. *International Journal of Life Science Pharma Research, 2*, 77–82.

Souza Filho, A. P. S., Santos, R. A., Santos, L. S., Guilhon, G. M. P., Santos, A. S., Arruda, M. S. P., ... Arruda, A. C. (2006). Potencial alelopático de *Myrcia guianensis*. *Planta Daninha, 24*, 649–656. http://dx.doi.org/10.1590/S0100-83582006000400005

Vyvyan, J. R. (2002). Allelochemicals as leads for new herbicides and agrochemicals. *Tetrahedron, 58*, 1631–1646. http://dx.doi.org/10.1016/S0040-4020(02)00052-2

Xuan, T. D., Shinkichi, T., Khanh, T. D., & Chung, I. M. (2005). Biological control of weeds and plant pathogens in paddy rice by exploiting plant allelopathy: an overview. *Crop Protection, 24*, 197–206. http://dx.doi.org/10.1016/j.cropro.2004.08.004

Zhou, Y. H., & Yu, J. Q. (2006). Allelochemicals and photosynthesis. In M. J. Reigosa, N. Pedrol, & L. Gonzalez (Eds.), *Allelopathy: A Physiological Process with Ecological Implications* (pp. 127–139). Dordrecht: Springer.

15

Vicia ervilia L. seeds newly explored biological activities

Mona M. Okba[1]*, Gehad A. Abdel Jaleel[2], Miriam F. Yousif[1,3], Kadriya S. El Deeb[1] and Fathy M. Soliman[1]

*Corresponding author: Mona M. Okba, Faculty of Pharmacy, Pharmacognosy Department, Cairo University, Kasr El-Ainy, 11562 Cairo, Egypt

E-mails: monamoradokba@gmail.com, mona.morad@pharma.cu.edu.eg

Reviewing editor: Sabrina Sabatini, Sapienza University of Rome, Italy

Additional information is available at the end of the article

Abstract: Within the global context of increasing poverty in the developing countries, natural products are important in devising new drugs. *Vicia ervilia* L. Willd., unlike several fabaceae seeds, is not used for human consumption till now. We aim to discover any possible medicinal use of the seed. Analgesic, anti-inflammatory, antiulcerogenic and antihyperglycemic activities were evaluated using hot plate, carrageenan-induced rat paw edema, ethanol-induced ulcer model and alloxan-induced diabetes methods, respectively. Antiviral activity was evaluated using Methylthiazol Tetrazolium assay. *V. ervilia* seeds ethanol (70%) extract had significant anti-inflammatory, analgesic, antiulcerogenic, antihyperglycemic and antiviral activities. It is of excellent choice for treatment of several illnesses in developing countries due to its diverse resource, easy accessibility, affordability and its newly proved significant wide range of biological activities.

Subjects: Environment & Agriculture; Food Science & Technology; Economics, Finance, Business & Industry; Health and Social Care

Keywords: antiviral; antihyperglycemic; antiulcerogenic; anti-inflammatory

1. Introduction
The use of herbal medicine is an age-old tradition worldwide. The recent progress in modern therapeutics has stimulated the use of natural products not only for its effectiveness but also for its

ABOUT THE AUTHORS
Dr Mona M. Okba, is a lecturer in the department of Pharmacognosy. Okba completed bachelor's degree at the Faculty of Pharmacy – Cairo University (FOPCU) in 2004. Okba completed master's and PhD in 2004–2014. Okba's research interests include Pharmacognosy and Phytochemistry. Okba's publications include (1) *Bull. Fac. Pharm. Cairo Univ.*, Vol. 47. No. 1, 87–96, (2009), (2) *Bull. Fac. Pharm. Cairo Univ.*, Vol. 45, No. 2 , 157- 179, (2007), (3) *Z. Naturforsch.*, 64 c, 611–614 (2009), (4) *Egypt. J. Biomed. Sci.*, Vol. 23, 121–134, (2007), (5) *Int. J. Pharm. Pharm. Sci.*, Vol 5, Suppl 3, 311–329, (2013), (6) *Int. J. Pharm. Pharm. Sci.*, Vol 6, Issue 1, 246–253, (2014), and (7) *Journal of Natural Sciences Research*, Vol. 6, No. 22, (2016). Okba's future is to explore other pharmacological activities and phytochemical content especially of *V. ervilia* L. to be incorporated in large-scale production of a dosage form that can be used for human consumption.

PUBLIC INTEREST STATEMENT
Within the global context of increasing poverty in the developing countries, natural products are important in devising new drugs. *Vicia ervilia* seeds are not used for human consumption till now although it is well known as an excellent sheep and cattle *feed* concentrate. This perspective article discovers some possible medicinal use of the seed. Analgesic, anti-inflammatory, antiulcerogenic, antihyperglycemic, and antiviral activities were evaluated. These treasure seeds are of excellent choice for treatment of several illnesses for low-income people in developing countries due to its diverse resource, easy accessibility, affordability and its newly explored wide range of biological activities.

relatively low prices and availability. Cultivation of high valued medicinal plants should be creating new dimension in the field of agriculture in developing countries (Amujoyegbe, Agbedahunsi, & Amujoyegbe, 2012). Egypt is a country with a developed economy according to the International Monetary Fund's World Economic Outlook report (2015).

Vicia ervilia L. Willd. (syn. *Ervium ervilia* L., Karsanah, kursene كرسنه), Family Fabaceae, is an annual herb distributed in the Western Mediterranean coastal region (Täckholm, 1974). The seeds are extensively used as stock feed in several countries (Haddad, 2006; Sadeghi, Pourreza, Samei, & Rahmani, 2009). However, to the best of the authors' knowledge, only one report concerning the medicinal importance of *V. ervilia* for human beings was traced (Fornstedt & Porath, 1975). The use of such cheap and easily cultivated seeds in treating several illnesses in developing country, like Egypt, will be of excellent value especially in the current state of increasing poverty. Antiulcerogenic and antihyperglycemic activities are among activities to be screened in this study due to high incidence of gastric ulcers (Hussein, 2010) and diabetes (National Center of Health & Population, 2004; Shaw, Sicree, & Zimmet, 2010) in Egypt. The seeds' anti-inflammatory and analgesic potentials are to be explored due to their major role in relieving hepatitis, cancer and rheumatic musculoskeletal disorders of high incidence in the country (Abdel-Tawab, Abdel-Nasser, & Darmawan, 2004; Ibrahim, Khaled, Mikhail, Baraka, & Kamel, 2014; Lotfi, Abdel-Nasser, & Hamdy, 2007; World Health Organization, 2011).

2. Experimental

2.1. Plant material
Samples of *V. ervilia* seeds were imported from Jordan in July 2012. They were cultivated in the Experimental Station of Medicinal Plants, Pharmacognosy Department, Faculty of Pharmacy, Cairo University, Giza. Photos of the cultivated plant were sent to Kew Garden, England to confirm their identity. Identification was studied by same authors in a previous publication (Okba, Yousif, El Deeb, & Soliman, 2014). Voucher samples were deposited at the Museum of the Pharmacognosy Department, Faculty of Pharmacy, Cairo University (herbarium No. 14.4.2013.2).

2.2. Extract preparation
Seeds were powdered and extracted with ethanol (70%) by percolation at room temperature for three consecutive days. The solvent was evaporated under vacuum to yield seeds ethanol (70%) extract (SEE).

2.3. Biological activities
Animals: were obtained from the animal house of the Laboratory National Research Center, Giza, Egypt. They were kept on standard laboratory diet. This study was conducted in accordance with ethical procedures and policies approved by Animal Care and Use Committee of Faculty of Pharmacy, Cairo University which follows the World Medical Association Declaration of Helsinki (WMA General Assembly 1964).

Determination of median lethal dose (LD50): was estimated according to (Kerber, 1931).

Acute anti-inflammatory activity: carrageenan-induced rat paw edema method was followed (Winter, Risley, & Nuss, 1962). Hind paw volume was measured by water displacement method using 7410, Ugo Basile, plythesmometer, Comerio, Italy (Chattopadhyay et al., 2002; Li, Hyun, Jeong, Kim, & Lee, 2003). Indomethacin (Epico, Egyptian Int. Pharmaceutical Industries Co.) was used as a positive control.

Analgesic activity: Hot plate method was carried out using (7280 Ugo Basile Biological Research Apparatus Company, Comerio, Italy) according to (Laviola & Alleva, 1990; Woolfe & McDonald, 1944). Indomethacin was used as a positive control.

Antiulcerogenic activity: ethanol-induced ulcer model was adopted (Morimoto, Shimohara, Oshima, & Sukamoto, 1991). Ranitidine was used as a reference drug. Lesions were examined under an illuminated magnifier (Adami, Marazzi-Uberti, & Turba, 1964).

Antihyperglycemic activity: The technique used by (Zhang, Huang, Hou, & Wang, 2006) was followed. Hyperglycemia was induced by alloxan monohydrate (Sigma Co., USA). BioMerieux kits were used for the assessment of blood glucose, triglycerides and cholesterol levels.

Antiviral activity: VERO cells (kidney epithelial cells of African green monkey) incubated into culture bottle were checked using inverted microscope. Healthy cells propagation, determination of extract cytotoxicity and MTT assay protocol were done according to (Alley, Scudiero, & Monks, 1988; Van de Loosdrecht, 1994).

3. Results and discussion

The LD_{50} of seed ethanol extract (SEE) was up to 5 g/kg b.wt. It is thus of high safety margin at the tested dose level (Buck, Osweiter, & Van Gelder, 1976). These reflect the possibility of the plant use for nutritional or medicinal purposes after elimination of its anti-nutritional factor canavanine (Berger, Robertson, & Cocks, 2003; Sadeghi, Samie, Pourreza, & Rahmani, 2004).

SEE showed a significant *anti-inflammatory* activity ($p < 0.05$). When indomethacin was administered the edema volume after 4 h of carrageenan injection was 24.5% of its original volume. The edema volume was 46.51% of its original volume when SEE was tested (500 mg/kg). The extract did not cause gastric ulcers, during the course of study, as usually happens with anti-inflammatory drugs. SEE showed a significant *analgesic activity.* SEE (500 mg/kg) analgesic activity was 80.21% of indomethacin (Table 1).

SEE showed a pronounced *antiulcerogenic* activity (Table 2). SEE (250 mg/kg) reduced ulcers number and severity by 80.6 and 82.32%, respectively. The antiulcerogenic activity at 500 mg/kg approached that of ranitidine.

SEE caused significant ($p < 0.05$) reduction in glucose, triglycerides, and cholesterol levels of *hyperglycemic* rats (Table 3). At 500 mg/kg b.wt SEE cause reduction in glucose level by 79.06%, a

Table 1. Anti-inflammatory and analgesic activities of *V. ervilia* seeds

		Control	Indomethacin	SEE	
				Dose (mg/kg)	
			10	250	500
Edema volume [a]	1 h	100.33 ± 7.5	59.71 ± 2.7*	60.32 ± 3.3*	55.53 ± 4*
	2 h	142.88 ± 7.7	67.22 ± 3.1*	75.52 ± 3*	65.69 ± 4.1*
	3 h	156.52 ± 9.5	40.46 ± 4.6*	68.80 ± 5.2*,**	54.16 ± 4.2*
	4 h	160.29 ± 10.8	24.50 ± 4.7*	50.70 ± 4.7*,**	46.53 ± 3.1*
Reaction time (min)	B.l.	6.84 ± 0.19	6.78 ± 0.23	6.02 ± 0.17	5.88 ± 0.13
	30	6.52 ± 0.18	7.94 ± 0.22*	7.02 ± 0.16	6.82 ± 0.16
	60	6.44 ± .09	9.46 ± 0.22*	8.48 ± 0.29*	7.92 ± 0.16*
	90	6.16 ± 0.07	11.22 ± 0.26*	9.62 ± 0.3*	9 ± 0.19*

Notes: Data represent the mean value ± S.E. of six rat per group and the percent changes vs. basal (zero min) values and 1, 2, 3 and 4 h post-carrageenan injection. Statistical analysis was done using one-way ANOVA Followed by LSD and Tukey for multiple comparisons.

Values represent the mean ± S.E. of six animals for each group ($n = 6$). B.l., base line, SEE: seed ethanol (70%) extract.

[a]% change from baseline.

*Significantly different from control group at $p < 0.05$.

**Significantly different from indomethacin group at $p < 0.05$.

Table 2. Effect of *V. ervilia* seeds on the gastric ulcer number and severity

Treatment	Dose (mg/kg)	Ulcer number		Ulcer severity	
		Mean ± SE	%Reduction	Mean ± SE	%Reduction
Control		0	–	0	–
Ethanol 99.5%		13.4 ± 1.07	–	39.6 ± 2.97	–
SEE	250	2.6 ± 0.50*	80.6	7 ± 0.70*,**	82.32
	500	3.8 ± 0.22*	71.64	6.8 ± 0.31*,**	82.83
Ranitidine	50	2.4 ± 0.24*	82.09	3.8 ± 0.37*	90.4

Notes: Statistical analysis was done using K independent test followed by Kruskal-wallis for nonparametric test. SEE: seed ethanol (70%) extract.

Values are expressed as means ± SEM (n = 6).

*Significantly different from ethanol treated group at $p < 0.05$.

**Significantly different from standard ranitidine group at $p < 0.05$.

Table 3. Antihyperglycemic activity of *V. ervilia* seeds

Groups	Dose mg/kg	mg/dl		
		Glucose	Triglyceride	Cholesterol
Normal		81.26 ± 1.74	34.15 ± 1.97	31.26 ± 1.32
Hyperglycemic		361.11 ± 7.47*	220.31 ± 5.23*	228.50 ± 10.11*
Gliclazid	5	103.63 ± 1.33**	55.69 ± 2.77**	49.23 ± 2.63**
SEE	250	111.58 ± 8.43**	70.15 ± 5.36*,**	115.63 ± 6.09*,**,***
	500	75.63 ± 6.3**	57.54 ± 4.93*,**	77.57 ± 4.41*,**,***

Notes: Statistical analysis was done using one way ANOVA followed by Tukey for multiple comparisons. SEE: seed ethanol (70%) extract.

Values are expressed as means ± SEM (n = 6).

*Significantly different from control group at $p < 0.05$.

**Significantly different from hyperglycemic group at $p < 0.05$.

***Significantly different from standard gliclazid group at $p < 0.05$.

Table 4. Antiviral activity of *V. ervilia* seeds

Test sample	Selected dose (mg/ml)[MNTC]	Viability (%)[a]	Cytotoxicity (%)	Antiviral effect (%)
Control (VERO cell line)		100	0	0
Virus control		31	69	
SEE	0.1	57	43	26

Notes: MNTC: maximum non-toxic concentration; SEE: seed ethanol (70%) extract.

[a]Average of three determinations.

value which is higher than that caused by the standard gliclazid (71.3%). SEE (500 mg/kg b.wt) reduced triglycerides and cholesterol levels in diabetic rats by 73.89 and 66.06%.

The *Antiviral* activity was screened against *Coxsackie B4* virus. It is a serotype of *Enterovirus B* which can trigger an autoimmune reaction resulting in destruction of the pancreas insulin-producing beta cells, which is one of several different etiologies of diabetes mellitus (Ylipaasto et al., 2004). SEE had a mild activity against *Coxsackie B4* virus (Table 4). SEE proved to have significant hypoglycemic action in this study (Table 3). This suggests the use of *V. ervilia* seeds in treatment of diabetes as it may relieve the case through two different mechanisms, antiviral and hypoglycemic ones. Further studies on the ability of *V. ervilia* seeds to treat diabetes through other mechanisms as increased insulin release from pancreatic beta cells, insulin sparing effect and controlling lipid peroxidation are recommended.

This is the first report on anti-inflammatory, analgesic, antiulcerogenic, antihyperglycemic and antiviral activities of such promising seeds. We are conducting further studies on *V. ervilia* to explore other pharmacological activities and its phytochemical content.

4. Conclusions

It was found that *V. ervillia* seeds possess a wide spectrum of biological activities. The plant can be used to relieve different illness in Egypt. The country climatic conditions are suitable for cultivation of such cheap seeds. The only restriction for the medicinal use of *V. ervilia* seed is its canavanine content which can be easily removed by soaking in boiling water before use. The seed's low price, availability, ease of cultivation, and medicinal effectiveness make it a good choice especially for low-income people in developed countries.

Funding
The authors received no direct funding for this research.

Competing Interest
The authors declare no competing interests.

Author details
Mona M. Okba[1]
E-mails: monamoradokba@gmail.com,
mona.morad@pharma.cu.edu.eg
Gehad A. Abdel Jaleel[2]
E-mail: gehad_abougharam@yahoo.com
ORCID ID: http://orcid.org/0000-0003-2445-3314
Miriam F. Yousif[1,3]
E-mail: miriamyousif@hotmail.com
Kadriya S. El Deeb[1]
E-mail: Ka_el_deeb@hotmail.com
Fathy M. Soliman[1]
E-mail: fathysoliman232@yahoo.com
[1] Faculty of Pharmacy, Pharmacognosy Department, Cairo University, Kasr El-Ainy, 11562 Cairo, Egypt.
[2] Pharmacology Department, National Research Center, Tahrir Street, Dokki, Cairo, Egypt.
[3] Faculty of Pharmaceutical Sciences and Pharmaceutical Industries, Pharmacognosy Department, Future University, Al Tagamoa Al Khames, 11528 New Cairo, Egypt.

References
Abdel-Tawab, R. R., Abdel-Nasser, A. M., & Darmawan, J. (2004). *The prevalence of rheumatic diseases in rural Egypt COPCORD-Egypt*. 11thAPLAR Congress, Jeju, Korea.

Adami, E., Marazzi-Uberti, E., & Turba, C. (1964). Pharmacological research on gefarnate, a new synthetic isoprenoid with an anti-ulcer action. *Archives Internationales de Pharmacodynamie et de Thérapie, 147*, 113–145.

Alley, M. C., Scudiero, D. A., & Monks, A. (1988). Feasibility of drug screening with panels of human tumor cell lines using a microculture tetrazolium assay. *Cancer Research, 48*, 589–601.

Amujoyegbe, B. J., Agbedahunsi, J. M., & Amujoyegbe, O. O. (2012). Cultivation of medicinal plants in developing nations: Means of conservation and poverty alleviation. *International Journal of Medicinal and Aromatic Plants, 2*, 345–353. ISSN 2249 – 4340.

Berger, J. D., Robertson, L. D., & Cocks, P. S. (2003). Agricultural potential of Mediterranean grain and forage legumes: 2) Anti-nutritional factor concentrations in the genus *Vicia*. *Genetic Resources and Crop Evolution, 50*, 201–212. http://dx.doi.org/10.1023/A:1022954232533

Buck, W. B., Osweiter, G. D., & Van Gelder, A. G. (1976). *Clinical and diagnostic veterinary toxicology* (2nd ed., p. 5201). Iowa: Kendall/Hunt Publishing Company.

Chattopadhyay, D., Arunachalam, G., Mandal, A. B., Sur, T. K., Mandal, S. C., & Bhattacharya, S. K. (2002). Antimicrobial and anti-inflammatory activity of folklore: *Mallotus peltatus* leaf extract. *Journal of Ethnopharmacology, 82*, 229–237. http://dx.doi.org/10.1016/S0378-8741(02)00165-4

Fornstedt, N., & Porath, J. (1975). Characterization studies on a new lectin found in seeds of Vicia ervilia. *FEBS Letters, 57*, 187–191. http://dx.doi.org/10.1016/0014-5793(75)80713-7

Haddad, S. G. (2006). Bitter vetch grains as a substitute for soybean meal for growing lambs. *Livestock Science, 99*, 221–225. http://dx.doi.org/10.1016/j.livprodsci.2005.06.014

Hussein, N. R. (2010). Helicobacter pylori and gastric cancer in the Middle East: A new enigma? *World Journal of Gastroenterology, 16*, 3226–3234. Retrieved from http://www.wjgnet.com/1007-9327/16/3226.pdf http://dx.doi.org/10.3748/wjg.v16.i26.3226

Ibrahim, A. S., Khaled, H. M., Mikhail, N. H., Baraka, H., & Kamel, H. (2014). Cancer Incidence in Egypt: Results of the national population-based cancer registry program. *Journal of Cancer Epidemiology*, Article ID 437971, 18 p. doi:10.1155/2014/437971

Kerber, J. (1931). Determination of LD_{50}. *J. Arch. Exp. Path. Pharmacol.*, 162–480.

Laviola, G., & Alleva, E. (1990). Ontogeny of muscimol effects on locomotor activity, habituation, and pain reactivity in mice. *Psychopharmacology, 102*, 41–48. http://dx.doi.org/10.1007/BF02245742

Li, D. W., Hyun, J. E., Jeong, C. S., Kim, Y. S., & Lee, E. B. (2003). Antiinflammatory activity of α-hederin methyl ester from the alkaline hydrolysate of the butanol fraction of kalopanax pictus bark extract. *Biological & Pharmaceutical Bulletin, 26*, 429–433. http://dx.doi.org/10.1248/bpb.26.429

Lotfi, A., Abdel-Nasser, A. M., & Hamdy, A. (2007). Hypovitaminosis D in female patients with chronic low back pain. *Clinical Rheumatology, 26*, 1895–1901. http://dx.doi.org/10.1007/s10067-007-0603-4

Morimoto, Y., Shimohara, K., Oshima, S., & Sukamoto, T. (1991). Effects of the new anti-ulcer agent KB-5492 on experimental gastric mucosal lesions and gastric mucosal defensive. *Journal of Pharmacology, 57*, 495–505.

National Center of Health and Population. (2004). *The burden of disease and injury in Egypt (mortality and morbidity)*.

Okba, M. M., Yousif, M. F., El Deeb, K. S., & Soliman, F. M. (2014). Botanical study, DNA fingerprinting, total protein profiling, nutritional values and certain proximates of *V. ervilia* L. *International Journal of Pharmacy and Pharmaceutical Sciences, 6*, 246–253.

Sadeghi, G. H., Pourreza, J., Samei, A., & Rahmani, H. (2009). Chemical composition and some anti-nutrient content of raw and processed bitter vetch (*Vicia ervilia*) seed for use as feeding stuff in poultry diet. *Tropical Animal Health and Production, 41*, 85–93. http://dx.doi.org/10.1007/s11250-008-9159-9

Sadeghi, G., Samie, A., Pourreza, J., & Rahmani, H. R. (2004). Canavanine content and toxicity of raw and treated bitter vetch (*Vicia ervilia*) seeds for broiler chicken. *International Journal of Poultry Science, 3*, 522–529.

Shaw, J. E., Sicree, R. A., & Zimmet, P. Z. (2010). Global estimates of the prevalence of diabetes for 2010 and 2030. *Diabetes Research and Clinical Practice, 87*, 4–14. http://dx.doi.org/10.1016/j.diabres.2009.10.007

Täckholm, V. (1974). *Students' flora of Egypt* (2nd ed., p. 245, 276). Cairo: Cairo University.

Van de Loosdrecht, A. A. (1994). A tetrazolium-based colorimetric MTT assay to quantitate human monocyte mediated cytotoxicity against leukemic cells from cell lines and patients with acute myeloid leukemia. *Journal of Immunological Methods, 174*, 311–320. http://dx.doi.org/10.1016/0022-1759(94)90034-5

Winter, C. A., Risley, E. A., & Nuss, G. W. (1962). Carrageenin-induced edema in hind paw of the rat as an assay for antiinflammatory drugs. *Experimental Biology and Medicine, 111*, 544–547. http://dx.doi.org/10.3181/00379727-111-27849

Woolfe, G., & McDonald, A. D. (1944). The evaluation of the analgesic action of pethidine hydrochloride (Demerol). *J Pharmacol Exper Ther, 80*, 300–307.

World Economic Outlook. (2015, April). *WEO groups and aggregates information* (pp. 150–153). Washington, DC: International Monetary Fund.

World Health Organization. (2011). *Hepatitis C*. Retrieved from https://en.wikipedia.org/wiki/Hepatitis_C

Ylipaasto, P., Klingel, K., Lindberg, A. M., Otonkoski, T., Kandolf, R., Hovi, T., & Roivainen, M. (2004). Enterovirus infection in human pancreatic islet cells, islet tropism *in vivo* and receptor involvement in cultured islet beta cells. *Diabetologia, 47*, 225–239. doi:10.1007/s00125-003-1297-z. ISSN 0012-186X

Zhang, J., Huang, Y., Hou, T., & Wang, Y. (2006). Hypoglycaemic effect of *Artemisia sphaerocephala* Krasch seed polysaccharide in alloxan-induced diabetic rats. *Swiss Medical Weekly, 136*, 529–532.

16

Protective effect of *Morinda citrifolia* L. (fruit extract) on methotrexate-induced toxicities —hematological and biochemical studies

Bhakti A. Mhatre[1] and Thankamani Marar[1]*

*Corresponding author: Thankamani Marar, School of Biotechnology & Bioinformatics, D.Y. Patil University, Sector 15, CBD Belapur, Navi Mumbai 400614, Maharashtra, India
E-mail: dr.marar@yahoo.com
Reviewing editor: Tsai-Ching Hsu, Chung Shan Medical University, Taiwan
Additional information is available at the end of the article

Abstract: Methotrexate (MTX) has been widely used as an anticancer drug. It acts as a folic acid analog, inhibits purine and pyrimidine synthesis, which accounts for its efficacy in the therapy of cancer as well as for some of its toxicities. The present study is an attempt of modulating MTX-induced toxicities using aqueous extract of *Morinda citrifolia* L. (Noni) as a nutritional supplement. Hematological parameters such as RBC, WBC, and platelet count that decreased ($p < 0.05$) after methotrexate injection was found to have been restored in rats co-treated with Noni. Enhanced levels of lipid peroxides ($p < 0.05$) in animals administered with MTX showed significant revision after co-administration of Noni. Alterations of other biochemical constituents in blood like glucose, urea, uric acid, and cholesterol ($p < 0.05$) were also reversed to near normal levels in animals co-treated with *Morinda citrifolia* L. The study provides preliminary evidence that *Morinda citrifolia* L. extract can ameliorate MTX mediated side effects.

Subjects: Biochemistry; Biophysics; Biotechnology

Keywords: methotrexate toxicity; *Morinda citrifolia* L.; Noni

ABOUT THE AUTHOR

Bhakti A. Mhatre is presently perusing PhD in Biotechnology from School of Biotechnology and Bioinformatics, D.Y. Patil University CBD Belapur, Navi Mumbai, Maharashtra, India. Mhatre is presently working on anticancer properties of *Morinda citrifolia* L. fruit extract. The use of chemotherapeutic agents in the treatment of cancer has hampered and complicated by toxic side effects. Many types of chemotherapy destroy cancer cells by generating free radicals which can cause cellular damage. Unfortunately, these free radicals are not discriminatory in their destructive action leading to undesirable side effects and sometimes even new cancers. Therefore, administration of antioxidants might reduce the side effects of camptothecin without compromising its efficacy. Further studies assessing the potential usefulness of *Morinda citrifolia* L. fruit extract treatment in methotrexate-induced toxicities on other organs and organ systems are warranted which may provide an effective way to improve their therapeutic efficacy.

PUBLIC INTEREST STATEMENT

Methotrexate (MTX) (4-amino-10-methylpteroylglutamic acid) is a potent antineoplastic agent used to treat choriocarcinoma, leukemia, osteosarcoma, non-Hodgkins lymphoma, breast cancer, and lung cancer. It is also involved in the treatment of non-cancerous conditions such as rheumatoid arthritis, psoriasis, immunological abnormalities, and systemic inflammation. Low-moderate to high doses of MTX causes various side effects and may lead to conditions such as liver cirrhosis or fibrosis. It has also been shown that MTX administration has severe side effects on the hematopoietic system. This work is to highlight the importance of *Morinda citrifolia* L. fruit extract supplementation as an antioxidant in chemotherapy to re-establish the levels of antioxidant and to strike a balance between the oxidant and antioxidant levels thus preventing the enormous toxicity observed due to this drug. Combinational chemotherapy gives an insight for an effective treatment to cancer patients helping them exhibit minimum levels of the deleterious drug-induced side effects.

1. Introduction

Methotrexate (MTX), a structural analog of folic acid, a potent inhibitor of enzyme dihydrofolate reductase is widely used as a chemotherapeutic agent for leukemia and other malignancies (Faten, Ibrahim, & Khaled, 2013). Nowadays, it is also used for sarcoidosis, inflammatory bowel diseases, vasculitis, arthritis, and severe refractory asthma (Patel & Ghodasara, 2014). However, its use is limited due to high incidence of serious dose-dependent toxicity, including hepatotoxicity, renal damage, bone marrow suppression, and gastrointestinal mucosal inflammation (Mohamed Akram, 2006; Ramadan, 2008).

Low-moderate to high doses of MTX causes various side effects and may lead to conditions such as liver cirrhosis or fibrosis (Katherine, Anjali, & Judith, 2010). It has also been shown that MTX administration has severe side effects on the hematopoietic system. MTX induces oxidative stress by increasing ROS production which is implicated in tissue injury (Viswa, Premila, & Bina, 2007). Further, associated increase in oxidative stress may play an important role in the pathophysiology of drug-induced side effects (Rakesh & Neeta, 2013).

Fruits of *Morinda citrifolia* L., also known as Noni, have been used as herbal medicines by ancient Hawaiians as remedies for various ailments. It has been reported to have a broad range of therapeutic effects, including antibacterial, antiviral, antifungal, antitumor, analgesic, hypotensive, anti-inflammatory, and immune enhancing effects (Khuntia & Panda, 2010). Empirical use of medicine derived from plants has been widely disseminated since ancient times to treat a wide range of diseases. Noni juice in particular has been used in the treatment of diabetes, heart disease, high blood pressure, and kidney and bladder disorders (Ajadi, Adenubi, & Thomas, 2011). A number of major chemical compounds and antioxidants like, alkaloids, flavonoids, tannins, and phenols have been identified in the leaves, roots, and fruits of Noni plant. Daily intake of Noni has reported to reduce free radical-induced oxidative damage and the consequent lipid peroxidation, and it improves the quality of life of patients undergoing chemotherapy. It has been reported to have a broad range of health benefits for cancer, infection, arthritis, diabetes, asthma, hypertension, and pain (Wang, West, & Jenesen, 2002). Concurrent use of *Morinda citrifolia* L. extract with chemotherapy can reduce the toxicity and increase the efficacy of the drug. Increased serum levels of lipid peroxides have been reported to decrease after noni administration and protects fall in leukocyte count, hemoglobin level, and mean osmotic fragility of erythrocytes (Bhakti & Thankamani, 2016). This study is a preliminary evaluation of the hematological and biochemical profile of rats treated with methotrexate and to determine the extent to which of *Morinda citrifolia* L., treatment could ameliorate the detrimental effects induced by MTX.

2. Materials and methods

Drugs and chemicals: methotrexate injection Folitrax-15 IP (15 mg/ml) was purchased from Ipca Pharmaceuticals. The fruits of *Morinda citrifolia* were purchased from Abirami Botanical, Tamil Nadu, India. Specimen was authenticated from St. Xaviers College, Department of Botany, Mumbai as *Morinda citrifolia* L. belonging to family Rubiaceae with Blatter Herbarium specimen number 108. The fruits were air-dried for 2 days and ground to powder. All other chemicals were of high analytical grade from SRL, Mumbai, India and Merck India Ltd, Mumbai, India and solvents were of Qualigen finechemicals, Mumbai, India.

The aqueous extract of Noni was prepared by cold maceration of 250 g of the shade-dried fruit powder in 500 ml of distilled water which was allowed to stand overnight. It was boiled for 5–10 min till the volume was reduced to half its original volume. The solution was then cooled, filtered through eight-layered muslin cloth, concentrated, Whatman, dried in vacuum (yield 36 g) and the residue stored in a refrigerator at 2–8°C for subsequent use (Bhakti & Thankamani, 2015).

Animal model: adult male albino rats of Wistar strain (100 ± 20 g) were obtained from Bharat Serum Pvt. Ltd, Thane, Navi Mumbai, India. The animals were maintained under standard conditions

Table 1. Design of treatment protocol	
Groups	**Treatment (n = 6)**
Group I (control)	Vehicle control (saline)
Group II	1 mg/kg/bw of methotrexate in saline (ip) twice weekly for 30 days
Group III	5 mg/kg/bw of concentrated Noni extract orally for 30 days
Group IV	5 mg/kg/bw of concentrated Noni extract orally + 1 mg/kg/bw of methotrexate in saline (ip) twice weekly for 30 days simultaneously

of temperature (25 ± 2°C), light (12 h light/12 h dark), and humidity. They were divided into groups as given in Table 1. They were fed standard rat pelleted diet obtained from Lipolin, India and water *ad libitum*. Experimental animals were handled according to the Institutional legislation, regulated by the Committee for the Purpose of Control and Supervision of Experiments on Animals (CPCSEA).

Experimental Design: following the acclimatization period of one week, a pilot study was conducted with 5 mg/kg/bw of methotrexate (ip) diluted in autoclaved distilled water twice weekly for 30 days. It was found to be highly toxic as there was extensive weight loss, loss of appetite, and diarrhea. Animals died within two weeks, hence the dose of methotrexate was modified to 1 mg/kg/bw of methotrexate ip twice weekly for 30 days and 5 mg/kg/bw of concentrated Noni extract diluted in distilled water. Rats were randomly divided into four groups consisting of six animals each (Table 1).

At the end of the experimental period, the animals were killed by cervical decapitation. Blood was collected in EDTA for plasma and without anticoagulant for serum. Blood was processed further for RBC and WBC counts (Raghuramulu, Madhavan, & Kalyansundaram, 1983) differential count (Dacie & Lewis, 1984), platelet count (Samuel, 1986), hemoglobin, (Varley, 2005), glucose (Trinder, 1969) and urea (Varley, 2005). The serum levels of protein (Reinhold, 1953), lipid peroxides (Yagi, 1984), uric acid, creatinine, bilirubin, and cholesterol (Varley, 2005) were also determined.

Statistical analysis: the results are expressed as mean ± standard deviation (SD) for six animals in each group. The statistical evaluation of all data was done using analysis of variance (ANOVA). p-value < 0.05 was considered statistically significant. It was complemented with unpaired Student's t-test. Data were analyzed by SPSS software, version 19 (Chicago, IL, USA).

3. Results and discussion
There were prominent, distinctive clinical signs, mortality, or morbidity observed in Group II during the experimental period. Control rats receiving *Morinda citrifolia* L. alone (Group III) did not show any significant change when compared with control rats (Group I), indicating that *Morinda citrifolia* L. does not have any adverse effect. Table 2 shows the results of the complete blood count. There is a significant decrease in number of RBCs ($p < 0.05$) and WBCs ($p < 0.01$) accompanied with a decrease in the levels of hemoglobin. Among the WBCs, the significant decrease is in the number of neutrophils. It is well known that myelosuppression and pancytopenia are characteristic features of chemotherapy. Actively proliferating progenitor cells in the bone marrow are sensitive to anticancer agents and are subjected to an irreversible removal process. According to clinical studies (Patel & Ghodasara, 2014) of methotrexate, the principal dose-limiting toxicities are neutropenia, thrombocytopenia, and anemia. In the present study, hematological effects of MTX including decreased RBC, WBC, and platelets are in agreement with the results of above clinical studies. Premature death of RBCs as a result of oxidative injury can also contribute to the reduction in RBC count and a decrease in hemoglobin. Pretreatment with *Morinda citrifolia* L. showed beneficial effect by restoring the levels of RBCs, WBCs, and hemoglobin suggesting the ameliorative effect of *Morinda citrifolia* L. in preventing MTX-induced bone marrow suppression. Glucose levels in blood of MTX-treated rats increased by 25 percent (Table 3). Insulin is essential for maintaining blood sugar levels and decreased availability

of insulin due to MTX-induced toxicity to the pancreas may contribute to the observed hyperglycemia. This hypoinsulinemia leads to less utilization of glucose by the cells and hence an increase in blood sugar (Singh & Marar, 2011).

Hepatotoxicities induced by MTX may also lead to increase in glucose levels since liver is also actively involved in glucose metabolism (Mohamed Akram, 2006).

MTX also causes increase in lipid peroxidation which has been reported to inactivate enzymes involved in glycolysis. *Morinda citrifolia* L. co-treated animals did not show hyperglycemia. Dietary supplementation of *Morinda citrifolia* L. has been reported to reduce blood sugar levels (Ajadi et al., 2011). Further protection rendered by *Morinda citrifolia* L. against lipid peroxidation could help in restoring activities of the enzymes (Table 3).

A significant decrease in serum cholesterol levels was seen in MTX-treated rats as shown in Figure 1. This might be correlated to malabsorption, maldigestion, and diarrhea observed in animals intoxicated with MTX. Further exocrine pancreatic insufficiency and hepatotoxicosis induced by MTX may also be a contributory factor. Supplementation of *Morinda citrifolia* L. in the diet could render protection to the gastrointestinal system and thus normalize the levels of cholesterol in serum.

There is an increase in the serum levels of creatinine, bilirubin, urea, and uric acid in rats treated with MTX (Table 3). An increase in the levels of urea ($p < 0.05$) and uric acid ($p < 0.01$) was also reported in animals injected with MTX (Kevin & Moder, 1995). Similarly, serum creatinine and urea levels were considerably increased ($p < 0.01$) than normal which indicated nephrotoxicity (Patel & Ghodasara, 2014). Moderate rise in urea is an indicator of increased protein breakdown associated with cytotoxicity of drugs. Uric acid is the end product of purine metabolism and it increases consistently with cell death induced by MTX. Our results reveal that blood creatinine, bilirubin, protein, urea, and uric acid returned to near-normal levels in animals supplemented with *Morinda citrifolia* L. These results further confirm the hepatoprotective nature of Noni (Seshachary & Satyavati, 2014).

Table 2. Effect of methotrexate and *Morinda citrifolia* L. fruit extract on complete blood count

Parameters	Group I control	Group II MTX (1 mg/kg/bw)	Group III Noni (5 mg/kg/bw)	Group IV Noni (5 mg/kg/bw) + MTX (1 mg/kg/bw)	ANOVA p-value
Hemoglobin (mg/dl)	11.7 ± 1.4	11.35 ± 1.1[aNS]	11.03 ± 0.6[bNS]	11.33 ± 1.3[cNS]	0.03
RBC (million/mm³)	7.7 ± 1.4	6.17 ± 0.7[aNS]	7.5 ± 0.2[bNS]	7.15 ± 0.3[cNS]	0.013
WBC (per mm³)	9,583.3 ± 69	8,483.3 ± 20[a***]	10,300 ± 34[b***]	9,083.3 ± 50[c***]	0.141
Platelets (×10³ per mm³)	1,229.1 ± 48	924 ± 14[a***]	1,233.5 ± 34[bNS]	1,231.33 ± 52[c***]	0.001
Neutrophils (%)	25.8 ± 26	17.7 ± 7.5[aNS]	21.83 ± 11[bNS]	15 ± 6[cNS]	0.638
Eosinophils (%)	69.5 ± 27	75.3 ± 17[aNS]	68 ± 1 0[bNS]	84.5 ± 5.9[cNS]	0.526

Notes: values are expressed as mean ± SD for six rats in a group. Statistical significance were tested with ANOVA and complemented with student's *t*-test; values carrying different superscripts are significantly different. NS: non-significant. One-way analysis revealed hemoglobin, RBC, and platelets show significance with $p < 0.05$, whereas p-values of WBC, neutrophils, and eosinophils are non-significant.

Comparisons are expressed as:

[a]Group I and Group II.

[b]Group I and Group III.

[c]Group II and Group IV.

Statistical significance represented as:

*$p < 0.05$.

**$p < 0.01$.

***$p < 0.001$.

Table 3. Effect of methotrexate and *Morinda citrifolia* L. fruit extract treatment on serum biochemical parameters of rats

Serological parameters	Group I control	Group II MTX (1 mg/kg/bw)	Group III Noni (5 mg/kg/bw)	Group IV Noni (5 mg/kg/bw) + MTX (1 mg/kg/bw)	ANOVA p-values
Glucose (mg/dl)	115.9 ± 3.6	82.55 ± 8.13[a]***	114.94 ± 2.6[bNS]	92.39 ± 1.3[c]*	0.001
Urea (mg/dl)	32.7 ± 1.3	46.97 ± 2.6[a]***	34.56 ± 0.9[b]*	40.83 ± 2.6[c]**	0.001
Uric acid (mg/dl)	3.93 ± 0.3	4.81 ± 0.4[a]*	3.56 ± 0.6[bNS]	4.10 ± 0.5[c]*	0.029
Total bilirubin (mg/dl)	0.43 ± 0.020	0.55 ± 0.018 [a]***	0.43 ± 0.010[bNS]	0.44 ± 0.030[c]***	0.001
Creatinine (mg/dl)	1.54 ± 0.01	2.56 ± 0.01[a]***	1.66 ± 0.03 [b]***	1.67 ± 0.03[c]***	0.001
Protein (g/dl)	8.50 ± 2.17	7.40 ± 2.6[a]***	8.84 ± 2.8[b]*	7.56 ± 1.8[cNS]	0.001
Serum lipid peroxides (U/ml)	65.25 ± 1.9	108.04 ± 4.4[a]***	66.65 ± 0.9[cNS]	75.23 ± 4[c]***	0.001

Notes: Values are expressed as mean ± SD for six rats in a group. Statistical significance were tested with ANOVA and complemented with Student's *t*-test; values carrying different superscripts are significantly different. NS: non significant. One way analysis revealed that all the serological parameters are significant with $p < 0.05$.

Comparisons are expressed as:

[a]Group I and Group II.

[b]Group I and Group III.

[c]Group II and Group IV.

Statistical significance represented as:

*$p < 0.05$.

**$p < 0.01$.

***$p < 0.001$.

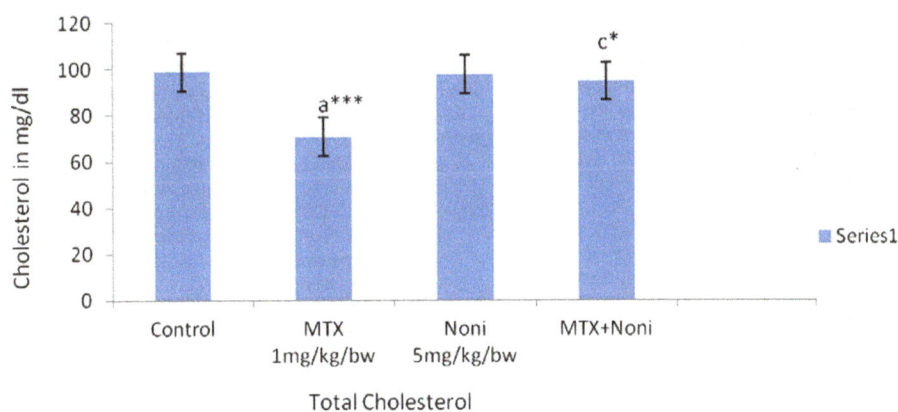

Figure 1. Effect of methoterexate and *Morinda citrifolia* L. fruit extract on levels of serum cholesterol.

Notes: Values are expressed as mean ± SD for six rats in a group. Statistical significance were tested with ANOVA and complemented with Student's *t*-test; values carrying different superscripts are significantly different. Comparisons are expressed as "a" Group I and Group II, "b" Group I and Group III, "c" Group II and Group IV. Statistical significance represented as: *$p < 0.05$, ***$p < 0.001$.

Membrane lipids succumb easily to deleterious actions of reactive oxygen species. In present study, the increased levels of lipid peroxides in serum of Group II animals reflect the oxidative stress (Table 3). Treatment with *Morinda citrifolia* L. protects the cells through attenuation of lipid peroxidation, as evident from the decreased levels of serum lipid peroxides in Group IV animals. Noni has been reported to possess antioxidant activity (Katherine et al., 2010) and its co-administration may have helped decrease oxidative stress.

4. Conclusion

The present investigation reveals that *Morinda citrifolia* L. fruit juice can ameliorate toxicities induced by MTX. Often, argument concerned with supplementation of antioxidants in chemotherapy arises. The basis of this disagreement is that many of the chemotherapeutic drugs induce the formation of oxygen-derived free radicals, the effect of which would be blocked by antioxidants. The anticancer effect of methotrexate however, does not depend on the formation of free radicals. Therefore, administration of antioxidants might reduce the side effects of methotrexate without compromising its efficacy. Further studies assessing the potential usefulness of *Morinda citrifolia* L. treatment in MTX-induced toxicities on other organs and organ systems are required which may provide an effective way to improve their therapeutic efficacy.

Funding
The authors received no direct funding for this research.

Competing Interests
The authors declare no competing interest.

Author details
Bhakti A. Mhatre[1]
E-mail: bhaktiamhatre@gmail.com
Thankamani Marar[1]
E-mail: dr.marar@yahoo.com
[1] School of Biotechnology & Bioinformatics, D.Y. Patil University, Sector 15, CBD Belapur, Navi Mumbai 400614, Maharashtra, India.

References
Ajadi, T., Adenubi, T., & Thomas, C. (2011). Protective effect of Tahitian Noni juice on the reproductive functions of Male Wistar Rats traeted with cyclophosphamide. *Iranian Journal of Pharmacology and Therapuetics, 10*, 39–43.

Bhakti, M., & Thankamani, M. (2015). Vetting and comparative analysis of antioxidants and phytochemicals from methanolic, aqueous and a popular commercial fruit juice (Noni) *Morinda Citrifolia* L. *Research Journal of Pharmaceutical, Biological and Chemical Sciences, 6*, 884–890.

Bhakti, M., & Thankamani, M. (2016). Ameliorativeeffect of *Morinda citrifolia* L. extracton Imatinib induced toxicities: In vitro studies on human RBC's. *Journal of Cell and Tissue Research, 16*, 5507–5510.

Dacie, J., & Lewis, S. (1984). *Practical Hematology* (pp. 84–86). NewYork, NY: Churchill-Lingstone Edinburgh.

Faten, R. A., Ibrahim, A. E., & Khaled, A. E. (2013). Protective and modulatory effects of Curcumin and L-Carnitine against Methotrexate-induced Oxidative stress in albino rats. *Research Journal of Pharmaceutical, Biological and Chemical Sciences, 4*, 744–754.

Katherine, S., Anjali, G., & Judith, A. S. (2010). Evaluation of the hepatic metabolism and antitumor activity of Noni juice (*Morinda citrifolia* L.) in combination with chemotherapy. *Journal of the Society for Integrative Oncology, 8*, 89–94.

Kevin, G., & Moder, M. (1995). Malignancies and the use of Methotrexate in rheumatoid arthritis. *America Journal of Medicines Hematologic , 99*, 276–281.

Khuntia, T., & Panda, D. (2010). Evaluation of antibacterial, antifungal and anthelmintic activity of *Morinda citrifolia* L. (Noni). *International Journal of PharmTech Research, 2*, 1030–1032.

Mohamed Akram, A. M. (2006). Histological and histochemical studies on the effects of methotrexate on the liver of adult male albino rat. *International Journal of Morphology, 24*, 417–422.

Patel, N. N., Ghodasara, D. (2014). Subacute toxicopathological studies of methotrexate in Wistar rats. *Dave Veterinary World, 223*, 489–495.

Raghuramulu, N., Madharan, N., Kalyansundaram, S. (1983). *A manual of laboratory techniques* (Vol. 78, pp. 257–258). Hyderabad: Silver prints.

Rakesh, K., & Neeta, S. (2013). *Morinda citrifolia* (Noni) alters oxidative stress marker and antioxidant activity in cervical cancer cell lines. *Asian Pacific Journal of Cancer Prevention, 14*, 4603–4606.

Ramadan, A. M. (2008). Curcumin attenuates methotraxate-induced hepatic oxidative damage in Rats. *Journal of the Egyptian National Cancer Institute, 20*, 141–148.

Reinhold, J. (1953). *Standard methods of clinical chemistry* (Vol. 1, pp. 88–97). M. Reiner (Ed.), NewYork, NY: Academic Press.

Samuel, K. M. (1986). *Notes on clinical lab techniques (168)*. Madras: MGK Iyyer and Sons Publishers.

Seshachary, A. K., & Satyavati, D. (2014). Chemoprotective effect of ethanolic extract of *Morinda citrifolia* against Cisplatin induced nephrotoxicity. *The Pharma Innovation , 3*, 84–91.

Singh, K., & Marar, T. (2011). Acute toxicity of Camptothecin and influence of α-Tocopherol on haematological and biochemical parameters. *Journal of Cell and Tissue Research, 11*, 2833–2837.

Trinder, P. (1969). *Analytical Clinical Biochemistry, 6*, 24–25.

Varley, H. (2005). *Practical Clinical Biochemistry* (4th ed.). New Delhi: CBC Publisher and distributor.

Viswa, K. K., Premila, A., & Bina, I. (2007). Alteration in antioxidant defense mechanisms in the small intestines of methotrexate treated rat may contribute to its gastrointestinal toxicity. *Cancer Therapy, 5*, 501–510

Wang, M.-Y., West, B. J., & Jenesen, C. J. (2002). *Morinda citrifolia* (Noni): A literature review andrecent advances in Noni research. *Act a Pharmacolog ica Si nic, 23*, 1127–2141.

Yagi, K. (1984). Assay for plasma lipid peroxides. *Methods in Enzymology, 109*, 328–331.

17

Comparison of visual observation and emission intensity of resazurin for antimicrobial properties of hexane, dichloromethane, methanol and water extracts from a brown alga, *Turbinaria ornata*

Kar-Yee Tye[1]*, Sook-Yee Gan[1], Swee-Hua Erin Lim[2], Sou-Eei Tan[1], Chen-Ai Chen[1] and Siew-Moi Phang[3]

*Corresponding author: Kar-Yee Tye, School of Pharmacy, International Medical University, 126 Jalan Jalil Perkasa 19, Bukit Jalil 57000, Kuala Lumpur, Malaysia
E-mail: tye.dragon@gmail.com
Reviewing editor: Jurg Bahler, University College London, UK
Additional information is available at the end of the article

Abstract: Marine natural products have been increasingly found to be a promising source of drug candidates for fighting human diseases. The present study was carried out to assess the antimicrobial properties of a brown alga, *Turbinaria ornata*. Hexane, dichloromethane, methanol, and water extracts were tested against 23 micro-organisms including Gram-positive and negative bacteria, yeasts, and fungi. The disk diffusion method was employed followed by modified resazurin microtitre assay (REMA). The results obtained from modified REMA using both methods of colorimetric and fluorometric were compared. The best antimicrobial activity was recorded in dichloromethane extract for disk diffusion. Further, modified REMA showed inhibition in *Bacillus cereus, Bacillus subtilis, Staphylococcus aureus* ATCC 25923, *S. aureus, Staphylococcus epidermidis, Staphylococcus saprophyticus, Enterococcus faecalis, Pseudomonas aeruginosa* ATCC 27853, *Candida parapsilosis* ATCC 22019, *Candida guilliermondii* ATCC 6260, and *Saccharomyces cerevisiae*. Both methods of modified REMA were substantially in agreement with each other based on Cohen's kappa statistical analysis (κ value = 0.712; $p < 0.0005$). Our findings suggested that *T. ornata* dichloromethane extract has the potential to be used as a source of antimicrobial compounds.

ABOUT THE AUTHORS

The main focus of our research is the exploration of natural resources for drugs and natural products. Natural products have played important roles in drug development for medicine and health. We are involved in sourcing plant and algal bioactive compounds with properties such as antioxidative, anti-neuroinflammatory, antifungal, antimicrobial as well as neuroprotective. Our research scopes include toxicology and neurodegenerative disorders. In addition, we are also involved in various molecular studies to understand the biological effects of the compounds using both animal models and *in vitro* cell-based assays.

PUBLIC INTEREST STATEMENT

Micro-organisms constantly evolve to adapt to new environments. As such, antimicrobial resistance is the ability of micro-organisms to grow in the presence of a chemical (drug) that would normally kill them or limit their growth. As a result of antimicrobial resistance, existing antibiotics become less and less effective in eliminating infections caused by these micro-organisms. One of the ways to address this issue is to develop new antibiotics. This paper discusses the potential of sourcing antimicrobial compounds from a marine alga, *Turbinaria ornata*. To test the antimicrobial properties of the marine alga's extract, a chemical indicator, resazurin was used. There are two methods of using resazurin to indicate antimicrobial properties; the authors used both methods and tried to analyze if the two methods were agreeable to each other.

Subject: Drug Discovery; Marine Biology; Microbiology; Natural Products

Keywords: antibacterial; antifungal; antimicrobial; seaweed; *Turbinaria ornata*

1. Introduction

Great interest has been developed in sourcing natural products from the marine environment due to the ocean's unique biodiversity (Baker, Chu, Oza, & Rajgarhia, 2007). Algae biosynthesize compounds called secondary metabolites in order to adapt to surrounding environment and to protect themselves against predators or pathogens (Lane et al., 2009). In addition, algae were reported to possess various biological properties such as antioxidant, anti-inflammatory, antitumor, antimicrobial, and anticoagulant (Ananthi et al., 2010; Aravindan, Delma, Thirugnanasambandan, Herman, & Aravindan, 2013; Manivannan, Karthikai Devi, Anantharaman, & Balasubramanian, 2011; Wang, Zhang, Zhang, Hou, & Zhang, 2011).

Antimicrobial resistance is a worldwide hazard and new resistance mechanisms have been emerging and spreading globally. Antimicrobial resistance results in prolonged illness, higher health costs, and higher risks of death (World Health Organization, 2015). One of the ways to address the issue of antimicrobial resistance is to actively encourage innovations and development of new antimicrobials. There is a great potential for sourcing antimicrobials from brown algae such as *Colpomenia sinuosa, Dictyota dichotoma, D. dichotoma* var. *implexa, Petalonia fascia,* and *Scytosiphon lomentaria* which possessed antimicrobial properties against *Bacillus subtilis* and *Staphylococcus aureus.* Compounds identified from these macroalgae included hydrocarbons, terpenes, acids, phenols, sulfur-containing compounds, and aldehydes (Demirel, Yilmaz-Koz, Karabay-Yavasoglu, Ozdemir, & Sukatar, 2009). Compounds such as phytol, neophytadiene, fucosterol, palmitoleic, and oleic acids had been identified in *Himanthalia elongate* and *Synechocystis* sp. These compounds were active against *Escherichia coli* and *S. aureus* (Plaza et al., 2010).

Turbinaria ornata is a brown alga under the family Sargassaceae. It is known to produce phenolic compounds such as phlorotannins (Girija, Hemalatha, Saranya, Parthiban, & Anantharaman, 2013) which were active against *S. aureus*, methicillin-resistant *S. aureus* (MRSA), *Salmonella* spp. *and E. coli* (Eom, Kim, & Kim, 2012). Using agar well diffusion method, Jeyaseelan (2012) reported that direct ethanol extract and sequentially extracted acetone and ethanol extracts of *T. ornata* were found to inhibit growth of *E. coli* and *S. aureus.* Furthermore, methanolic extract of *T. ornata* contained phenolic compounds which exhibited activities against nine Gram-positive and negative bacteria based on agar well diffusion test (Vijayabaskar & Shiyamala, 2011).

Conversely, it is also possible to detect antimicrobial activity in lipophilic extracts of algae. Metabolites with high lipophilicity have the advantage of being able to remain potent for longer periods as they diffuse slowly into seawater (Cortés et al., 2014). Therefore, four different solvents (dichloromethane, hexane, methanol, and water) with the potential to extract both hydrophilic and lipophilic compounds were employed in the extraction of *T. ornata*. These extracts were tested for antimicrobial activity against a total of 23 micro-organisms consisting of Gram-positive, Gram-negative bacteria, yeasts and fungi. This study utilized the modified resazurin microtitre assay (REMA) where methods of visual observation (colorimetric) and emission intensity (fluorometric) were compared.

2. Results and discussion

2.1. Results for disk diffusion assay

Both hexane and dichloromethane extracts showed growth inhibition in all seven strains of Gram-positive bacteria tested (Table 1). Dichloromethane extract had stronger antimicrobial activity in *Bacillus cereus, B. subtilis, S. aureus* ATCC 25923, and *Staphylococcus epidermidis* compared to hexane extract. For dichloromethane extract, except for *S. aureus* ATCC 25923 and *S. aureus*, in which inhibition started at 10 mg/mL, the rest of the strains were inhibited at the lowest extract

Table 1. Antimicrobial activity of T. ornata extracts against Gram-positive bacteria by disk diffusion method

Extract	mg/mL	Zone of inhibition (mm)						
		B. cereus	B. subtilis	S. aureus	S. aureus ATCC 25923	S. epidermidis	S. saprophyticus	E. faecalis
T. ornata hexane	3	–	–	12.00 ± 0.00	–	–	9.70 ± 0.58	–
	5	8.00 ± 0.00	–	14.00 ± 1.00	–	–	11.00 ± 1.00	–
	10	8.30 ± 1.15	–	15.30 ± 0.58	–	–	14.00 ± 0.00	–
	15	11.60 ± 1.15	–	15.60 ± 1.15	7.30 ± 0.58	–	15.00 ± 0.00	–
	25	13.00 ± 1.41	–	16.00 ± 1.00	7.70 ± 0.58	–	16.30 ± 0.58	–
	50	15.60 ± 0.58	8.50 ± 0.71	17.70 ± 0.58	9.00 ± 1.34	8.30 ± 0.58	18.00 ± 0.00	8.00 ± 0.00
	100	16.00 ± 1.00	9.60 ± 2.08	19.30 ± 0.58	9.70 ± 0.58	9.30 ± 0.58	20.60 ± 0.58	9.70 ± 0.00
T. ornata dichloromethane	3	7.67 ± 0.58	7.33 ± 0.58	–	–	14.33 ± 0.58	12.00 ± 0.00	6.33 ± 0.58
	5	10.00 ± 0.00	8.33 ± 0.58	–	–	17.00 ± 0.00	13.67 ± 0.00	6.67 ± 0.58
	10	11.33 ± 0.58	11.00 ± 0.00	13.00 ± 0.00	7.67 ± 0.58	19.00 ± 0.00	15.67 ± 0.00	8.00 ± 0.00
	15	14.00 ± 0.00	12.67 ± 1.53	16.67 ± 0.58	13.00 ± 0.00	21.00 ± 0.00	16.00 ± 0.00	8.00 ± 0.00
	25	15.33 ± 0.58	17.00 ± 0.00	17.67 ± 0.58	16.33 ± 0.58	24.00 ± 1.00	17.33 ± 0.00	8.00 ± 0.00
	50	16.00 ± 0.00	18.67 ± 1.53	19.00 ± 0.00	18.67 ± 0.58	26.00 ± 0.00	19.00 ± 0.00	9.00 ± 0.00
	100	16.33 ± 0.58	20.00 ± 1.73	20.67 ± 0.58	19.67 ± 0.58	28.00 ± 0.00	20.00 ± 0.00	9.67 ± 0.58
T. ornata methanol	3	–	–	–	–	–	–	–
	5	–	–	–	–	–	–	–
	10	–	–	–	–	–	–	–
	15	–	–	–	–	–	–	–
	25	–	–	–	–	–	–	–
	50	8.00 ± 0.00	7.70 ± 0.58	–	7.00 ± 0.00	–	8.00 ± 0.00	–
	100	10.30 ± 0.58	9.00 ± 0.00	8.00 ± 0.00	10.00 ± 0.00	–	8.80 ± 0.29	–
Gentamicin	10 µg/disk	30.80 ± 0.41	25.00 ± 0.00	23.00 ± 0.00	27.00 ± 1.10	31.00 ± 0.89	29.30 ± 0.52	12.30 ± 0.52

Notes: "–" indicates no antimicrobial activity. Gentamicin was used as positive control. 99.5% DMSO was used as negative control where no inhibition was observed. Disk diffusion test was performed in at least duplicates and diameter of zone of inhibitions was expressed as mean ± standard deviation in mm. Zones of inhibition include 6 mm for disk diameter.

Table 2. Antimicrobial activity of _T. ornata_ extracts against Gram-negative bacteria by disk diffusion method

Extract	mg/mL	Zone of Inhibition (mm)	
		P. aeruginosa ATCC 27853	_Enterobacter_ spp.
T. ornata dichloromethane	3	–	–
	5	–	–
	10	–	–
	15	–	–
	25	–	–
	50	7.00 ± 0.00	–
	100	7.00 ± 0.00	–
T. ornata methanol	3	–	–
	5	–	–
	10	–	–
	15	–	–
	25	–	–
	50	–	–
	100	–	8.33 ± 0.58
Gentamicin	10 ug/disk	18.20 ± 0.41	20.80 ± 0.41

Notes: "–" indicates no antimicrobial activity. Gentamicin was used as positive control. 99.5% DMSO was used as negative control where no inhibition was observed. Disk diffusion test was performed in at least duplicates and diameter of zone of inhibitions was expressed as mean ± standard deviation in mm. Zones of inhibition include 6 mm for disk diameter.

concentration of 3 mg/mL. Methanol extract only exhibited moderate zones of inhibition at higher extract concentrations of 50 mg/mL in strains of _B. cereus_, _B. subtilis_, _S. aureus_ ATCC 25923, and _Staphylococcus saprophyticus_ and 100 mg/mL in _S. aureus_. As for water extract, no antimicrobial activity was detected in all strains of Gram-positive bacteria.

Out of the six Gram-negative bacteria tested, _Pseudomonas aeruginosa_ was the only Gram-negative bacteria inhibited by dichloromethane extract starting at concentration of 50 mg/mL with zone of inhibition 7.00 ± 0.00 mm (Table 2). Methanol extract only inhibited growth of _Enterobacter_ spp. at 100 mg/mL with zone of inhibition recorded as 8.33 ± 0.58 mm. There was no inhibition observed in all Gram-negative bacteria using the hexane extract as well as water extract.

Dichloromethane extract caused growth inhibition in three strains of yeasts, namely _Candida guilliermondii_ ATCC 6262, _Candida parapsilosis_ ATCC 22019, and _Saccharomyces cerevisiae_ (Table 3). Hexane extract inhibited _S. cerevisiae_ as well, but only at higher extract concentrations (≥25 mg/mL) and smaller zones of inhibition were observed as compared to dichloromethane extract. _C. parapsilosis_ ATCC 22019 was the only yeast inhibited by methanol extract, again at higher extract concentrations (≥50 mg/mL) and smaller zones of inhibition compared to dichloromethane extract. The growth of all strains showing positive results was inhibited by the extracts in a concentration-dependent manner. All four extracts were not active against the three strains of fungi tested (_Aspergillus niger_ ATCC 46404, _Aspergillus fumigatus_, and _Penicillium chrysogenum_).

2.2. Results for modified resazurin microtitre assay
The results obtained in REMA are tabulated in Table 4. The organisms tested were most susceptible to dichloromethane extract. Percentage viability of microbial cells obtained from fluorometric REMA for each extract of _T. ornata_ is illustrated in Figures 1–3.

Table 3. Antimicrobial activity of *T. ornata* extracts against yeasts by disk diffusion method

Extract	mg/mL	Zone of inhibition (mm)		
		C. guilliermondii ATCC 6260	*C. parapsilosis* ATCC 22019	*S. cerevisiae*
T. ornata hexane	3	–	–	–
	5	–	–	–
	10	–	–	–
	15	–	–	–
	25	–	–	9.00 ± 1.41
	50	–	–	9.30 ± 0.58
	100	–	–	10.00 ± 0.00
T. ornata dichloromethane	3	–	–	–
	5	7.00 ± 0.00	8.00 ± 0.00	10.00 ± 0.00
	10	9.00 ± 0.00	9.00 ± 0.00	13.00 ± 0.00
	15	11.00 ± 0.00	9.55 ± 0.64	17.67 ± 0.58
	25	14.67 ± 1.15	10.00 ± 0.00	21.33 ± 1.15
	50	18.67 ± 2.52	10.67 ± 0.58	29.00 ± 1.41
	100	20.33 ± 2.08	11.33 ± 0.58	32.00 ± 0.00
T. ornata methanol	3	–	–	–
	5	–	–	–
	10	–	–	–
	15	–	–	–
	25	–	–	–
	50	–	8.00 ± 0.00	–
	100	–	8.50 ± 0.71	–
Amphotericin B	10 ug/disk	32.50 ± 0.84	20.00 ± 0.63	29.70 ± 0.52

Notes: "–" indicates no antimicrobial activity. Amphotericin B was used as positive control. 99.5% DMSO was used as negative control where no inhibition was observed. Disk diffusion test was performed in at least duplicates and diameter of zone of inhibitions was expressed as mean ± standard deviation in mm. Zones of inhibition include 6 mm for disk diameter.

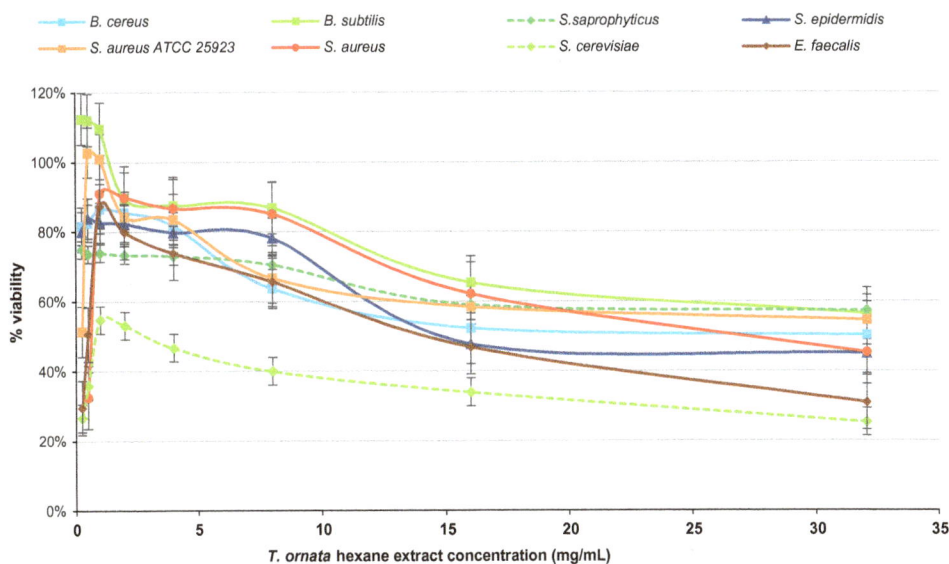

Figure 1. Percentage viability of microbial cells tested using different concentrations of *Turbinaria ornata* hexane extract in fluorometric REMA.

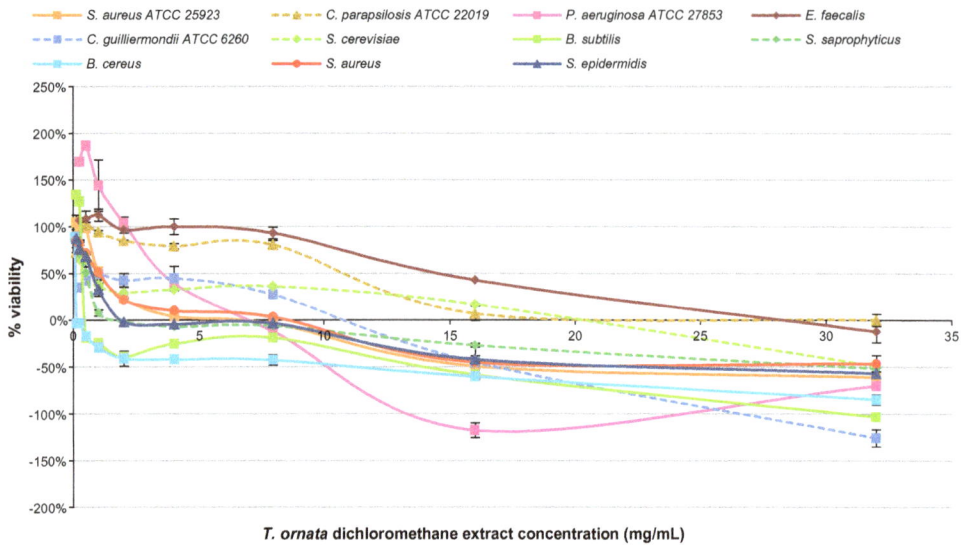

Figure 2. Percentage viability of microbial cells tested using different concentrations of *Turbinaria ornata* dichloromethane extract in fluorometric REMA.

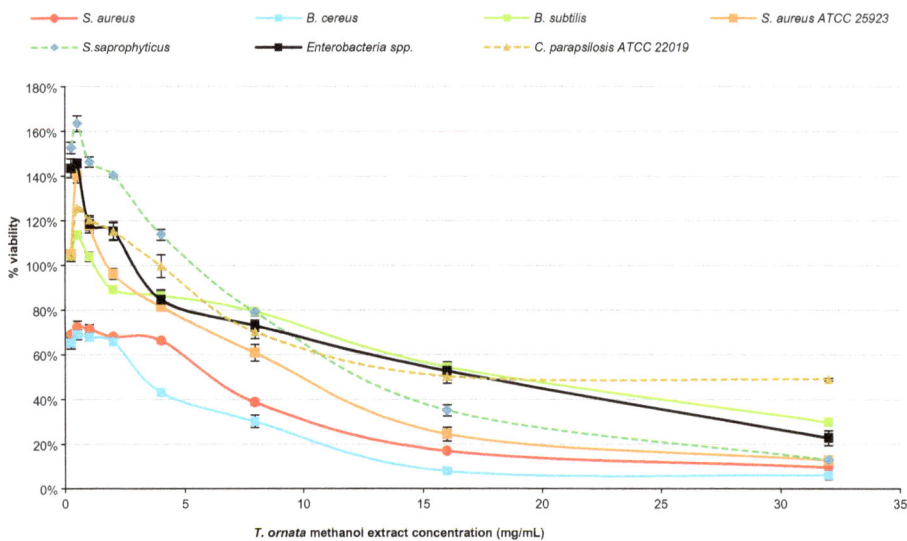

Figure 3. Percentage viability of microbial cells tested using different concentrations of *Turbinaria ornata* methanol extract in fluorometric REMA.

Cohen's κ was conducted using SPSS (IBM Corporation) to evaluate the extent of agreement between readings obtained from the experiment using the methods of colorimetric and fluorometric REMA. The κ value obtained was 0.712 with $p < 0.0005$, indicating there was substantial agreement between the two methods following the suggestions for kappa-statistic interpretation by Landis and Koch (1977). Since $p < 0.0005$, the κ coefficient was statistically significantly different from zero.

2.3. Discussion

In this study, it was observed that dichloromethane was the best solvent for extracting the effective antimicrobial compounds from *T. ornata* when this extract was tested against the selected strains of microbes compared to hexane, methanol, and water extracts. Demirel et al. (2009) utilized the three solvents (hexane, dichloromethane, methanol) in extraction of five brown algae; the dichloromethane extracts were found to have more potent antimicrobial activity then the hexane and methanol extracts at extract concentrations of 1.0 and 1.5 mg/disk. However, it was found in the present study that at extract concentrations of 1.0 mg/disk, *T. ornata* dichloromethane extract possessed superior activity than that of the dichloromethane extracts of brown algae reported by Demirel et al. *T. ornata*

Table 4. Antimicrobial activity of T. ornata extracts against bacteria and yeasts using REMA

Micro-organisms	Hexane (mg/mL)		Dichloromethane (mg/mL)		Methanol (mg/mL)		Gentamicin (µg/mL)	Amphotericin B (µg/mL)
	Colorimetric	Fluorometric	Colorimetric	Fluorometric	Colorimetric	Fluorometric	Colorimetric	Colorimetric
B. cereus	>32 ± 0	>32 ± 0	0.125 ± 0	0.125 ± 0	>32 ± 0	16 ± 0	0.5 ± 0	NA
B. subtilis	>32 ± 0	>32 ± 0	0.5 ± 0	0.5 ± 0	>32 ± 0	>32 ± 0	1 ± 0	NA
S. aureus ATCC 25923	>32 ± 0	>32 ± 0	2 ± 0	4 ± 0	>32 ± 0	>32 ± 0	0.67 ± 0.3	NA
S. aureus	>32 ± 0	>32 ± 0	4 ± 0	4 ± 0	>32 ± 0	32 ± 0	0.25 ± 0	NA
S. saprophyticus	>32 ± 0	>32 ± 0	1 ± 0	1 ± 0	>32 ± 0	>32 ± 0	0.06 ± 0	NA
S. epidermidis	>32 ± 0	>32 ± 0	2 ± 0	2 ± 0	NT	NT	0.25 ± 0	NA
E. faecalis	>32 ± 0	>32 ± 0	32 ± 0	32 ± 0	NT	NT	0.5 ± 0	NA
Enterobacter spp.	NT	NT	NT	NT	>32 ± 0	>32 ± 0	0.67 ± 0.3	NA
P. aeruginosa ATCC 27853	NT	NT	32 ± 0	8 ± 0	NT	NT	0.5 ± 0	NA
C. parapsilosis ATCC 22019	NT	NT	>32 ± 0	16 ± 0	>32 ± 0	>32 ± 0	NA	1 ± 0
C. guilliermondii ATCC 6260	NT	NT	>32 ± 0	16 ± 0	NT	NT	NA	0.25 ± 0
S. cerevisiae	>32 ± 0	>32 ± 0	0.5 ± 0	32 ± 0	NT	NT	NA	4 ± 0

Notes: NT indicates not tested. NA indicates not applicable. Gentamicin and Amphotericin B were used as positive controls for bacteria and yeasts/fungi respectively. Tween 80, PBS and DMSO (6.25%) were used as negative controls for hexane, dichloromethane and methanol extracts respectively. REMA was performed in at least duplicates and concentrations were expressed as mean ± standard deviation in mg/mL.

dichloromethane extract exhibited zone of inhibition of 18.67 mm in *B. subtilis*, while Demirel et al. recorded zone of inhibition of only 6.5–7.5 mm in *B. subtilis* ATCC 6633.

Positive results in Gram-positive bacteria were expected as they lack an outer membrane (lipopolysaccharide layer) which is present in Gram-negative bacteria (Nikaido, 2003). The limited susceptibility in Gram-negative bacteria could thus be attributed to limited outer membrane permeability and presence of porins in the membrane which narrows penetration of the extract (Delcour, 2009; Nikaido, 2003). Interestingly, dichloromethane extract exhibited positive effect against *P. aeruginosa* which lacks general diffusion porins and possesses "slow porins" instead (Nikaido, 2003). *P. aeruginosa* has an outer membrane with decreased permeability and has a more effective drug efflux mechanism compared to the other Gram-negative bacteria (Nikaido, 2003). In fact, *P. aeruginosa* is less susceptible to most antibiotics than *Enterobacteriaceae* (Nikaido, 2003).

In the present study, *T. ornata* water extract did not show any antimicrobial activity although inhibitory effect against *S. aureus* was reported by Zubia, Payri, and Deslandes (2008). In addition, Vijayabaskar and Shiyamala (2011) reported that methanol extract of *T. ornata* inhibited four Gram-negative bacteria, namely *E. coli*, *K. pneumoniae*, *P. vulgaris*, and *P. aeruginosa*. However, this was not observed in the present study; the only Gram-negative bacterium inhibited was *Enterobacter* spp. These discrepancies might be attributed to the different source of *T. ornata* used. Seasonal and geographical variations may alter algal production of antibacterial substances resulting in different antimicrobial activities (Moreau, Pesando, Bernard, Caram, & Pionnat, 1988; Stirk, Reinecke, & van Staden, 2007; Vidyavathi & Sridhar, 1991). Discrepancies might also result from the different assays used; agar well diffusion method (Berghe & Vlietinck, 1991) was used in Vijayabaskar and Shiyamala's (2011) study. The strains of the particular bacteria species employed were different too. Adherence to established guidelines such as that of CLSI might help to reduce conflicting reports on antimicrobial activities in the future.

The inhibition observed in selected strains of yeasts suggested that hexane, dichloromethane, and methanol extracts of *T. ornata* contained antifungal substances. Brown algae were known to have antifungal activities against *Candida* species (Khaled, Hiba, & Asma, 2012) and *S. cerevisiae* (Sridhar, Kumar, Babu, Aruna, & Mansuya, 2010). The underlying antifungal mechanisms are yet to be completely elucidated. However, purified phlorotannins from brown seaweeds were suggested to have an effect on ergosterol and respiration in yeasts (Lopes, Pinto, Andrade, & Valentão, 2013).

The use of REMA is preferable over agar disk diffusion method since it is more sensitive and accurate. Although RPMI 1640 medium is recommended by CLSI for susceptibility testing of yeast and fungi, Mueller Hinton broth (Oxoid, England) was used in this study and had comparable results as evidenced by the achievement of the same MIC value for positive antibiotic control for the yeast control strain *C. parapsilosis* ATCC 22019 (for amphotericin B, MIC value obtained in this study was 1 ± 0 µg/mL which was within CLSI standard's range of 0.5–4 µg/mL for 48 h incubation). In dichloromethane extract, MIC values obtained from colorimetric and fluorometric REMA were the same (Table 4), with exception for the strains *S. aureus* ATCC 25923, *P. aeruginosa* ATCC 27853, *C. guilliermondii* ATCC 6260, *C. parapsilosis* ATCC 22019, and *S. cerevisiae*. For the case of *S. aureus* ATCC 25923, colorimetric measurement with the unaided eye might be a shortcoming as the color of the well could not be distinguished clearly between blue and purple. A purple color would be regarded as a trailing result, where some metabolic activities and a longer incubation time would cause purple color to change to pink (Monteiro et al., 2012).

In some of the MIC values obtained, fluorometric REMA readings were lower than colorimetric REMA readings. In dichloromethane extract, for the case of *P. aeruginosa* ATCC 27853, the colorimetric method revealed an MIC value of 32 mg/mL. However, percent viability calculations showed an MIC value of 8 mg/mL. This could be perhaps explained by the drawback of using the modified REMA. In modified REMA, the antibiotic (extract), strain, and resazurin were added at the same time before a certain period of incubation. It was possible that in modified REMA, living microbial cells produced enough reduced resofurin during the incubation time to give a highly fluorescent pink color, although most of the cells were dead at the time of reading (O'Brien, Wilson, Orton, & Pognan, 2000). This would mean an overestimation of survival. In order to improve the accuracy of the assay, the classic REMA method where resazurin was added after the incubation time could be utilized. Alternatively, it might be of use to employ a slower reacting indicator to allow ample time for the extract to react with the bacterial cells (Gabrielson et al., 2002).

For the yeast strain *S. cerevisiae*, there was a discrepancy between the colorimetric and fluorometric measurements using dichloromethane extract. MIC value was taken at 0.5 mg/mL during visual observation. However, it was not expected that fluorometric measurements revealed an MIC value at 32 mg/mL. MIC is a quantitative endpoint measurement and there are various factors controlling the ultimate endpoint (Othman et al., 2011). Further validation of the MIC value could be done by performing the minimum bactericidal concentration (MBC) test.

No prior literature has made a comparison between colorimetric and fluorometric modified REMA. The use of resazurin as a colorimetric indicator in addition to CLSI's protocol of microdilution assay was first described by Tiballi, He, Zarins, Revankar, and Kauffman (1995) and was found to be simple, sensitive, rapid, robust, and reliable in testing antimicrobial properties of natural products (Sarker, Nahar, & Kumarasamy, 2007). This study is the first to compare between colorimetric and fluorometric REMA utilizing kappa statistics. Based on the results obtained, the colorimetric method is reliable and could be used in resource-limited laboratories where a microplate reader is not available.

3. Experimental

3.1. Solvent extraction
T. ornata (Turner) J. Agardh was collected from Pulau Kerindingan, Semporna on 4 October 2012 and transported to the IMU Research laboratory. The seaweed sample was authenticated by Prof. Phang Siew Moi, University of Malaya (Voucher No.: PSM12862). The samples were cleaned, rinsed with sterile distilled water, oven-dried at 4°C, and powdered in a mixer grinder. Samples were sequentially extracted by soaking in various solvent (Merck, USA) systems started with hexane, followed by dichloromethane and methanol. Each solvent extraction process was conducted for three days. The extracts were then concentrated using rotary evaporator. For the water extraction, the samples were soaked with ultrapure water for three days prior to freeze-drying (Labconco, USA). All extracts were kept in a desiccator with silica gel until use.

3.2. Test micro-organisms
All micro-organisms were obtained from the culture collections of Institute of Medical Research, Kuala Lumpur, Malaysia and the American Type Culture Collection (ATCC, USA). In this study, the seven Gram-positive bacteria screened were *Bacillus cereus*, *B. subtilis*, *S. aureus* ATCC 25923, *S. aureus*, *S. epidermidis*, *Staphylococcus saprophyticus*, and *Enterococcus faecalis*. Six Gram-negative bacteria used were *E. coli* ATCC 25922, *Enterobacter* spp., *Klebsiella pneumoniae*, *P. aeruginosa* ATCC 27853, *Proteus vulgaris*, and *Proteus mirabilis*. The yeast strains employed were *Candida albicans* ATCC 60193, *Candida glabrata* ATCC 2001, *Candida parapsilosis* ATCC 22019, *Candida tropicalis* ATCC 201380, *C. guilliermondii* ATCC 6260, *Candida lusitaniae* ATCC 34449, and *S. cerevisiae*. Three fungi *A. niger* ATCC 46404, *A. fumigatus*, *P. chrysogenum* were also tested. All bacteria were maintained on Tryptone Soy Agar (Oxoid, UK) and incubated for 18 h while yeasts and fungi were cultured on Malt Extract Agar (Oxoid, UK) and incubated for 48 h. Overnight fresh microbial cultures were used to prepare inoculum suspensions for the following antimicrobial studies.

3.3. Disk diffusion antimicrobial assay
All four extracts were preliminarily screened for antimicrobial activity using the Kirby-Bauer disk diffusion method according to the Clinical and Laboratory Standards Institute (CLSI) M02-A11 guidelines (CLSI, 2012a). Gentamicin (10 µg/disk) (Oxoid, England) was used as positive control for bacterial strains while amphotericin B (10 µg/disk) (Sigma–Aldrich, USA) was used as positive control

for yeasts and fungi. Dimethylsulfoxide (DMSO) 99.5% (Sigma-Aldrich, USA) which was used to dissolve the hexane, dichloromethane, and methanol extracts served as negative control. Ultrapure water served as negative control in the case of water extract. Mueller-Hinton agar (MHA) was used for bacteria while MHA supplemented with 0.5 µg/mL methylene blue dye and 2% glucose was used for yeasts and fungi for a more defined zone of inhibition. The agar plates were incubated for 18 h at 37°C for bacteria and 48 h at 37°C for yeasts and fungi. The antimicrobial activity was assessed by measuring the diameter of zone of inhibition around sterile filter paper disks (6 mm diameter; Oxiod, England) impregnated with 20 µL of algal extracts at concentrations of 100, 50, 25, 15, 10, 5, and 3 mg/mL. All tests were performed in at least duplicates.

3.4. Modified resazurin microtitre assay
The strains of microbes found to have inhibition in the disk diffusion method were subjected to broth microdilution test to determine the minimum inhibitory concentration (MIC). The MIC assay was performed in accordance to CLSI guidelines (M07-A9 guideline for bacteria; M27-A3 guideline for yeasts; M38-A2 guideline for filamentous fungi) with modification by the addition of resazurin dye (Sigma-Aldrich, USA) at a final concentration of 0.002% (CLSI, 2008a, 2008b, 2012b). Mueller-Hinton broth was used in replacement of Roswell Park Memorial Institute (RPMI) 1640 medium.

The hexane extract was dissolved in Tween 80 (Sigma-Aldrich, USA). The dichloromethane extract was dissolved well in phosphate buffered saline (PBS) (pH7.2–7.6) (MP Biomedicals, USA) with sonication for more than 30 min with continuous stirring prior to use. DMSO with a final well concentration of 6.25% was used to dissolve the methanol extract. The following were placed in each well of 96-well plates: 50 µL of test algal extract for final well concentrations ranging from 0.25–32 mg/mL, 40 µL of microbial cells suspended in 2.5X Mueller-Hinton broth adjusted to 0.5 MacFarland to achieve a final bacteria concentration of 5×10^5 cfu/mL or final yeast/fungi concentration of 0.5–2.5×10^3 cfu/mL, and 10 µL of sterile resazurin of final concentration 0.002% (w/v). Gentamicin (Bioplus, USA) served as positive control for bacteria while for yeasts and fungi, amphotericin B (Sigma-Aldrich, USA) was used. Tween 80, PBS and DMSO (6.25%) were used as negative controls for hexane, dichloromethane, and methanol extracts, respectively. All experiments were performed in at least duplicates. Plates were incubated at 37°C for 18 h (bacteria) and 48 h (yeasts and fungi). The MICs were determined by colorimetric and fluorometric methods. By visual inspection, the lowest concentration at which the color of resazurin changed from blue to pink was taken as the MIC value. The MIC value was also determined by measuring fluorescence at wavelength of emission 590 nm and excitation 560 nm using a microplate reader (Tecan Infinite 200 PRO). Percentage viability for microbial cells was then calculated after background fluorescence correction with MIC value taken as the well yielding % viability of less than 10% (IC_{90}). The formula to calculate % viability was:

$$\% \text{ variability} = \frac{\text{extract mean} - \text{colour control mean}}{\text{growth control mean} - \text{sterlity control mean}} \times 100\%$$

where extract: extract, bacterial/yeast/fungi suspension, resazurin; color control: extract, Mueller Hinton broth, resazurin; growth control: Tween 80/PBS/ DMSO, bacterial/yeast/fungi suspension, resazurin; sterility control: Tween 80/PBS/DMSO, Mueller-Hinton broth, resazurin

3.5. Statistical analysis
Cohen's κ was conducted using SPSS (IBM Corporation) to evaluate the extent of agreement between readings obtained from the experiment using the methods of colorimetric and fluorometric REMA, following the suggestions for kappa-statistic interpretation by Landis and Koch (1977).

4. Conclusion
The results reported herein indicated that there was presence of antimicrobial activity in *T. ornata* hexane, dichloromethane, and methanol extracts. In particular, dichloromethane extract had the most potent antimicrobial activity, suggesting the extract's potential as an antimicrobial agent. However, this study only served as preliminary screening. In the sequence of this work, active compounds could be isolated from dichloromethane *T. ornata* extract. The MICs of the active compounds

could then be identified to provide lead in drug discovery. Both colorimetric and fluorometric REMA were compared and the two methods were found to be substantially agreeable to each other (κ value = 0.712 with $p < 0.0005$).

Non-standard abbreviations

ATCC   American type culture collection

CLSI   Clinical and Laboratory Standards Institute

IC90   concentration giving 90% inhibition

MBC   minimum bactericidal concentration

MHA   Mueller-Hinton agar

MIC   minimum inhibitory concentration

MRSA   Methicillin-resistant *S. aureus*

REMA   modified resazurin microtitre assay

RPMI   Roswell Park Memorial Institute

Acknowledgements
The authors would also like to thank Mr. Lee Ser Yong for helping with the water extraction of *T. ornata* and assays. We also appreciate the kindness of Dr Chong Chun Wie and Ms. Tsen Min Tze for providing assistance in statistical analysis.

Funding
This work was supported by the International Medical University, Malaysia [grant number BP1–01/11(14)2014].

Competing Interests
The authors declare no competing interest.

Author details
Kar-Yee Tye[1]
E-mail: tye.dragon@gmail.com
Sook-Yee Gan[1]
E-mail: sookyee_gan@imu.edu.my
ORCID ID: http://orcid.org/0000-0002-4826-9143
Swee-Hua Erin Lim[2]
E-mail: erinlim@perdanauniversity.edu.my
ORCID ID: http://orcid.org/0000-0001-5177-0257
Sou-Eei Tan[1]
E-mail: josephine_tse92@hotmail.com
Chen-Ai Chen[1]
E-mail: chenchenai10@gmail.com
ORCID ID: http://orcid.org/0000-0002-8578-1410
Siew-Moi Phang[3]
E-mail: phang@um.edu.my

[1] School of Pharmacy, International Medical University, 126 Jalan Jalil Perkasa 19, Bukit Jalil 57000, Kuala Lumpur, Malaysia.
[2] Perdana University-Royal College of Surgeons in Ireland (PU-RCSI), Jalan MAEPS Perdana, Perdana University, Block B & D Aras 1, MAEPS Building, MARDI Complex, Serdang 43400, Selangor DE, Malaysia.
[3] Institute of Ocean and Earth Sciences and Institute of Biological Sciences, University of Malaya, Kuala Lumpur, Malaysia.

Cover image
Source: Authors.

References
Ananthi, S., Raghavendran, H. R., Sunil, A. G., Gayathri, V., Ramakrishnan, G., & Vasanthi, H. R. (2010). In vitro antioxidant and *in vivo* anti-inflammatory potential of crude polysaccharide from *Turbinaria ornata* (marine brown alga). *Food and Chemical Toxicology, 48*, 187–192. http://dx.doi.org/10.1016/j.fct.2009.09.036

Aravindan, S., Delma, C. R., Thirugnanasambandan, S. S., Herman, T. S., & Aravindan, N. (2013). Anti-pancreatic cancer deliverables from sea: First-hand evidence on the efficacy, molecular targets and mode of action for multifarious polyphenols from five different brown-algae. *PLoS One, 8*, e61977. http://dx.doi.org/10.1371/journal.pone.0061977

Baker, D. D., Chu, M., Oza, U., & Rajgarhia, V. (2007). The value of natural products to future pharmaceutical discovery. *Natural Product Reports, 24*, 1225–1244. http://dx.doi.org/10.1039/b602241n

Berghe, V. A., & Vlietinck, A. J. (1991). Screening methods for anti-bacterial and antiviral agents from higher plants. *Methods in Plant Biochemistry, 6*, 47–68.

Clinical Laboratory Standards Institute. (2008a). *Reference method for broth dilution antifungal susceptibility testing of yeasts; approved standard, CLSI document M27-A3.* ISBN 1-56238-666-2.

Clinical Laboratory Standards Institute. (2008b). *Reference method for broth dilution antifungal susceptibility testing of filamentous fungi; approved standard, CLSI document M38-A2.* ISBN 1-56238-668-9.

Clinical and Laboratory Standards Institute. (2012a). *Performance standards for antimicrobial disk susceptibility tests; approved standard, CLSI document M02-A11.* ISBN 1-56238-781-2.

Clinical and Laboratory Standards Institute. (2012b). *Methods for dilution antimicrobial susceptibility tests for bacteria that grow aerobically; approved standard* (p. 32). ISBN 1-56238-783-9.

Cortés, Y., Hormazábal, E., Leal, H., Urzúa, A., Mutis, A., Parra, L., & Quiroz, A. (2014). Novel antimicrobial activity of a dichloromethane extract obtained from red seaweed *Ceramium rubrum* (Hudson) (Rhodophyta: Florideophyceae) against *Yersinia ruckeri* and *Saprolegnia parasitica*, agents that cause diseases in salmonids. *Electronic Journal of Biotechnology, 17*, 126–131. http://dx.doi.org/10.1016/j.ejbt.2014.04.005

Delcour, A. H. (2009). Outer membrane permeability and antibiotic resistance. *Biochimica et Biophysica Acta (BBA)- Proteins and Proteomics, 1794*, 808–816. http://dx.doi.org/10.1016/j.bbapap.2008.11.005

Demirel, Z., Yilmaz-Koz, F. F., Karabay-Yavasoglu, U. N., Ozdemir, G., & Sukatar, A. (2009). Antimicrobial and antioxidant activity of brown algae from the Aegean sea. *Journal of the Serbian Chemical Society, 74*, 619–628. http://dx.doi.org/10.2298/JSC0906619D

Eom, S. H., Kim, Y. M., & Kim, S. K. (2012). Antimicrobial effect of phlorotannins from marine brown algae. *Food and Chemical Toxicology, 50*, 3251–3255. http://dx.doi.org/10.1016/j.fct.2012.06.028

Gabrielson, J., Hart, M., Jarelöv, A., Kühn, I., McKenzie, D., & Möllby, R. (2002). Evaluation of redox indicators and the use of digital scanners and spectrophotometer for quantification of microbial growth in microplates. *Journal of Microbiological Methods, 50*, 63–73. http://dx.doi.org/10.1016/S0167-7012(02)00011-8

Girija, K., Hemalatha, A., Saranya, C., Parthiban, C., & Anantharaman, P. (2013). Extraction and isolation of phlorotannins from brown seaweed *Turbinaria ornata* (Turner) J. Agardh and its antioxidant activity. *International Journal of Bioassays, 2*, 1185–1189.

Jeyaseelan, E. (2012). Antibacterial activity of some selected algae present in the costal lines of Jaffna peninsula. *International Journal of Pharmaceutical & Biological Archive, 3*, 352–356.

Khaled, N., Hiba, M., & Asma, C. (2012). Antioxidant and antifungal activities of *Padina pavonica* and *Sargassum vulgare* from the Lebanese Mediterranean coast. *Advances in Environmental Biology, 6*, 42–48.

Landis, J. R., & Koch, G. G. (1977). The measurement of observer agreement for categorical data. *Biometrics, 33*, 159–174. http://dx.doi.org/10.2307/2529310

Lane, A. L., Nyadong, L., Galhena, A. S., Shearer, T. L., Stout, E. P., Parry, R. M., ... Kubanek, J. (2009). Desorption electrospray ionization mass spectrometry reveals surface-mediated antifungal chemical defense of a tropical seaweed. *Proceedings of the National Academy of Sciences, 106*, 7314–7319. http://dx.doi.org/10.1073/pnas.0812020106

Lopes, G., Pinto, E., Andrade, P. B., & Valentão, P. (2013). Antifungal activity of phlorotannins against dermatophytes and yeasts: Approaches to the mechanism of action and influence on *Candida albicans* virulence factor. *PLoS One, 8*, e72203. http://dx.doi.org/10.1371/journal.pone.0072203

Manivannan, K., Karthikai Devi, G., Anantharaman, P., & Balasubramanian, T. (2011). Antimicrobial potential of selected brown seaweeds from Vedalai coastal waters, Gulf of Mannar. *Asian Pacific Journal of Tropical Biomedicine, 1*, 114–120. http://dx.doi.org/10.1016/S2221-1691(11)60007-5

Monteiro, M. C., de la Cruz, M., Cantizani, J., Moreno, C., Tormo, J. R., Mellado, E., ... Vicente, F. (2012). A new approach to drug discovery: High-throughput screening of microbial natural extracts against *Aspergillus fumigatus* using resazurin. *Journal of Biomolecular Screening, 17*, 542–549. http://dx.doi.org/10.1177/1087057111433459

Moreau, J., Pesando, D., Bernard, P., Caram, B., & Pionnat, J. (1988). Seasonal variations in the production of antifungal substances by some dictyotales (brown algae) from the French Mediterranean coast. *Hydrobiologia, 162*, 157–162. http://dx.doi.org/10.1007/BF00014538

Nikaido, H. (2003). Molecular basis of bacterial outer membrane permeability revisited. *Microbiology and Molecular Biology Reviews, 67*, 593–656. http://dx.doi.org/10.1128/MMBR.67.4.593-656.2003

O'Brien, J., Wilson, I., Orton, T., & Pognan, F. (2000). Investigation of the Alamar Blue (resazurin) fluorescent dye for the assessment of mammalian cell cytotoxicity. *European Journal of Biochemistry, 267*, 5421–5426. http://dx.doi.org/10.1046/j.1432-1327.2000.01606.x

Othman, M., Loh, H., Wiart, C., Khoo, T. J., Lim, K. H., & Ting, K. N. (2011). Optimal methods for evaluating antimicrobial activities from plant extracts. *Journal of Microbiological Methods, 84*, 161–166. http://dx.doi.org/10.1016/j.mimet.2010.11.008

Plaza, M., Santoyo, S., Jaime, L., García-Blairsy Reina, G. G., Herrero, M., Señoráns, F. J., & Ibáñez, E. (2010). Screening for bioactive compounds from algae. *Journal of Pharmaceutical and Biomedical Analysis, 51*, 450–455. http://dx.doi.org/10.1016/j.jpba.2009.03.016

Sarker, S. D., Nahar, L., & Kumarasamy, Y. (2007). Microtitre plate-based antibacterial assay incorporating resazurin as an indicator of cell growth, and its application in the *in vitro* antibacterial screening of phytochemicals. *Methods, 42*, 321–324. http://dx.doi.org/10.1016/j.ymeth.2007.01.006

Sridhar, S., Kumar, J. S., Babu, S., Aruna, P., & Mansuya, P. (2010). Pharmacognostical and antifungal activity of selected seaweeds from Gulf of Mannar region. *Recent Research in Science and Technology, 2*, 115–119.

Stirk, W. A., Reinecke, D. L., & van Staden, J. (2007). Seasonal variation in antifungal, antibacterial and acetylcholinesterase activity in seven South African seaweeds. *Journal of Applied Phycology, 19*, 271–276. http://dx.doi.org/10.1007/s10811-006-9134-7

Tiballi, R. N., He, X., Zarins, L. T., Revankar, S. G., & Kauffman, C. A. (1995). Use of a colorimetric system for yeast susceptibility testing. *Journal of Clinical Microbiology, 33*, 915–917.

Vidyavathi, N., & Sridhar, K. (1991). Seasonal and geographical variations in the antimicrobial activity of seaweeds from the Mangalore coast of India. *Botanica Marina, 34*, 279–284.

Vijayabaskar, P., & Shiyamala, V. (2011). Antibacterial activities of brown marine algae (*Sargassum wightii* and *Turbinaria ornata*) from the Gulf of Mannar biosphere reserve. *Advances in Biological Research, 5*, 99–102.

Wang, J., Zhang, Q., Zhang, Z., Hou, Y., & Zhang, H. (2011). In-vitro anticoagulant activity of fucoidan derivatives from brown seaweed *Laminaria japonica*. *Chinese Journal of Oceanology and Limnology, 29*, 679–685. http://dx.doi.org/10.1007/s00343-011-0181-9

World Health Organization. (2015). *Antimicrobial resistance fact sheet N 194.*

Zubia, M., Payri, C., & Deslandes, E. (2008). Alginate, mannitol, phenolic compounds and biological activities of two range-extending brown algae, *Sargassum mangarevense* and *Turbinaria ornata* (Phaeophyta: Fucales), from Tahiti (French Polynesia). *Journal of Applied Phycology, 20*, 1033–1043. http://dx.doi.org/10.1007/s10811-007-9303-3

Appendix

Table A1. Cross-tabulation table of output for Cohen's kappa generated by SPSS

cREMA_coded * fREMA_coded Crosstabulation
Count

cREMA_coded		fREMA_coded							Total
		0.125	0.5	1	2	4	32	>32	
	0.125	1	0	0	0	0	0	0	1
	0.5	0	1	0	0	0	1	0	2
	1	0	0	1	0	0	0	0	1
	2	0	0	0	1	1	0	0	2
	4	0	0	0	0	1	0	0	1
	32	0	0	0	0	0	1	1	2
	>32	0	0	0	0	0	1	13	14
Total		1	1	1	1	2	3	14	23

Notes: cREMA = colorimetric REMA, fREMA = fluorometric REMAs.

Table A2. Symmetric Measures table of output for Cohen's kappa generated by SPSS

	Value	Asymp. Std. Error[a]	Approx. T[b]	Approx. Sig.
Measure of Kappa Agreement	.712	.123	6.533	.000
N of Valid Cases	23			

[a]Not assuming the null hypothesis,

[b]Using the asymptotic standard error assuming the null hypothesis.

18

Anti-inflammatory activities of extracts from *Oliva sp.*, *Patella rustica*, and *Littorina littorea* collected from Ghana's coastal shorelines

Lawrence Sheringham Borquaye[1,2]*, Godfred Darko[2], Michael Konney Laryea[2], Victor Roberts[2], Richmond Boateng[2] and Edward Ntim Gasu[1]

*Corresponding author: Lawrence Sheringham Borquaye, Central Laboratory, Kwame Nkrumah University of Science and Technology, Kumasi, Ghana; Department of Chemistry, Kwame Nkrumah University of Science and Technology, Kumasi, Ghana

E-mails: lsborquaye.sci@gmail.com, lsborquaye.sci@knust.edu.gh

Reviewing editor: Laura Giamperi, University of Urbino, Italy

Additional information is available at the end of the article

Abstract: Inflammation is one of the means the human body uses to defend itself in the event of infection, trauma, or exposure to toxic substances and it is closely associated with a number of disease symptoms. Steroidal and non-steroidal anti-inflammatory agents have been the drugs of choice for managing inflammation. However, reports of unpleasant side effects have necessitated a search for new anti-inflammatory agents which have minimal side effects. Marine-derived natural products continue to make significant contributions in the pharmaceutical, nutra-ceutical, and cosmeceutical industries and a number of extracts and compounds from marine origin have shown promise as anti-inflammatory agents. In Ghana, extracts that have been screened for their potential anti-inflammatory effects have almost exclusively come from plants. In this work, the anti-inflammatory activi-ties of extracts from three different marine mollusks (*Oliva sp.*, *Patella rustica*, and

ABOUT THE AUTHORS

Lawrence Sheringham Borquaye is a bioorganic chemist with a research focus on science at the interface of chemistry and biology. He's been exploring biologically active natural products from marine organisms (mollusks, bacteria, and fungi) and from plant sources. Other research interests include the development of methods for the analysis of pharmaceutical and personal care products in the environment and the characterization of volatiles (essential oils) from plants.

Godfred Darko's research interests spans environmental chemistry (www.sheathe.org), analytical chemistry, nanochemistry, and natural products.

Michael Konney Laryea is a graduate student in the Borquaye Lab working on the isolation of biologically active natural products from plant and marine sources.

Richmond Boateng and Victor Roberts are undergraduate researchers in the Borquaye Lab. Visit us at www.borquayelab.com for detailed profiles of all authors.

Edward Ntim Gasu is a research assistant at the KNUST Central Laboratory investigating the unique properties of antimicrobial peptides.

PUBLIC INTEREST STATEMENT

The marine environment is emerging as an important contributor in providing lead compounds for various therapeutic areas in many drug development programs worldwide. In Ghana, attempts to utilize the sea in this regard have been few and far-fetched. Interestingly, less than 5% of non-fish organisms in Ghana's marine ecosystem have been properly characterized. In this work, extracts from three mollusks commonly found along the shorelines of Ghana were tested for their ability to reduce inflammation. Current inflammation drugs have unwanted side effects that make their use unpleasant, especially over long periods. The results indicated that these mollusks harbor compounds that could potentially be exploited for pharmaceutical, nutraceutical, and cosmeceutical purposes.

Littorina littorea) were evaluated. Extracts were obtained by cold maceration. The carrageenan-induced paw edema model in seven-day old chicks was used to evaluate anti-inflammatory potentials. Of the extracts tested, the ethyl acetate fraction of *Oliva sp.* was the most potent, with an ED_{50} of 10.16 mg/kg. The ethanol extract of *L. littorea* proved to be least effective in reducing inflammation, with an ED_{50} value of 119.80 mg/kg. When compared, extracts from *Oliva sp.* seemed to possess greater anti-inflammatory potentials than either *P. rustica* or *L. littorea* counterparts. The ethyl acetate fraction of *Oliva sp.* was a potent and promising anti-inflammatory agent and could be explored for anti-inflammatory lead compounds.

Subjects: Biochemistry; Nutrition; Pharmacology; Marine Biology; Medicinal & Pharmaceutical Chemistry; Natural Products

Keywords: inflammation; marine mollusks; bioactivity; carrageenan; chick edema

1. Background

It is widely known that inflammation is associated with most diseased conditions. Inflammation is usually characterized by redness, heat, pain, swellings, and disruption in normal physiological functions. The body uses inflammation as a protective response to infection, injury (or trauma), or chemical exposure. Inflammation is used as a mechanism by the body to inactivate invading organisms, remove irritants, and initiate tissue repair procedures (Mitchell & Cotran, 2003). Asthma, colitis, hepatitis, and arthritis, which are all inflammatory diseases, have been implicated as a leading cause of permanent disability and ultimately death in people worldwide (Emery, 2006). Usually, steroidal and non-steroidal anti-inflammatory agents are used to manage inflammation cases. Reported side effects associated with the use of these drugs have, however, been unpleasant. The side effects include gastric ulceration, bleeding, abnormal kidney function, hypertension, and immunosuppression (Chandra, Chatterjee, Dey, & Bhattacharya, 2012). The long term use of steroidal and non-steroidal anti-inflammatory drugs is understood to lead to drug-induced toxic effects and/ or secondary adverse effects (Beg, Hasan, Hussain, Swain, & Barkat, 2011; Drew & Myers, 1997). The search for new anti-inflammatory agents with improved efficacy and limited side effects has therefore become the focus of global scientific research, with natural products leading the way.

The ocean is the largest biosphere of the earth and houses a colossal number of diverse organisms. The marine environment is made up of intricate ecologies, with many benthic and sessile organisms employing potent, biologically active molecules as the mainstay of their basic defense mechanisms (Thakur, Thakur, & Müller, 2005). The defense systems of these organisms are highly complex and toxic and the bioactive compounds play significant roles in these systems. Chemicals released into the aquatic ecosystem undergo rapid dilution. This means that such chemicals must be of high potency to be effective for use in chemical defense systems. Additionally, such chemicals must be water soluble and should be reasonably specific for the task required (Jimeno, Faircloth, Fernández Sousa-Faro, Scheuer, & Rinehart, 2004; Newman & Cragg, 2004). All these are good characteristics of potential drug candidates.

Marine invertebrates have proven to be valuable sources of pharmacologically active compounds. Many crude extracts and compounds isolated from the marine organisms have been found to possess antitumor, antifouling, antibacterial, antifungal, antiviral, and anti-inflammatory effects (Blunt et al., 2014; Cragg & Newman, 2013; Faulkner, 2000; Fusetani, 2011; Martins, Vieira, Gaspar, & Santos, 2014). Most compounds isolated have been classified as polyketides, terpenes, steroidal or triterpene saponins, carbohydrates, aliphatic compounds, amino acids, alkaloids, peptides, lipopeptides, and proteins. Of the many compounds isolated thus far, a number of them are either on the market as approved drugs or in various phases of clinical trials. In 2015, trabectedin (Pharmamar, trademark: Yondelis), an alkaloid that was isolated from a tunicate, was approved for use in the management of soft tissue sarcoma and ovarian cancer (Allavena et al., 2005; National Cancer Institute, 2015; U.S. Food & Drug Administration, 2015). Debromohymenialdisine, a secondary metabolite isolated from a marine sponge has been explored for use in the management of rheumatoid arthritis and osteoarthritis (Mayer & Lehmann, 2001). Other compounds whose anti-inflammatory properties have been investigated include manoalide, pseudopterosins, topsentins, and scytonemin (Cragg & Newman, 2013; Martins et al., 2014; Mayer, Rodriguezn, Berlinck, & Hamann, 2010). Secondary metabolites from marine sources therefore provide a promising avenue to prospect for new anti-inflammatory agents.

In this work, the anti-inflammatory activity of crude extracts from *Oliva sp, Patella rustica,* and *Littorina littorea* were investigated *in vivo* using the carrageenan-induced foot edema model of inflammation in chicks. *Oliva sp.,* is a sand-borer and usually found in the beach sand, whereas *P. rustica* and *L. littorea* are usually found on rocks in the sea. *P. rustica, Oliva sp.,* and *L. littorea* are gastropods belonging to the families Patellidae, Olividae, and Littorinidae, respectively. Our investigations indicate that the extracts of these marine mollusks possess significant anti-inflammatory activities. The active principles could therefore be isolated and explored as potential leads in drug discovery programs.

2. Methods

2.1. Sample collection
Samples of *Oliva sp.* were collected from Eikwe in the Western Region, while *P. rustica* and *L. littorea* specimens were collected from Labadi in the Greater Accra Region, all in Ghana. All samples were collected between November 2015 and February 2016. Samples were transported on ice to the laboratory. The body tissues were removed from the shell and washed with copious amount of distilled water, then stored at 0°C until use. Body tissues were used within 24 h of removal. A sample of each mollusk was sent to the Department of Fisheries and Marine Sciences, Kwame Nkrumah University of Science and Technology (KNUST), Kumasi, Ghana for authentication.

2.2. Solvent extraction
The body tissues were washed with distilled water and homogenized with a blender. To 100 g of blended body tissue was added 100 mL of appropriate solvent (methanol, 70% ethanol, or ethyl acetate) and kept for 24 h at room temperature with intermittent shaking. The mixture was centrifuged at 5000 rpm (SciSpin ONE, UK) for 15 min. The supernatant was transferred into a round-bottom flask and concentrated *in vacuo* (Cole Parmer SB-1200, Shanghai, China). Extracts were stored in vials at 4°C until use. Extracts obtained were labeled **OSE** (*Oliva sp.,* 70% ethanol extract), **OSM** (*Oliva sp.,* Methanol extract), **OSEA** (*Oliva sp.,* ethyl acetate extract), **PRE** (*P. rustica* 70% ethanol extract), **PREA** (*P. rustica* ethyl acetate extract), **LLE** (*L. littorea* 70% ethanol extract), **LLM** (*L. littorea* methanol extract), and **LLEA** (*L. littorea* ethyl acetate extract).

2.3. Anti-inflammatory assay

2.3.1. Animals
One-day post-hatch cockerels (*Gallus gallus,* strain Shaver 5790) were acquired from Akate Farms (Kumasi, Ghana) and maintained in the Animal House of the Department of Pharmacology, KNUST in stainless steel cages (34 × 57 × 18 cm^3) at a population density of 12–13 chicks per cage. The

temperature in the cages was maintained between 26 and 29°C, with a 12-h light–dark cycle. Food and water was available *ad libitum* via 1-quart gravity-fed feeders and water troughs. Cages were maintained daily during the first quarter of the light cycle. Seven-day old chicks were used for the assay. Group sample sizes of 4–5 were employed throughout the study. Handling of the chicks and experimental protocols were in accordance with the National Institute of Health guidelines for the care and use of laboratory animals.

2.4. Carrageenan-induced foot edema in chicks

The anti-inflammatory activity of the extracts of *Oliva sp., P. rustica,* and *L. littorea* was evaluated using the carrageenan-induced foot edema model of inflammation in the seven-day old chicks with slight modifications (Roach & Sufka, 2003; Woode et al., 2007). Diclofenac (25 mg/kg) was used as the positive control. The various extracts were suspended in sterile distilled water. Two percent of tween-20 was added to enhance the suspension of the extracts. Carrageenan (10 μL of a 2% suspension in saline) was injected sub-plantar into the right footpads of the chicks to induce edema. Extracts were orally administered at different concentrations (50, 100, and 150 mg/kg) 1 h after edema induction. A digital Vernier Caliper was used to measure the foot volume before injection and at various time points after injection. The control animals received only tween-20 in sterile water, serving as the negative control. All drugs and extracts were orally administered in volumes not exceeding 100 mL/kg.

2.5. Data analysis

The raw scores for foot volume increase at each time interval for each chick was normalized as the percentage difference from the initial foot volume at time zero and then averaged for each treatment group. Increase in foot volume was computed using the equation below

$$\% \text{ Increase of foot volume} = \frac{\text{Foot volume at time } t - \text{Foot volume at time zero}}{\text{Foot volume at time } t} \times 100$$

Total foot volume for each treatment group was calculated in arbitrary unit as the area under the curve (AUC). Percentage inhibition of edema for each treatment group was then determined as follows:

$$\% \text{ Inhibition of edema} = \frac{\text{AUC control} - \text{AUC treatment}}{\text{AUC control}} \times 100$$

One-way analysis of variance (ANOVA) followed by Holm–Sidak's *post hoc* test was used to analyze differences in AUCs. ED_{50} values (dose responsible for 50% of maximal effect) for each extract were determined using an iterative computer least square method with the following nonlinear regression (three-parameter logistics) equation.

$$Y = \frac{a + (b - a)}{1 + 10^{(\log ED50 - X)}}$$

where X is the logarithm of dose and Y is the response. Y starts at a (the bottom) and goes to b (the top) with a sigmoid shape. Graphpad Prism for windows version 7.01 (GraphPad Software, San Diego, CA, USA) was used for all statistical analyses and ED_{50} determinations. $p < 0.05$ was considered statistically significant.

3. Results

To evaluate the anti-inflammatory activities of the extracts, acute inflammation was induced in chicks using carrageenan. Subcutaneous administration of carrageenan resulted in a noticeable increase in paw size of the chicks. In control chicks, maximum inflammation was observed after 30 min and this persisted until the 2-h mark. This reduced very slowly to the third hour, an indication of the body's own ability to combat inflammation (Figure 4(a), Control).

Oral administration of diclofenac (the positive control) or the extracts significantly reduced inflammation in the chicks and at a faster rate than the negative control group (Figures 1–4). In general, all extracts examined in this study significantly reduced inflammation in a dose-dependent manner. One-way ANOVA data treatment followed by Dunnett's *post hoc* test indicated that effect of extracts on edema were significant ($p < 0.05$). OSEA (at all concentrations tested; 50, 100, 150 mg/kg) had a significant effect in reducing inflammation ($p < 0.05$), similar to the standard drug, diclofenac.

When the total edema over the experimental period for all extracts were expressed as AUCs (in arbitrary units) over the time course curves, a dose-dependent effect in decreasing total edema was also observed. The highest % inhibition of edema (83.90%) was recorded again by OSEA at a dose of 150 mg/kg (Figure 2), whereas the least % inhibition of 33.42% was recorded by LLE for its 50 mg/kg dose (Figure 1). The dose-dependent effect was also evident in the % inhibitions of all the extracts studied. As an example, PRE had % inhibitions of 46.69, 57.58, and 79.26 for 50, 100, and 150 mg/kg doses, respectively (Figure 3). In comparison, the % inhibition of diclofenac at 25 mg/kg was 65.48%.

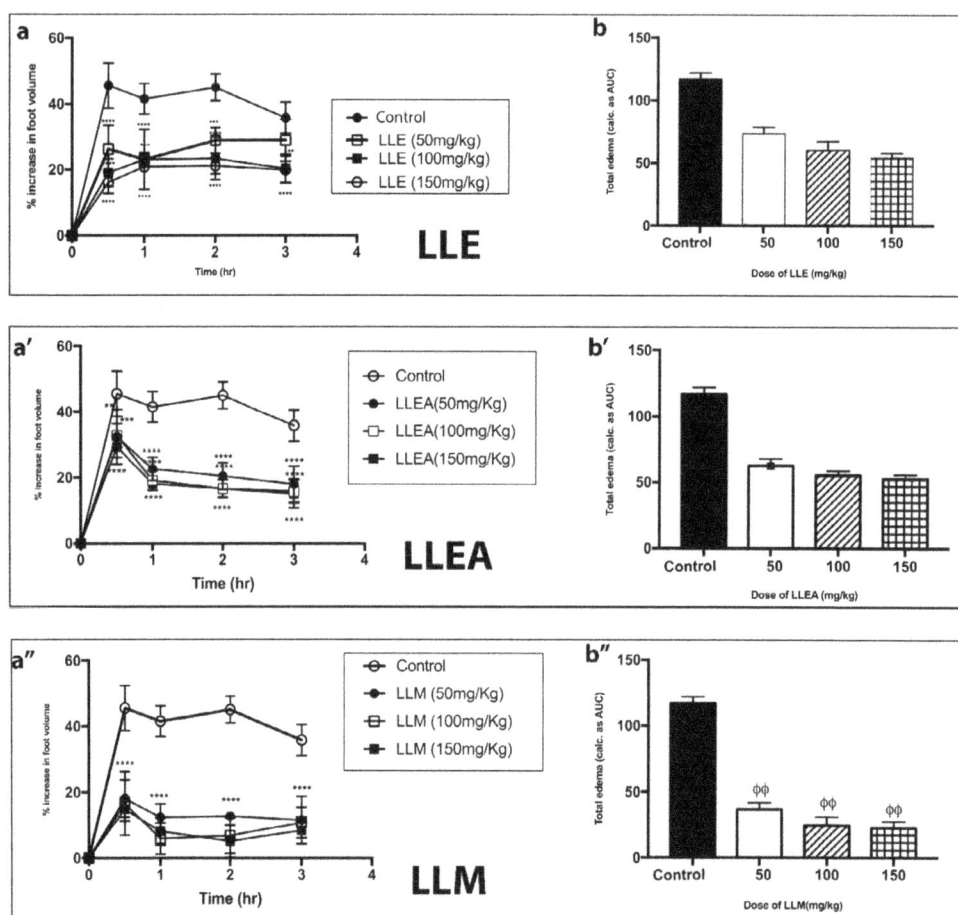

Figure 1. Effect of LLE, LLEA, and LLM (50–150 mg kg⁻¹) on time course curve (a, a′, & a″) and the total edema response in carrageenan-induced foot edema in chicks (b, b′, & b″). Values are means ± SEM ($n = 4$). $p > 0.05$ (ns), $p < 0.05$ (*), $p < 0.01$ (**), $p < 0.001$ (***), $p < 0.0001$ (****) compared to vehicle-treated group. (One-way ANOVA followed by Dunnett *post hoc* test). $p > 0.05$(ns), $p < 0.05$ (φ), $p < 0.01$ (φφ), $p < 0.001$ (φφφ), $p < 0.0001$ (φφφφ) compared to vehicle-treated group (one-way ANOVA followed by Holm–Sidak's *post hoc* test). LLE (*Litorrina littorea* extract), LLEA

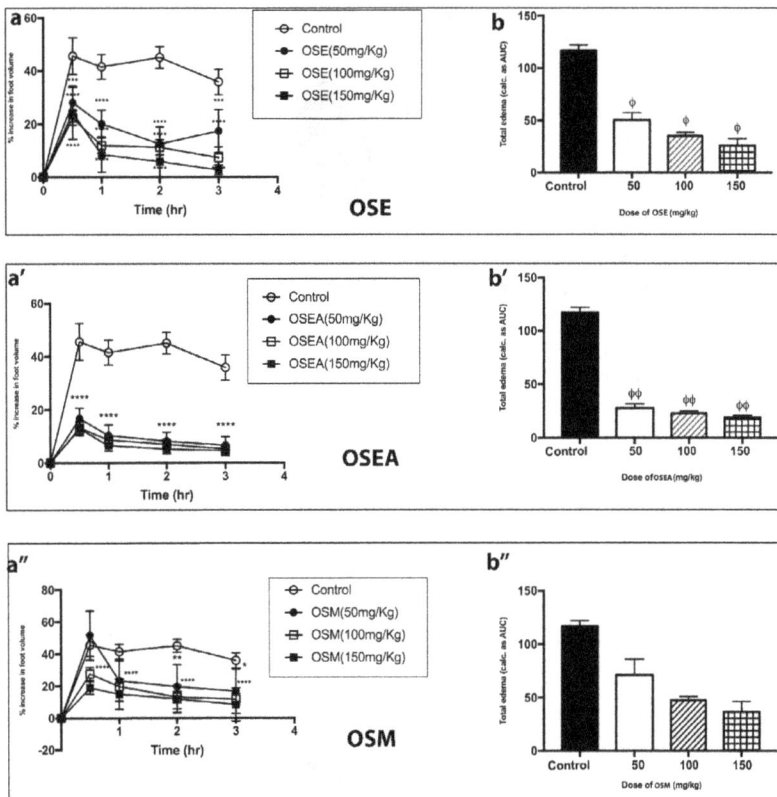

Figure 2. Effect of OSE, OSEA, and OSM (50–150 mg kg⁻¹) on time course curve (c, c′, & c″) and the total edema response in carrageenan-induced foot edema in chicks (d, d′,& d″). Values are means ± SEM (n = 4). $p > 0.05$ (ns), $p < 0.05$ (*), $p < 0.01$ (**), $p < 0.001$ (***), $p < 0.0001$ (****) compared to vehicle-treated group. (One-way ANOVA followed by Dunnett *post hoc* test). $p > 0.05$(ns), $p < 0.01$ (φ), $p < 0.001$ (φφ), $p < 0.0002$ (φφφ), $p < 0.0001$ (φφφφ) compared to vehicle-treated group (one-way ANOVA followed by Holm–Sidak's *post hoc* test). OSE (*Oliva sp.*, 70% ethanol extract), OSEA (*Oliva sp.*, ethyl acetate extract), and OSM (*Oliva sp.*, Methanol extract).

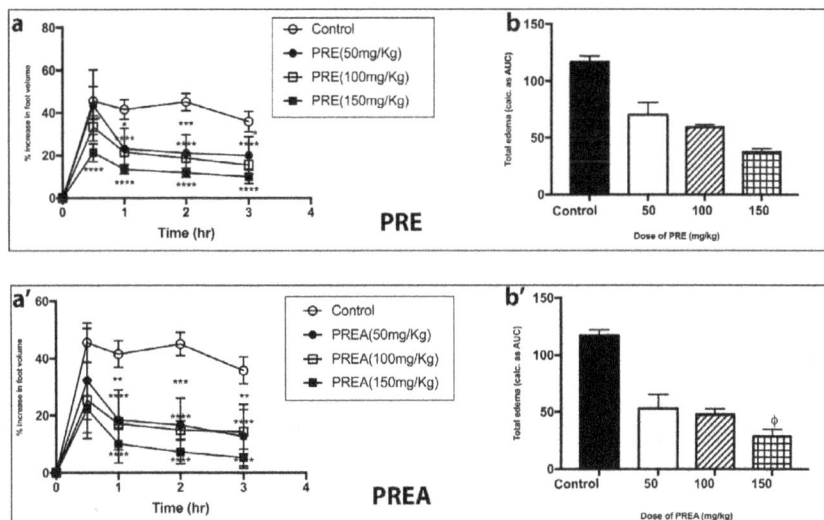

Figure 3. Effect of PRE and PREA (50–150 mg kg⁻¹) on time course curve (e & e′) and the total edema response in carrageenan-induced foot edema in chicks (f & f′). Values are means ± SEM (n = 4). $p > 0.05$ (ns), $p < 0.05$ (*), $p < 0.01$ (**), $p < 0.001$ (***), $p < 0.0001$ (****) compared to vehicle-treated group. (One-way ANOVA followed by Dunnett's *post hoc* test). $p > 0.05$ (ns), $p < 0.05$ (φ), $p < 0.01$ (φφ), $p < 0.004$ (φφφ), $p < 0.0001$ (φφφφ) compared to vehicle-treated group (one-way ANOVA followed by Holm–Sidak's *post hoc* test). PRE (*Patella rustica* ethanol extract), PREA (*Patella rustica* ethyl acetate extract).

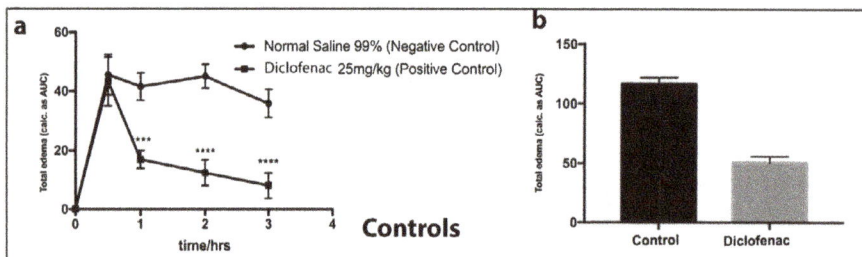

Figure 4. Effect of Normal saline 99% and Diclofenac (25 mg/kg) on time course curve (a) and the total edema response in carrageenan-induced foot edema in chicks (b). Values are means ± SEM ($n = 4$). $p > 0.05$ (ns), $p < 0.05$ (*), $p < 0.01$ (**), $p < 0.001$ (***), $p < 0.0001$ (****) compared to vehicle-treated group. (One-way ANOVA followed by Dunnett's *post hoc* test).

The ED_{50} values were computed via the three-parameter logistics equation. The most potent extract from this study was OSEA with an ED_{50} of 10.16 mg/kg. This was over 10 times better than LLE, whose ED_{50} value of 119.80 mg/kg was the highest and hence least potent (Table 1). To compare how potent extracts from a specific mollusk was against another, the ED_{50} value of each extract from that particular mollusk were added and the mean computed (Figure 5). *Oliva sp.* extracts were the most potent with a mean ED_{50} of 40.15 ± 18.35. *L. littorea* extracts were the least potent. One-way ANOVA treatment of data followed by Holm–Sidak's *post hoc* test, however, revealed no statistically significant difference in potency between the extracts obtained from these three organisms ($p > 0.05$).

Table 1. Effect of extracts on carrageenan-induced edema in seven-day old chicks	
Extract	**ED_{50} (mg/kg) ± SEM**
LLE	119.80 ± 1.15
LLEA	80.72 ± 1.18
LLM	16.61 ± 1.70
OSE	36.83 ± 1.30
OSEA	10.16 ± 1.47
OSM	73.46 ± 1.19
PRE	82.70 ± 1.15
PREA	43.25 ± 1.49

Notes: LLE (*Litorrina littorea* 70% ethanol extract); LLEA (*Litorrina littorea* ethyl acetate extract); LLM (*Litorrina littorea* methanol extract); OSE (*Oliva sp.*, 70% ethanol extract); OSEA (*Oliva sp.*, ethyl acetate extract); OSM (*Oliva sp.*, Methanol extract); PRE (*Patella rustica* 70% ethanol extract); PREA (*Patella rustica* ethyl acetate extract).

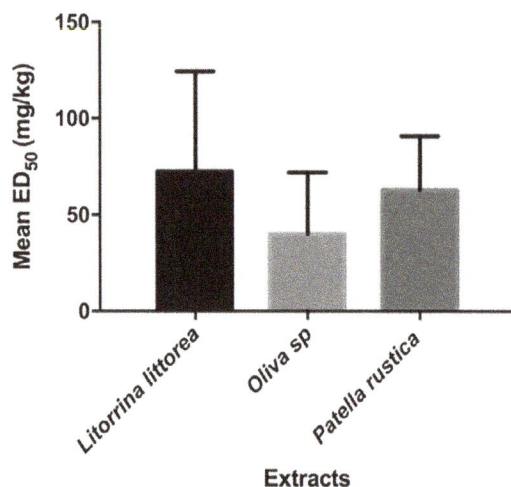

Figure 5. Comparative effect of extracts from mollusks (mean ED_{50}) on carrageenan-induced edema in seven-day old chicks (one-way ANOVA followed by Holm–Sidak's *post hoc* test; $p > 0.05$; no significant difference between mean ED_{50} of each mollusk extracts).

4. Discussion

Two distinct phases have been associated with the carrageenan-induced paw edema model. The first phase (0–60 min) is mediated by mast cell degranulation with histamine and serotonin release followed by a late phase where inflammatory mediators such as prostaglandins, proteases, and lysosomes are produced (Eddouks, Chattopadhyay, & Zeggwagh, 2012; Posadas et al., 2004; Vinegar, Schreiber, & Hugo, 1969). It has also been suggested that prostaglandin release may be attributed to the cyclooxygenase (COX) induction in tissues (Di Rosa, 1972). All extracts used in this study profoundly inhibited carrageenan-induced acute edema in chick foot. The ethyl acetate extracts of *Oliva sp.* (OSEA), in particular, exhibited impressive potency among all extracts examined.

Even though the exact mechanism by which the extracts inhibited carrageenan-induced inflammation was not explored in this study, it is possible that reduction in inflammation occurs via the inhibition of the many inflammatory mediators released during the carrageenan-induced acute inflammation. For most of the extracts, reduction of inflammation was prominent after 30 min and seemed to plateau from 60 min till the end of the experiment (Figures 2 and 3). This suggests that the extracts were probably interacting with the inflammatory mediators that are released in the first phase of carrageenan-induced paw edema (i.e. histamine and serotonin). LLE did not seem to follow this route (Figure 1(a)), with no discernable reduction in inflammation after the first hour. It is, however, interesting to note that the highest level of inflammation which was achieved after 30 min for LLE is significantly lower than that of the control group (Figure 1(a)), indicating some level of potency. It has been suggested that the late phase are particularly susceptible to most anti-inflammatory agents (Di Rosa & Willoughby, 1971; Vinegar et al., 1969). The results obtained for diclofenac (Figure 4(a)) supports this notion. OSEA, in particular, had a time course profile (Figure 2(a′)) similar to that of the standard drug, diclofenac (Figure 4(a)).

The aquatic environment has proven to be a viable source of biologically active extracts and compounds and results from this study corroborates this notion. Many compounds isolated from the marine environment have shown impressive activities in various therapeutic areas. Antimicrobial and antitumor compounds accounts for most of these activities. However, others with antimalarial, antiviral, anti-inflammatory, anti-helminthic, and antioxidant potentials have also being unearthed (Blunt et al., 2014; Borquaye, Darko, Oklu, Anson-Yevu, & Ababio, 2016; Mayer et al., 2010).

The anti-inflammatory potentials of solvent extracts of some marine invertebrates were investigated by Herencia et al. (1998). Extracts of *Coscinasterias tenuispina* and *Holothuria tubulosa* were shown to dose dependently inhibit edema, with lower elastase activity and decreased PGE2 levels measured in homogenates from inflamed paws, without affecting the levels of this prostanoid present in stomach homogenates. The sesquiterpenoid derivatives, avarol and avarone isolated from the Mediterranean sponge, *Dysidea avara*, have been shown to potently inhibit paw edema induced by carrageenan in mice (Ferrándiz et al., 1994). Other marine-derived compounds such as bolinaquinone and petrosiaspongiolide M have been shown to inhibit neutrophilic infiltration, interleukin-1β, prostaglandin E_2 levels, and cyclooxygenase 2 (COX2) protein expression *in vivo*. This has led to plans to further develop these compounds for "protective strategies" against intestinal inflammatory diseases (Busserolles, Payá, & D'Auria, 2005). These examples, and many others, are indicative of the huge potential of marine natural products.

In Ghana, many crude extracts with substantial anti-inflammatory activities have been recorded (Abotsi, Ainooson, & Woode, 2012; Fleischer, Annan, Dickson, Mensah, & Sarpong, 2013; Mensah, Donkor, & Fleischer, 2011; Mensah, Mireku, & Okwuonu, 2014; Woode et al., 2009). The anti-inflammatory effects of *Ficus exasperata* (sandpaper tree) extracts have been evaluated to confirm their traditional use in pain management (Fleischer et al., 2013; Woode et al., 2009). However, investigations into the use of extracts from marine sources in such studies are negligible. We have shown earlier that extracts from *P. rustica* and *L. littorea* possess significant antimicrobial and antioxidant activities (Borquaye, Darko, Ocansey, & Ankomah, 2015; Borquaye et al., 2016). This work further shows that these mollusks harbor important compounds that could be beneficial in our search for new therapeutic agents.

5. Conclusion

In this study, the solvent extracts of *Oliva sp, P. rustica,* and *L. littorea* were shown to reduce inflammation in carrageenan-induced paw edema in chicks in a dose-dependent fashion. *Oliva sp.* extracts were most potent. Isolation and characterization of the pure compounds that elicited these anti-inflammatory activities is currently in progress in our laboratories. This work shows the immense utility of the marine environment as an important source of bioactive compounds for various therapeutic uses.

Author contributions

LSB and GD conceived the study and helped draft the manuscript. All experiments were designed by LSB, GD, and MKL. Samples were collected by RB, VR, MKL, and ENG. VR, RB, and ENG carried out all the experiments. Data analysis was by LSB, GD, and MKL. All authors read and approved the final manuscript.

Acknowledgment

The authors are grateful to the Departments of Chemistry and Pharmacology of the Kwame Nkrumah University of Science and Technology, Kumasi, for the use of its facilities for this study. We are also grateful to Dr Nathaniel Owusu Boadi, Department of Chemistry for helpful discussions as well as Mr Godwin Darku of the Animal House, Department of Pharmacology for technical support.

Funding

The authors received no direct funding for this research.

Competing Interests

The authors declare no competing interest.

Author details

Lawrence Sheringham Borquaye[1,2]
E-mails: lsborquaye.sci@gmail.com,
lsborquaye.sci@knust.edu.gh
ORCID ID: http://orcid.org/0000-0002-5037-0777
Godfred Darko[2]
E-mail: godfreddarko@yahoo.com
ORCID ID: http://orcid.org/0000-0001-7157-646X
Michael Konney Laryea[2]
E-mail: michaellaryeakonney@yahoo.com
Victor Roberts[2]
E-mail: vick.berts@gmail.com
Richmond Boateng[2]
E-mail: richboattosky@gmail.com
Edward Ntim Gasu[1]
E-mail: gasuhuckel66@gmail.com
[1] Central Laboratory, Kwame Nkrumah University of Science and Technology, Kumasi, Ghana.
[2] Department of Chemistry, Kwame Nkrumah University of Science and Technology, Kumasi, Ghana.

Cover image

Source: Authors.

References

Abotsi, W. K. M., Ainooson, G. K., & Woode, E. (2012). Anti-inflammatory and antioxidant effects of an ethanolic extract of the aerial parts of *Hilleria latifolia* (Lam.) H. Walt. (Phytolaccaceae). *African Journal of Traditional, Complementary and Alternative Medicines, 9*, 138–152.

Allavena, P., Signorelli, M., Chieppa, M., Erba, E., Bianchi, G., Marchesi, F., … D 'incalci, M. (2005). Anti-inflammatory properties of the novel antitumor agent yondelis (trabectedin): Inhibition of macrophage differentiation and cytokine production. *Cancer Research, 65*, 2964–2971. https://doi.org/10.1158/0008-5472.CAN-04-4037

Beg, S., Hasan, H., Hussain, M. S., Swain, S., & Barkat, Ma (2011). Systematic review of herbals as potential anti-inflammatory agents: Recent advances, current clinical status and future perspectives. *Pharmacognosy Reviews, 5*, 120. https://doi.org/10.4103/0973-7847.91102

Blunt, J. W., Copp, B. R., Keyzers, R. A., Munro, M. H. G., Prinsep, M. R., Zatylny-Gaudin, C., … Molgó, J. (2014). Marine natural products. *Natural Product Reports, 31*, 160. https://doi.org/10.1039/c3np70117d

Borquaye, L. S., Darko, G., Ocansey, E., & Ankomah, E. (2015). Antimicrobial and antioxidant properties of the crude peptide extracts of *Galatea paradoxa* and *Patella rustica*. *SpringerPlus, 4*, 1. https://doi.org/10.1186/s40064-015-1266-2

Borquaye, L. S., Darko, G., Oklu, N., Anson-Yevu, C., & Ababio, A. (2016). Antimicrobial and antioxidant activities of ethyl acetate and methanol extracts of *Littorina littorea* and *Galatea paradoxa*. *Cogent Chemistry, 2*, 1161865.

Busserolles, J., Payá, M., & D'Auria, M. V. (2005). Protection against 2, 4, 6-trinitrobenzenesulphonic acid-induced colonic inflammation in mice by the marine products bolinaquinone and petrosaspongiolide M. *Biochemical Pharmacology, 69*, 1433–1440. https://doi.org/10.1016/j.bcp.2005.01.020

Chandra, S., Chatterjee, P., Dey, P., & Bhattacharya, S. (2012). Evaluation of *in vitro* anti-inflammatory activity of coffee against the denaturation of protein. *Asian Pacific Journal of Tropical Biomedicine, 2*, S178–S180. https://doi.org/10.1016/S2221-1691(12)60154-3

Cragg, G. M., & Newman, D. J. (2013). Natural products: A continuing source of novel drug leads. *Biochimica et Biophysica Acta (BBA) - General Subjects, 1830*, 3670–3695. https://doi.org/10.1016/j.bbagen.2013.02.008

Di Rosa, M. (1972). Biological properties of carrageenan. *Journal of Pharmacy and Pharmacology, 24*, 89–102. https://doi.org/10.1111/jphp.1972.24.issue-2

Di Rosa, M., & Willoughby, D. A. (1971). Screens for anti-inflammatory drugs. *Journal of Pharmacy and Pharmacology, 23*, 297–298. https://doi.org/10.1111/jphp.1971.23.issue-4

Drew, A. K., & Myers, S. P. (1997). Safety issues in herbal medicine: Implications for the health professions. *The Medical Journal of Australia, 166*, 538–541.

Eddouks, M., Chattopadhyay, D., & Zeggwagh, N. A. (2012). Animal models as tools to investigate antidiabetic and anti-inflammatory plants. *Evidence-Based Complementary and Alternative Medicine, 2012*, 1–14.

Emery, P. (2006). Treatment of rheumatoid arthritis. *BMJ: British Medical Journal, 332*, 152–155. https://doi.org/10.1136/bmj.332.7534.152

Faulkner, D. J. (2000). Highlights of marine natural products chemistry (1972–1999). *Natural Product Reports, 17*, 1–6. https://doi.org/10.1039/a909113k

Ferrándiz, M. L., Sanz, M. J., Bustos, G., Payá, M., Alcaraz, M. J., & De Rosa, S. (1994). Avarol and avarone, two new anti-inflammatory agents of marine origin. *European Journal of Pharmacology, 253*, 75–82. https://doi.org/10.1016/0014-2999(94)90759-5

Fleischer, T. C., Annan, K., Dickson, R. A., Mensah, A. Y., & Sarpong, F. M. (2013). Anti-inflammatory, antioxidant and antimicrobial activity of the stem bark extract and fractions of Ficus exasperata Vahl. (Moraceae). Journal of Pharmacognosy and Phytochemistry, 2, 38–44.

Fusetani, N. (2011). Antifouling marine natural products. Natural Product Reports, 28, 400–410. https://doi.org/10.1039/C0NP00034E

Herencia, F., Ubeda, A., Ferrándiz, M. L., Terencio, M. C., Alcaraz, M. J., García-Carrascosa, M., ... Payá, M. (1998). Anti-inflammatory activity in mice of extracts from Mediterranean marine invertebrates. Life Sciences, 62, 115–120.

Jimeno, J., Faircloth, G., Fernández Sousa-Faro, J., Scheuer, P., & Rinehart, K. (2004). New Marine Derived Anticancer Therapeutics — A Journey from the Sea to Clinical Trials. Marine Drugs, 2, 14–29. https://doi.org/10.3390/md201014

Martins, A., Vieira, H., Gaspar, H., & Santos, S. (2014). Marketed marine natural products in the pharmaceutical and cosmeceutical industries: Tips for success. Marine Drugs, 12, 1066–1101. https://doi.org/10.3390/md12021066

Mayer, A. M. S., & Lehmann, V. K. B. (2001). Marine pharmacology in 1999: Antitumor and cytotoxic compounds. Anticancer Research, 21, 2489–2500.

Mayer, A. M. S., Rodriguezn, A. D., Berlinck, R. G. S., & Hamann, M. T. (2010). Marine pharmacology in 2005–6: Marine compounds with anthelmintic, antibacterial, anticoagulant, antifungal, anti-inflammatory, antimalarial, antiprotozoal, antituberculosis, and antiviral activities; affecting the cardiovascular, immune and nervous syste. Biochimica et Biophysica Acta (BBA) - General Subjects, 1790, 283–308.

Mensah, A. Y., Donkor, P. O., & Fleischer, T. C. (2011). Anti-inflammatory and antioxidant activities of the leaves of Wissadula amplissima var Rostrata. African Journal of Traditional, Complementary and Alternative Medicines, 8, 185–195.

Mensah, A., Mireku, E., & Okwuonu, V. (2014). Anti-inflammatory and anti-oxidant activities of Secamone afzelii (Rhoem) Ascleipiadaceae. Journal of Medical and Biomedical Sciences, 3, 23–30. https://doi.org/10.4314/jmbs.v3i1.4

Mitchell, R., & Cotran, R. (2003). Acute and chronic inflammation. In R. L. Johnson (Ed.), Robbins basic pathology(Vol. 7, pp. 33–59). Philadelphia, PA: Saunders.

National Cancer Institute. (2015). FDA approves trabectedin for sarcoma. Author. Retrieved March 12, 2017, from https://www.cancer.gov/news-events/cancer-currents-blog/2015/fda-trabectedin-sarcoma

Newman, D. J., & Cragg, G. M. (2004). Marine natural products and related compounds in clinical and advanced preclinical trials. Journal of Natural Products, 67, 1216–1238. https://doi.org/10.1021/np040031y

Posadas, I., Bucci, M., Roviezzo, F., Rossi, A., Parente, L., Sautebin, L., & Cirino, G. (2004). Carrageenan-induced mouse paw oedema is biphasic, age-weight dependent and displays differential nitric oxide cyclooxygenase-2 expression. British Journal of Pharmacology, 142, 331–338. https://doi.org/10.1038/sj.bjp.0705650

Roach, J., & Sufka, K. (2003). Characterization of the chick carrageenan response. Brain Research, 994, 216–225. https://doi.org/10.1016/j.brainres.2003.09.038

Thakur, N. L., Thakur, A. N., & Müller, W. E. G. (2005). Marine natural products in drug discovery. Natural Product Radiance, 4, 471–477.

U.S. Food and Drug Administration. (2015). Press announcements - FDA approves new therapy for certain types of advanced soft tissue sarcoma. Retrieved March 12, 2017, from https://www.fda.gov/NewsEvents/Newsroom/PressAnnouncements/ucm468832.htm

Vinegar, R., Schreiber, W., & Hugo, R. (1969). Biphasic development of carrageenin edema in rats. Journal of Pharmacology and Experimental Therapeutics, 166, 96–103.

Woode, E., Ansah, C., Ainooson, G. K., Abotsi, W. M. K., Mensah, A. Y., & Duwiejua, M. (2007). Anti-inflammatory and antioxidant properties of the root extract of Carissa edulis (Forsk.) Vahl (Apocynaceae. Journal of Science and Technology, 27, 5–15.

Woode, E., Poku, R. A., Ainooson, G. K., Boakye-Gya, E., Abotsi, W. K. M., Mensah, T. L., & Amoh-Barim, A. K. (2009). An evaluation of the anti-inflammatory, antipyretic and antinociceptive effects of Ficus exasperata (Vahl) Leaf Extract. Journal of Pharmacology and Toxicology, 4, 138–151. https://doi.org/10.3923/jpt.2009.138.151

Permissions

List of Contributors

Ashutosh Pathak, Shashi Kant Shukla, Rajesh Kumar, Madhu Pandey, Manisha Pandey, Afifa Qidwai and Anupam Dikshit
Biological Product Lab, Department of Botany, University of Allahabad, Allahabad 211012, Uttar Pradesh, India

Rohit K. Mishra
CMDR, Motilal Nehru National Institute of Technology, Allahabad 211004, Uttar Pradesh, India

Komguem Tagne Pélagie Michelin and Aghofack-Nguemezi Jean
Laboratoire de Botanique Appliquée, Faculté des Sciences, Université de Dschang, BP 67, Dschang, Cameroon

Gatsing Donatien, Lacmata Tamekou Stephen and Kuiate Jules-Roger
Laboratory of Microbiology and Antimicrobial Substances, Faculty of Science, Department of Biochemistry, University of Dschang, P.O. Box 67, Dschang, Cameroon

Lunga Paul Keilah
Laboratory of Phytobiochemistry and Medicinal Plants Study, Faculty of Science, Department of Biochemistry, University of Yaoundé 1, P.O. Box 812, Yaoundé, Cameroon

Amina Ghalem and Rachid Tarik Bouhraoua
Faculty of Science, Laboratory n 31, Department of Agro- Forestry, Management Conservatory Water, Soil and Forests, University of Tlemcen, P.O. Box 119, Tlemcen, Algeria

Inês Barbosa
Center for Environmental and Sustainability Research (CENSE), Nova University of Lisbon, Campus de Caparica, 2829-516, Caparica, Portugal

Augusta Costa
Center for Environmental and Sustainability Research (CENSE), Nova University of Lisbon, Campus de Caparica, 2829-516, Caparica, Portugal Instituto Nacional de Investigação Agrária e Veterinária, I.P.,Quinta do Marquês, Av. da República, 2780-159 Oeiras, Portugal

Ememobong Gideon Asuquo and Chinweizu Ejikeme Udobi
Faculty of Pharmacy, Department of Pharmaceutics and Pharmaceutical Technology, Pharmaceutical Microbiology Unit, University of Uyo, Uyo, Nigeria

Yogish Somayaji T, Vidya V and Vishakh R
Central Research Laboratory, K S Hegde Medical Academy, Nitte University, Mangaluru, Karnataka, India

Jayarama Shetty and Alex John Peter
Department of Oncology, Nitte Leela Narayan Shetty Memorial Institute of Oncology, Mangaluru, Karnataka, India

Suchetha Kumari N
Department of Biochemistry, K S Hegde Medical Academy, Nitte University, Mangaluru, Karnataka, India

Md Shamsuddin Sultan Khan and Amin Malik Shah Abdul Majid
School of Pharmaceutical Sciences, University of Science Malaysia, Penang, Malaysia

Muhammad Adnan Iqbal
School of Chemical Sciences, University of Science Malaysia, Penang, Malaysia

Hui Wu, Wen-Ying Li, Lei Wu, Ling-Yun Zhu, Er Meng and Dong-Yi Zhang
Research Center of Biological Information, College of Science, National University of Defense Technology, Changsha, Hunan 410073, China

Theophine Chinwuba Akunne, Peter A. Akah, Ifeoma A. Nwabunike, Chukwuemeka S. Nworu, Emeka K. Okereke, Nelson C. Okereke and Francis C. Okeke
Faculty of Pharmaceutical Sciences, Department of Pharmacology and Toxicology, University of Nigeria, Nsukka 410001, Enugu State, Nigeria

Sutapa Joardar
Department of Biotechnology, Neotia Institute of Technology, Management and Science, Jhinga, Diamond Harbour Road, South 24 Parganas, Amira 743368, India.

Department of Food Technology and Bio-Chemical Engineering, Jadavpur University, 188, Raja S. C. Mallick Road, Kolkata 700 032, India

Paramita Bhattacharjee
Department of Food Technology and Bio-Chemical Engineering, Jadavpur University, 188, Raja S. C. Mallick Road, Kolkata 700 032, India

Shounak Ray and Suvendu Samanta
Department of Chemistry, Indian Institute of Engineering Science and Technology, Shibpur, Howrah 711 103, India

Zakaria M. Sawan
Agricultural Research Center, Cotton Research Institute, Ministry of Agriculture and Land Reclamation, 9 Gamaa street, 12619 Giza, Egypt

Zakaria M. Sawan
Cotton Research Institute, Agricultural Research Center, Ministry of Agriculture and Land Reclamation, 9 Gamaa Street, 12619, Giza, Egypt

Moffat K. Njoroge
Department of Crops, Ministry of Agriculture and Livestock, P.O. Box 16, Kitui, Kenya
Department of Plant and Crop Protection, University of Nairobi, P.O. Box 30197, Nairobi, Kenya

D.W. Miano and D.C. Kilalo
Department of Plant and Crop Protection, University of Nairobi, P.O. Box 30197, Nairobi, Kenya

D.L. Mutisya
Department of Crop Health, KALRO-Katumani, P.O. Box 340, Machakos, Kenya

Rowida S. Baeshen
Faculty of Science, Department of Biology, University of Tabuk, P.O. Box 741, Tabuk 71491, Saudi Arabia

Wafaa M. Hikal
Faculty of Science, Department of Biology, University of Tabuk, P.O. Box 741, Tabuk 71491, Saudi Arabia
Parasitology Laboratory, Water Pollution Researches Department, National Research Center, 33 El-Bohouth St., Dokki, Giza 12622, Egypt

Hussein A.H. Said-Al Ahl
Medicinal and Aromatic Plants Researches Department, National Research Centre, 33 El-Bohouth St., Dokki, Giza 12622, Egypt

Joham Sarfraz Ali, Attarad Ali and Muhammad Zia
Department of Biotechnology, Quaid-i-Azam University Islamabad, Islamabad 45320, Pakistan

Ihsan ul Haq and Madiha Ahmed
Department of Pharmacy, Quaid-i-Azam University Islamabad, Islamabad 45320, Pakistan

Mona M. Okba, Kadriya S. El Deeb and Fathy M. Soliman
Faculty of Pharmacy, Pharmacognosy Department, Cairo University, Kasr El-Ainy, 11562 Cairo, Egypt

Gehad A. Abdel Jaleel
Pharmacology Department, National Research Center, Tahrir Street, Dokki, Cairo, Egypt

Miriam F. Yousif
Faculty of Pharmacy, Pharmacognosy Department, Cairo University, Kasr El-Ainy, 11562 Cairo, Egypt
Faculty of Pharmaceutical Sciences and Pharmaceutical Industries, Pharmacognosy Department, Future University, Al Tagamoa Al Khames, 11528 New Cairo, Egypt

Bhakti A. Mhatre and Thankamani Marar
School of Biotechnology & Bioinformatics, D.Y. Patil University, Sector 15, CBD Belapur, Navi Mumbai 400614, Maharashtra, India

Kar-Yee Tye, Sook-Yee Gan, Sou-Eei Tan and Chen-Ai Chen
School of Pharmacy, International Medical University, 126 Jalan Jalil Perkasa 19, Bukit Jalil 57000, Kuala Lumpur, Malaysia

Swee-Hua Erin Lim
Perdana University-Royal College of Surgeons in Ireland (PU-RCSI), Jalan MAEPS Perdana, Perdana University, Block B & D Aras 1, MAEPS Building, MARDI Complex, Serdang 43400, Selangor DE, Malaysia

Siew-Moi Phang
Institute of Ocean and Earth Sciences and Institute of Biological Sciences, University of Malaya, Kuala Lumpur, Malaysia

Edward Ntim Gasu
Central Laboratory, Kwame Nkrumah University of Science and Technology, Kumasi, Ghana

Lawrence Sheringham Borquaye
Central Laboratory, Kwame Nkrumah University of Science and Technology, Kumasi, Ghana
Department of Chemistry, Kwame Nkrumah University of Science and Technology, Kumasi, Ghana

Godfred Darko, Michael Konney Laryea, Victor Roberts and Richmond Boateng
Department of Chemistry, Kwame Nkrumah University of Science and Technology, Kumasi, Ghana

Index

www.ingramcontent.com/pod-product-compliance
Lightning Source LLC
Chambersburg PA
CBHW082015190326
41458CB00010B/3193